Science for Governing Japan's Population

Twenty-first-century Japan is known for the world's most aged population. Faced with this challenge, Japan has been a pioneer in using science to find ways of managing a declining birth rate. *Science for Governing Japan's Population* considers the question of why these population phenomena have been seen as problematic. What roles have population experts played in turning this demographic trend into a government concern? Aya Homei examines the medico-scientific fields around the notion of population that developed in Japan from the 1860s to the 1960s, analyzing the role of the population experts in the government's effort to manage its population. She argues that the formation of population sciences in modern Japan had a symbiotic relationship with the development of the neologism, "population" (*jinkō*), and with the transformation of Japan into a modern sovereign power. Through this history, Homei unpacks assumptions about links between population, sovereignty, and science. This title is also available as Open Access.

Aya Homei is Lecturer in Japanese Studies at the University of Manchester.

SCIENCE IN HISTORY

Series Editors
Simon J. Schaffer, University of Cambridge
James A. Secord, University of Cambridge

Science in History is a major series of ambitious books on the history of the sciences from the mid-eighteenth century through the mid-twentieth century, highlighting work that interprets the sciences from perspectives drawn from across the discipline of history. The focus on the major epoch of global economic, industrial and social transformations is intended to encourage the use of sophisticated historical models to make sense of the ways in which the sciences have developed and changed. The series encourages the exploration of a wide range of scientific traditions and the interrelations between them. It particularly welcomes work that takes seriously the material practices of the sciences and is broad in geographical scope.

Science for Governing
Japan's Population

Aya Homei

University of Manchester

CAMBRIDGE
UNIVERSITY PRESS

CAMBRIDGE
UNIVERSITY PRESS

University Printing House, Cambridge CB2 8BS, United Kingdom

One Liberty Plaza, 20th Floor, New York, NY 10006, USA

477 Williamstown Road, Port Melbourne, VIC 3207, Australia

314–321, 3rd Floor, Plot 3, Splendor Forum, Jasola District Centre, New Delhi – 110025, India

103 Penang Road, #05–06/07, Visioncrest Commercial, Singapore 238467

Cambridge University Press is part of the University of Cambridge.

It furthers the University's mission by disseminating knowledge in the pursuit of education, learning, and research at the highest international levels of excellence.

www.cambridge.org
Information on this title: www.cambridge.org/9781009186834
DOI: 10.1017/9781009186827

First published 2023

A catalogue record for this publication is available from the British Library.

ISBN 978-1-009-18683-4 Hardback

Cambridge University Press has no responsibility for the persistence or accuracy of URLs for external or third-party internet websites referred to in this publication and does not guarantee that any content on such websites is, or will remain, accurate or appropriate.

The source of the cover image: *Jinkō mondai kenkyū*, 3, no. 6 (June 1942): 33, https://www.ipss.go.jp/syoushika/bunken/data/pdf/14193908.pdf

This book is dedicated to the four magnificent historians of medicine and science whom I have been lucky to know thus far in my academic life: Roberta Bivins, Elizabeth Toon, Michael Worboys, and the late John Pickstone. Without their guidance and support, I would not be who I am today.

Contents

Figures

Acknowledgments

Many around me would know exactly what I mean when I say this book is a product of much juggling. Between when I began writing this book and now, some from the first cohort of students I taught at the University of Manchester have received their PhDs, others have got fabulous jobs, and others are married. My children have graduated from their primary school, my mother has had and survived cancer, and I have had two research leaves and been through an Achilles tendon rupture and COVID-19. Throughout all this drama, at work and in life, many supported me. At work, the guidance and friendship of four historians of medicine and science I respect dearly – Roberta Bivins, Elizabeth Toon, Michael Worboys, and the late John Pickstone – shaped me profoundly as a historian of science and medicine. I would also like to acknowledge the generosity and kindness of my colleagues in Japanese Studies at the University of Manchester. Outside Manchester, Professor Yoko Matsubara gave me invaluable guidance through our research collaboration, which I am happy to say lasts to this date. Specifically for this monograph, I would first like to express my sincere thanks to the Wellcome Trust (085926/Z/08/Z), JSPS London (JSPS London Symposium Scheme FY16), and the Great Britain Sasakawa Foundation (No. 5073), whose financial support gave me opportunities to shape this book's ideas. An earlier version of Chapter 5 appeared in *Japan Forum* ("Birth Control Survey Research, Technical Bureaucrats and the Imagining of Japan's Population, 1945–60," published online on July 16, 2020, doi.org/10 .1080/09555803.2020.1750450), and I would like to thank the journal's editors and reviewers who helped me refine the chapter's argument. Dr. Reiko Hayashi at the National Institute of Population and Social Security Research generously granted me access to the Tachi Bunko archive, gave me research affiliate status at the institute in 2016–17, and read an earlier draft of Chapter 2; thank you for all the help. I would also like to send my special thanks to Professor Ryuzaburo Sato, who was so generous as to read the entire manuscript and always sent me words of encouragement. Professor Sato's insider knowledge as a renowned demographer has been extremely helpful. My thanks extend to colleagues across the world, specifically to Ruselle Meade, Reut Harari, Jaehwan Hyun, and John DiMoia, who commented on earlier drafts of this book. The members of the Hayama seminar group led by Dr. Kenji Ito at Sokendai gave me invaluable advice toward the end of the book project. Rebekah Zwanzig was more than a proofreader; over the years, she has become my writing partner. Lucy Rhymer, Rachel Blaifeder, Emily Plater, and Natasha Whelan of Cambridge University Press, and Jessica B. Murphy of the Center for the History of Medicine, Francis A. Countway Library, Harvard Medical School, have been fantastically friendly and efficient, making the book production process so much easier than it could have been. I have many more colleagues and friends to thank here but cannot due to the limited space, so please accept my sincere apologies for not listing all your names. Finally, I want to thank my families: My parents for looking after their grandchildren during summers when I was writing; my brother and sister-in-law for offering me shelter while I was doing research in Japan; my nephew Rintaro for having been my reliable research assistant; my husband, who helped me plough through this book manuscript; and my children for being their happy selves so I could concentrate on finishing this book.

Note on the Text

The romanization of Japanese words largely follows the modified Hepburn system. In regard to names of Japanese persons in the main text, surnames appear first, followed by the first name. The place of publication for Japanese language sources is Tokyo, unless specified otherwise.

Abbreviations

ACPP	Advisory Council on Population Problems
CBS	Cabinet Bureau of Statistics
CPB	Cabinet Planning Board
DPHD	Department of Public Health Demography
EPA	Economic Planning Agency
EPL	Eugenic Protection Law
GPL	General Plan to Establish National Land Planning
GPP	General Plan to Establish the National Population Policy
HHSG	Health and Hygiene Survey Group, Home Ministry
IC-PFP	Investigative Commission for Population and Food Problems
IHD	International Health Division, Rockefeller Foundation
IPH	The Institute of Public Health
IPP	Institute of Population Problems
IPPF	International Planned Parenthood Federation
IRPP	Foundation-Institute for Research of Population Problems
IRPP-CPM	Foundation-Institute for Research of Population Problems Committee on Population Measures
ISI	International Statistical Institute
JOICFP	Japanese Organization for International Cooperation in Family Planning
MHW	Ministry of Health and Welfare
NCMH	National Committee on Maternal Health
NRS	Natural Resources Section, SCAP-GHQ
OCMA	Osaka City Midwives' Association
OIPS	Osaka Infant Protection Society
OMA	Osaka Midwives Association
OPR	Office of Population Research at Princeton University
PAJ	Population Association of Japan
PC	Population Council
PH&W	Public Health and Welfare Section, SCAP-GHQ
SCAP	Supreme Commander for Allied Powers

SCAP-GHQ	Supreme Commander for Allied Powers General Headquarters
TFR	Total Fertility Rate
TTPSG	Temporary Taiwan Population Survey Group
WWI	World War I
WWII	World War II

Introduction

One day during the summer of 2017, I was in a city in Japan and entered a large bookshop to avoid the excruciating heat outside. In the section displaying bestsellers, I found a sensational caption on a book titled *The Future Chronology*.[1] Evoking a sense of crisis over the aging and shrinking population, the caption, written in bold gothic style with a red and white background, read:

2020: Half the women are over fifty years old
2024: One-third of the total population is sixty years old and over
2027: There is a shortage of blood for blood transfusions
2033: One in three households is empty
2039: There is a shortage of crematories
2040: Half of the local authorities are gone
2042: The population of the aged peaks at forty million

A similar image of Japan's dystopic demographic future is found in the 2016 version of the government's *White Paper of Health, Labour and Welfare*. In it, the then Minister of Health, Labour, and Welfare Shiozaki Yasuhisa stated: "Our country … is a 'low-fertility and aging society' … It is estimated that by 2060, two in five will be older people and the total population will go under 90 million."[2] Attached to the *White Paper* was the annually updated leaflet, "Japan as Seen from the Perspective of a Population of 100" (Figure 0.1).

Despite the cute characters surrounding the demographic figures and the upbeat tone typically associated with government publications of this kind, the message it conveyed was rather gloomy. It said that 12.7 persons out of the 100 people were fifteen years old or younger, while

[1] Kawai Masashi, *Mirai no nenpyō: Jinkō genshō nihon de korekara okiru koto* (Kōdansha, 2017).
[2] Shiozaki quoted in Kōseirōdōshō, *Heisei 28-nendo ban kōsei rōdō hakusho* (Kōseirōdōshō, 2016) (no page numbers are assigned for this reference).

Figure 0.1 Japan as seen from the perspective of a population of 100. Source: Ministry of Health, Labour and Welfare of the Government of Japan, "Jinkō 100-nin demita nihon," www.mhlw .go.jp/wp/hakusyo/kousei/16-3/dl/01.pdf, accessed July 2, 2020.

26.7 persons were aged sixty-five years or more. According to demographers Satō Ryuzaburō and Kaneko Ryūichi, these figures clearly indicate Japan is making headway as a country in a "post-demographic transition phase."[3] The phase is inundated with "many difficult problems," for instance rural communities disappearing due to the aging and contracting population.[4] Precisely for this reason, demographers and policymakers across the world are closely watching Japan's population trend and the government's response to it.

What is so obvious that we tend to overlook is that the public narrative of a population crisis is substantiated by numerical facts. The 2016 *White Paper* drew data from the population census conducted in 2015.[5] In turn, *The Future Chronology* was based on the results of the medium fertility projection presented in "Population Projections for Japan (2016–2065)," produced by the National Institute of Population and Social Security Research in 2017 (Figure 0.2).[6]

The results, based on the 2015 data, estimated the population of Japan would "decrease to around 110.92 million by 2040, fall below 100 million to 99.24 million by 2053, and drop to 88.08 million by 2065." The population aged sixty-five years and more was expected to grow from 33.87 million in 2015 to 37.16 million by 2030, and peak at 39.35 million by 2042.[7] Public life in Japan today is dominated by what historian Barbara J. Shapiro once called a "culture of fact," a firm consensus that facts, especially those represented in numbers, provide a credible perspective with which to view the natural and human world.[8]

What is more, we also take for granted that these demographic data should necessarily urge official or societal responses to the population crisis. *The Future Chronology* and *White Paper* are significant, not simply because they have contributed to constructing a public discourse about

[3] Ryuzaburo Sato and Ryuichi Kaneko, *Posuto jinkō tenkanki no nihon* (Hara shobo, 2016), 2–6.
[4] Ryuzaburo Sato and Ryuichi Kaneko, "Entering the Post-Demographic Transition Phase in Japan: Its Concept, Indicators and Implications," paper presented at the European Population Conference, Budapest, Hungary, June 25–28, 2014, accessed July 21, 2020, https://epc2014.princeton.edu/papers/140662; see also Sato and Kaneko, *Posuto jinkō tenkanki no nihon*.
[5] Sōmushō Tōkeikyoku, "Heisei 27-nen kokusei chōsa jinkō nado kihon shūkei kekka yōyaku" (Sōmushō Tōkeikyoku, December 16, 2015), www.stat.go.jp/data/kokusei/2015/kekka/kihon1/pdf/youyaku.pdf.
[6] Kawai, *Mirai no nenpyō*, 7–9.
[7] National Institute of Population and Social Security Research, "Population Projections for Japan (2016–2065): Summary Population Statistics," (2017), www.ipss.go.jp/pp-zenkoku/e/zenkoku_e2017/pp_zenkoku2017e.asp.
[8] Barbara J. Shapiro, *A Culture of Fact: England, 1550–1720* (Ithaca and London: Cornell University Press, 2000).

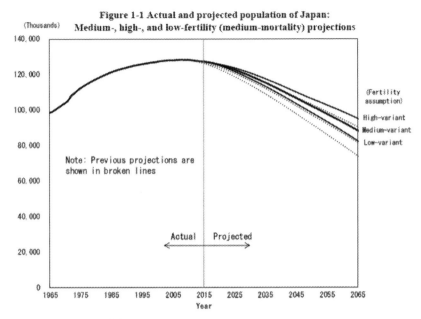

Figure 0.2 Actual and projected population of Japan: medium-, high-, and low-fertility.
Source: National Institute of Population and Social Security Research, 2017, www.ipss.go.jp/pp-zenkoku/e/zenkoku_e2017/g_images_e/g-tablese/pp29gg0101e.gif, accessed July 3, 2020.

the population crisis, but also because they illustrate how population facts are problematized as public issues and what solutions are considered viable. In fact, half of *The Future Chronology* is dedicated to the section, "10 Prescriptions for Saving Japan." For instance, one of these "prescriptions," following a proposal made in January 2017 by a working group of gerontologic scholars, is to raise the minimum age for the demographic category of "the elderly" (*kōreisha*) from the current sixty-five years old to seventy-five.[9] Using this new definition, Japan's aging population problem would appear less pressing.[10] The *White Paper*, with its overall theme of considering "a social model with which to overcome the population aging," contained an entire section mapping out policy suggestions.[11] Though less provocative than the "prescriptions" proposed

[9] Kawai, *Mirai no nenpyō*, 162–65.
[10] Ibid., 163–65.
[11] Shiozaki, quoted in Kōseirōdōshō, "*Heisei 28-nendoban*," 105–224.

by *The Future Chronology*, the *White Paper* also made recommendations, such as the creation of a system to encourage the employment of the elderly population.[12] Both publications clearly illustrate how the sense of a social crisis generated by the demographic facts drove social critics and the government to come up with suggestions for policy measures. They also display how making policy recommendations is self-evident in this type of publication.

However, historically speaking, this way of presenting, interpreting, and responding to population facts is relatively new. As I show in this book, the idea that humans could be represented in decimal numbers – a practice we regard as natural today – was literally foreign to many Japanese intellectuals in the early 1860s. The same was the case with the notion that the numerical presentation and analysis of a population could inform the characteristics and chronological trend of a society. Though many local rulers prior to the 1860s collected the details of people's life events (e.g., births and deaths), as well as their status within the family, for the purpose of religious control, corvée, or taxation, the practice of collecting this demographic information was based on neither the mathematical nor the sociological understanding of population we are familiar with today. Our assumptions about the inherent link between numerical data, population trends, and social phenomena, as well as the roles assigned to government and public intellectuals to come up with solutions to the population problems by means of policy, have been constructed gradually throughout Japan's history over the past 160 years or so.

This book illustrates how these assumptions and roles were normalized alongside the changing contours of Japan as a modern sovereignty by focusing on the critical, yet hitherto overlooked, role that science played in the Japanese state's attempts to govern its population for the sake of its sovereignty. It analyzes how discourses related to the Japanese population mobilized scientists to conduct policy-oriented population research and state administrative activities and how, in turn, their practices and knowledge shaped the mode of governance. It also considers how population scientists constructed medico-scientific disciplines and their own professional identities through policymaking, while the government's political agenda, which required the redefinition and management of its population – be it for nation-building, colonialism, war mobilization, or postwar reconstruction – shaped the contours of the scientific fields they wished to promote. This book demonstrates that the creation of the human and social science of the population, as well as the state sovereignty predicated on population management, had a symbiotic

---

[12] Ibid.

relationship, each driven by surrounding ideologies, institutional agendas, sociopolitical and material conditions, and personal motivations.

To elaborate on these points, this book draws on interrelated lines of inquiry in disciplines and subfields that have been particularly beneficial for my analysis. One is historical studies on science, technology, and medicine in modern Japan.[13] The studies have stressed the mutually exclusive relationship between the development of modern science, knowledge production, and the formation of Japan as a modern state and empire.[14] Within the field, works on sciences dealing with human subjects, for instance anthropology and medicine, have specifically considered the implications of scientific knowledge and practice for changing modes of governing people.[15] They have illustrated how scientific theories about people and race resulted from research activities provided foundations for the fundamental decisions Japan made about nation- and empire-building, for instance, where to draw the "boundaries of the Japanese," and who would need policy interventions.[16] They have also demonstrated how the knowledge production process reinforced, and was reinforced

[13] The recent outpouring of studies in this field has been so impressive that I can only offer samples here. Toru Sakano and Togo Tsukahara, eds., *Teikoku nihon no kagaku shisōshi* (Keiso shobo, 2018); Osamu Kanamori, ed., *Meiji, Taisho-ki no kagaku shisōshi* (Keiso shobo, 2017); Toru Sakano, *Teikoku wo shiraberu: Shokuminchi fīrudo wāku no kagakushi* (Keiso shobo, 2016); Osamu Kanamori, ed., *Showa zenki no kagaku shisōshi* (Keiso shobo, 2011); David G. Wittner and Philip C. Brown, eds., *Science, Technology, and Medicine in the Modern Japanese Empire* (London: Routledge, 2016); Morris Low, ed., *Building a Modern Japan: Science, Technology, and Medicine in the Meiji Era and Beyond* (New York: Palgrave Macmillan, 2005). See also Nihon Kagakushi Gakkai, ed., *Kagakushi jiten* (Maruzen, 2021); Shigeru Nakayama, Kunio Goto, and Hitoshi Yoshioka, eds., *A Social History of Science and Technology in Contemporary Japan*, vols. 1–4 (Melbourne: Trans Pacific Press, 2001–06).

[14] In addition to the works listed above, see the series Teikoku no gakuchi published by Iwanami Shoten in 2006–07.

[15] See Miriam Kingsberg Kadia, *Into the Field: Human Scientists of Transwar Japan* (Stanford: Stanford University Press, 2020); Jaehwan Hyun, "Racializing Chōsenjin: Science and Biological Speculations in Colonial Korea," *East Asian Science, Technology and Society* 13, no. 4 (December 2019): 489–510; Kristin A. Roebuck, "Japan Reborn: Mixed-Race Children, Eugenic Nationalism, and the Politics of Sex after World War II" (PhD diss., Columbia University, 2015); Arno Nanta, "Physical Anthropology and the Reconstruction of Japanese Identity in Postcolonial Japan," *Social Science Japan Journal* 11 (2008): 29–47; Toru Sakano, *Teikoku nihon to jinrui gakusha: 1884–1952 nen* (Keiso shobo, 2005); Tessa Morris-Suzuki, "Ethnic Engineering: Scientific Racism and Public Opinion Surveys in Midcentury Japan," *Positions: East Asia Cultures Critique* 8, no. 2 (November 2000): 499–529; Tessa Morris-Suzuki, "Debating Racial Science in Wartime Japan," *Osiris* 13, no. 1 (January 1998): 354–75; Eiji Oguma, "Tsumazuita junketsu shugi: Yūseigaku seiryoku no minzoku seisakuron," *Jyōkyō* 5, no. 11 (1994): 38–50.

[16] Eiji Oguma, *"Nihonjin" no kyōkai: Okinawa, Ainu, Taiwan, Chōsen, shokuminchi shihai kara fukki undō made* (Shin'yōsha, 1998).

by, preexisting socioeconomic conditions, institutional constraints, the political agenda affecting Japan's nation- and empire-building exercise, and the social norms and cultural work on the Japanese in relation to the marginalized Other.[17] At the same time, they have also shown that the interplay between the knowledge produced by human sciences, citizenship, and nationhood was often messy, in part because of the inconsistencies in how people as constituents of the Japanese nation, were articulated in the language with multiple yet overlapping expressions: *kokumin, shinmin, minzoku,* and *jinshu.*[18] The elephant in the room is "population" *(jinkō),* a concept coterminous with all these expressions. Like other categories that undergirded concepts of nationhood, the notion of population was omnipresent in the areas of public life that touched on the issues of citizenship, national/racial identification, sovereignty, and the disciplining of bodies throughout Japan's modern history. However, surprisingly few studies have focused on the concept of population and the sciences that engaged with it.[19]

---

[17] See Lawrence Yoshitaka Shimoji, *"Konketsu" to "nihonjin": Hāfu, daburu, mikkusu no shakaishi* (Seidosha, 2018); Christopher P. Hanscom and Dennis C. Washburn, eds., *The Affect of Difference: Representations of Race in East Asian Empire* (Honolulu: University of Hawai'i Press, 2016); Noriaki Hoshino, "Racial Contacts Across the Pacific and the Creation of Minzoku in the Japanese Empire," *Inter-Asia Cultural Studies* 17, no. 2 (April 2016): 186–205; Tessa Morris-Suzuki, *Borderline Japan: Foreigners and Frontier Controls in the Post-War Era* (Cambridge: Cambridge University Press, 2010); Tessa Morris-Suzuki, *Re-Inventing Japan: Time, Space, Nation* (Armonk, NY: M.E. Sharpe, 1998); Michael A. Weiner, *Japan's Minorities: The Illusion of Homogeneity* (London and New York: Routledge, 1997); Naoki Sakai, "Ethnicity and Species/Radical Philosophy," *Radical Philosophy* 95, no. 1 (June 1999), www .radicalphilosophy.com/article/ethnicity-and-species; Eiji Oguma, *Tan'itsu minzoku shinwa no kigen: "Nihonjin" no jigazō no fukei* (Shin'yōsha, 1995).

[18] The word *kokumin* can be conventioanlly translated as "citizens" and "nationals," *shinmin* "imperial subjects," *minzoku* "ethnic [group]," and *jinshu* "race," but the boundaries of these concepts were blurry precisely because biology, culture, and nationality were conflated in the articulation of national identity. Morris-Suzuki, *Re-Inventing Japan*, 79–109.

[19] The exception is several works that have emerged recently. Hiroshi Kojima and Kiyoshi Hiroshima, eds., *Jinkō seisaku no hikakushi: semegiau kazoku to gyōsei* (Nihon Keizai Hyouronsha, 2019); Sidney Xu Lu, *The Making of Japanese Settler Colonialism: Malthusianism and Trans-Pacific Migration, 1868–1961* (Cambridge: Cambridge University Press, 2019); Sujin Lee, "Problematizing Population: Politics of Birth Control and Eugenics in Interwar Japan" (PhD diss., Cornell University, 2017); Jinkyung Park, "Interrogating the 'Population Problem' of the Non-Western Empire: Japanese Colonialism, the Korean Peninsula, and the Global Geopolitics of Race," *Interventions* 19, no. 8 (November 2017): 1112–31; Aya Homei, "The Science of Population and Birth Control in Post-War Japan," in *Science, Technology, and Medicine in the Modern Japanese Empire*, eds. David G. Wittner and Philip C. Brown (London: Routledge, 2016), 227–43; Akiko Ishii, "Statistical Visions Of Humanity: Toward a Genealogy Of Liberal Governance in Modern Japan," (PhD diss., Cornell University, 2013).

This book is built on the wealth of knowledge created by scholarship and examines how the scientific communities or disciplines predicated on the modern concept of population were formed while engaging in activities and producing knowledge that contributed to the Japanese state's attempts to govern its people. In so doing, the book elaborates on the relationship between population science and modern governance, which relied on the state's population management effort.

Modern Governance and Issues with "Demography"

A great advantage to focusing on population science for studying modern governance is that its subject matter is thoroughly entangled with running the state as a modern, sovereign power. According to the theory of governmentality first elaborated by Michel Foucault, population was a central object of power that shaped the specific ways people as "species bodies" were governed in the modern era.[20] Inspired by Foucault's canonical theory, a number of works on modern Japanese history have also depicted how people's lives became subjected to modern power through the enhancement, disciplining, and management of bodies and health through the diffused network that prevailed in the government, schools, hospitals, and other nonstate and private institutions, at the specific historical moment that Japan was rising as a modern nation-state with imperial aspirations.[21] The population depicted in this literature is primarily carnal, made up of individuals with quotidian bodily needs, such as demands for better food, sex, and sleep, the desire or duty to stay healthy and have robust offspring, and yearnings for a better life in general.[22]

However, more recent literature has pointed out that the areas of entanglement between the population and modern governance were far

---

[20] Michel Foucault, *The Foucault Reader*, ed. Paul Pabinow, Penguin reprint (London: Penguin Books, 1991), 258–64.

[21] See Hideto Tsuboi, ed., *Sengo nihon wo yomikaeru*. Volume 4 of *Jendā to seiseiji* (Kyoto: Rinsen shoten, 2019); Jin-kyung Park, "Corporeal Colonialism: Medicine, Reproduction, and Race in Colonial Korea" (PhD diss., University of Illinois Urbana-Champaign, 2008); Hiroko Takeda, *The Political Economy of Reproduction in Japan: Between Nation-State and Everyday Life* (London and New York: RoutledgeCurzon, 2005); Yuki Terazawa, "The State, Midwives, and Reproductive Surveillance in Late Nineteenth – and Early Twentieth-Century Japan," *US-Japan Women's Journal* 24 (2003): 59–81.

[22] E.g., Rickie Solinger and Mie Nakachi, eds., *Reproductive States: Global Perspectives on the Invention and Implementation of Population Policy* (Oxford and New York: Oxford University Press, 2016).

more exhaustive, precisely because of the vast range of associations the idea of population evoked throughout modern history.[23] Population, in the words of historian Alison Bashford, "touched on almost everything: international relations; war and peace; food and agriculture; economy and ecology; race and sex; labor, migration, and standards of living."[24] This book shows that this applied to modern Japanese history, as evidenced by the representation of population as "national power" (*kokuryoku*), for example. This notion, on the one hand, embraced the Foucauldian formulation of the corporeal population, which stressed its capacity to expand or perish as "species bodies." The population shaped the "politics of life," affecting issues related to the workings of the human body – sex, race, food, health, etc. On the other hand, population as national power was also described in abstract terms, as itself constituting power in the sense of physics. Population imagined in this way was described with terms such as "military force," "workforce," "manpower," and "human resources." It supported the modern military and capitalist (and during the war, controlled) economy that the Japanese government endorsed in the process of nation- and empire-building. Population problems based on this conceptualization dovetailed with issues related to political economy (e.g., labor, urban-rural divide, poverty, migration, and security). Furthermore, the interpretation of population dynamics changed over time depending on the context. For instance, critics in the late 1930s celebrated a large population size as embodying the "racial power" that would bring economic and political prosperity to the nation-state-empire at war (Chapter 4).[25] However, in the decades prior to and after, the

---

[23] The Population Knowledge Network, *Twentieth Century Population Thinking: A Critical Reader of Primary Sources* (Abingdon, Oxon and New York: Routledge, 2015). For examples of works that follow this wide interpretation of population, see Alison Bashford, *Global Population: History, Geopolitics, and Life on Earth* (New York: Columbia University Press, 2014); Michelle Murphy, "Economization of Life: Calculative Infrastructure of Population and Economy," in *Relational Architectural Ecologies Architecture, Nature and Subjectivity*, ed. Peg Rawes (Florence: Taylor and Francis, 2013), 139–55; Thomas Robertson, *Malthusian Moment: Global Population Growth and the Birth of American Environmentalism* (New Brunswick: Rutgers University Press, 2012); Nick Cullather, *The Hungry World: America's Cold War Battle Against Poverty in Asia* (Cambridge, MA and London: Harvard University Press, 2010).

[24] Bashford, *Global Population*, 5.

[25] For understanding Japan as a modern sovereignty in the period leading up to 1945 when nation-building and empire-building efforts were often entangled with each other, Tomoko Akami's work has been particularly useful. According to Akami, a nation-state and an empire should be presented as a "unit of analysis" and an "actor in international politics" to foster historical studies that show a "mutually constitutive relationship between metropolitan centres and colonial peripheries." Tomoko Akami, "The Nation-State/Empire as a Unit of Analysis in the History

same phenomenon was stigmatized as "overpopulation" (*kajō jinkō*), a "surplus" that could disrupt economic, social, and political orders (Chapters 3, 5, and 6). Because of population's far-reaching implications and multiple meanings and, in the case of modern Japan, because these associations were matters of national importance at one time or another, the domains of people's lives that were subjugated to state intervention under the name of population were expansive. For historians, the comprehensive manner in which population was entwined with national affairs is what makes population a great subject for studying the specific mode of modern governance that Japan organized while yearning to become a "modern" sovereign power.

This book incorporates this expansive rendering of population within the study of the population science that developed in Japan. When doing this, I avoid using the established nomenclature, "demography" (*jinkōgaku*). Instead, I adopt a more extensive – perhaps to specialists somewhat unconventional – definition that also includes diverse scientific, medical, and healthcare fields and practices that are less immediately associated with population studies today. This statement might come as puzzling to those who know the field of demography well, because demography is actually one of the most inclusive academic disciplines. In fact, interdisciplinarity – or in the words of Henrich Hartmann and Corinna R. Unger, "transdisciplinary character" – is what has defined the field from the onset.[26] This is certainly the case in Japan, too, as evidenced by the vast array of disciplines introduced in the canonical publication of the Population Association of Japan (PAJ), *The Population Encyclopedia*.[27]

of International Relations: A Case Study in Northeast Asia, 1868–1933," in The Nation State and Beyond: Governing Globalization Processes in the Nineteenth and Early Twentieth Centuries, eds. Isabella Löhr and Roland Wenzlhuemer (Berlin and Heidelberg: Springer, 2013), 177–79. For this reason, I use the expression, "nation-state-empire" to refer to the Japanese sovereignty in the chapters dealing with the period leading up to 1945. For other chapters dealing with the period after 1945, which saw the demise of the Japanese Empire, the works of Toyomi Asano, Barak Kushner, and Sherzod Muminov have been helpful. These works offer a useful framework with which to see Japan as a modern sovereignty in relation to the shifting regional geopolitical dynamics. See also Barak Kushner and Sherzod Muminov, eds. *The Dismantling of Japan's Empire in East Asia: Deimperialization, Postwar Legitimation and Imperial Afterlife* (London: Routledge, 2017); Toyomi Asano, ed., *Sengo nihon no baishō mondai to higashi ajia chiiki saihen* (Tokyo: Jigakusha shuppan, 2013).

[26] Heinrich Hartmann and Corinna R. Unger, eds., *A World of Populations: Transnational Perspectives on Demography in the Twentieth Century* (New York: Berghahn Books, 2014), 2.

[27] Jinkō Daijiten Henshū Iin, ed., *Jinkō daijiten* (Heibonsha, 1957).

I argue that demography would be a less productive heuristic device for this book for two main reasons: The first is the fact that demography as a scientific discipline did not exist in Japan during most of the period covered in this book. If we apply the concept of scientific discipline offered by the sociology of science and regard factors such as the formation of organizational structures and the foundation of communication through standardized publication as markers of a disciplinary formation in modern science, in Japan, demography certainly did not exist as a scientific discipline until the middle of the twentieth century.[28] Yet, this did not mean communities dedicated to, and mobilized by, scientific inquiries into the concept of population were absent prior to this period. On the contrary, as the book shows, wide-ranging medico-scientific communities were formed through engagement with population politics from the 1860s onward. Furthermore, it was their various modes of engagement that provided a foundation for the rise of demography in the mid-twentieth century. Thus, rather paradoxically, by dodging the term demography, this book also historicizes the discipline of demography as it appears today.

Linked to the one above, the second reason behind the decision to stay away from the term "demography" is because it masks the complexities that shaped the interactions between medico-scientific activities and the state's population management efforts. As the book demonstrates, if I used "demography" to approach the book's subject, the convoluted history, which can be captured only through the analysis of practice, would be difficult to grasp. For instance, if I examine the vital statistics developed under the Meiji state only through the lens of demography, I would pass over the significant role medical midwifery played in this history. In turn, if we examine the day-to-day paperwork and regulatory activities involved in the collection of vital statistics – as I do in Chapter 2 – we can see medical midwifery's contributions to both the making of population statistics and to population statistics' role in the governing of Japan's population. This latter approach enables the book to clarify the mostly parallel, yet at times intertwined, relationship between the formation of medical midwifery, the modern administrative system, and population statistics; the part of the history critical for understanding the role of science in the Japanese state's engagement with population politics, which has been obscured thus far.

---

[28] Rudolf Stichweh, "The Sociology of Scientific Disciplines: On the Genesis and Stability of the Disciplinary Structure of Modern Science," *Science in Context* 5, no. 1 (1992): 3–15.

Instead of demography, I follow some critical studies of demography and use the expressions "population science" or the "science of population."[29] However, instead of simply referring to a disciplinary category, I also use these terms to depict the site where the sciences catalyzed by the modern concept of population intersected with the state-led population management endeavor. So, among the sciences represented by the terms are included more obvious disciplinary fields, such as population statistics, but because population's extensive links with the issues of labor, industrial production, the distribution of wealth, and migration were the subjects of national policy, policy-relevant debates on these topics, in which diverse scholar-advisors such as economists and social policy specialists participated, are also included in the definition of population science in this book. Finally, due to population's corporeal quality, activities related to reproductive medicine, public health, and social and welfare policy, triggered for the sake of population management, are also included. In fact, this take on population science makes it impossible for the book to cover every field and activity linked to the natural, human, or social sciences of population. Yet, it at least permits me to show the diverse modes of interaction between medico-scientific activities and the statecraft that affected the science of population in modern Japan.

But, what about science? How did science interact with statecraft? The following section explains the strands of inquiry that address these questions – to which this book is indebted.

Conceptual Frameworks
for Understanding "Science"

I draw from two bodies of research to explain how the science of population and state-led population-governing exercises mutually interacted in Japan.[30] The first is studies that clarified the instrumental role of population works vis-à-vis modern governance, and the second is the history of science that intersects with science and technology studies (STS).[31] The works in the first group, in part built on Foucault's work on governmentality, have long argued that population works – or, more

---

[29] See Minami Ryōsaburō's definition in the prologue of Jinkō Daijiten Henshū Iin, *Jinkō daijiten*; and Susan Greenhalgh, *Just One Child: Science and Policy in Deng's China* (Berkeley: University of California Press, 2008).

[30] This is similar to the approach adopted by Greenhalgh, *Just One Child*, 6–10.

[31] Peter Dear and Sheila Jasanoff, "Dismantling Boundaries in Science and Technology Studies," *Isis* 101, no. 4 (2010): 759–74; Lorraine Daston, "Science Studies and the History of Science," *Critical Inquiry* 35, no. 4 (January 2009): 798–813.

generally, systematic explorations of demographic facts – acted as an "instrument of modern governance" by providing rhetorical devices with which to capture individuals or groups as legally and socially contained population groups amenable to political interventions.[32] For instance, according to Tong Lam, the knowledge production accompanying the national census in Republican-era China "fundamentally transformed the nature of governance by making the complex human world appear to be knowable and manipulable in ways that were not possible before."[33] In a similar vein, population works in Japan in the nineteenth century normalized the statistical representation of individuals as closely related to the health and wealth of a nation at the specific moment when the new political elites were striving to construct a strong nation-state. In the 1930s, population works constructed images of enemy populations in a way that was legible to the Japanese nation-state-empire at war. Therefore, based on this scholarship, this book examines how population science shaped the narrative of the modern political subject and, in so doing, highlights the critical role population science played, not only as a technology of the nation-state and empire, but also in the process of nation- and empire-building.

The history of science that overlaps with STS is the second field to which this book is indebted.[34] Specifically, the following three frameworks have been beneficial for my analysis: the quantification of social facts, the coproduction of natural and social orders, and micropolitics. First, engaging with the sociology of quantification, historians of science have clarified how trust in numbers in legal and social transactions shaped epistemologies and methodologies in the modern science of statistics, while statistics verified social facts by quantification.[35] This book, by incorporating this framework, also examines how the formation of population science was predicated on the ways in which Japanese society and government conferred authority upon the act of counting numbers to recognize facts about human endeavors. In turn, it also depicts how the specific way of discerning the population's relationship with human endeavors mobilized

---

[32] Hartmann and Unger, *A World of Populations*, 4.

[33] Tong Lam, *A Passion for Facts: Social Surveys and the Construction of the Chinese Nation-State, 1900–1949* (Berkeley: University of California Press, 2011), 1.

[34] Those familiar with STS literature may want to skip this part of the introduction.

[35] Alain Desrosières and Camille Naish, *The Politics of Large Numbers: A History of Statistical Reasoning* (Cambridge, MA: Harvard University Press, 1998); Theodore M. Porter, *Trust in Numbers: The Pursuit of Objectivity in Science and Public Life* (Princeton: Princeton University Press, 1995); Theodore M. Porter, *The Rise of Statistical Thinking 1820–1900* (Princeton: University Press, 1986). For a recently published case study, see Arunabh Gosht, *Making It Count: Statistics and Statecraft in the Early People's Republic of China* (Princeton: Princeton University Press, 2020).

the government to act on the population. On this point, Ishii Akiko's work, which argues that what she called the "statistical vision" formed a foundation for liberal governance in Japan, is particularly useful.[36] While borrowing from Ishii's insightful work, I also show that the ways in which this "statistical vision" interacted with the mode of governance in Japan was not always constant, but was susceptible to change depending on the institutional and political context. The "statistical vision" and modern governance might have almost always been coconstituted, but the ways they were and the effects this had were contingent upon history.

Linked to this point, the book also builds on the second, STS framework mentioned above. Scientific knowledge, in the words of Sheila Jasanoff, "embeds and is embedded in ... all the building blocks of what we term the *social*" because "the ways in which we know and represent the world (both nature and society) are inseparable from the ways in which we choose to live in it."[37] More recently, the coproduction concept has been applied to the examination of the forms of "scientific sense-making" in the context of a political regime.[38] This book adopts the coproduction idiom to analyze how the science of population and the state's effort to govern the population established a mutually exclusive relationship in Japan. In so doing, I describe two specific ways in which natural and social/political orders were coproduced in the science of population. First, I show how knowledge about the naturalized concept of population, constructed through policy-relevant scientific research, was coproduced with vectors that consolidated existing social orders. Second, the book also depicts how population science was coproduced with the consolidation of state power in order to intervene in people's lives via policymaking. For instance, Chapters 5 and 6 describe how the population research after World War II that was accountable for the national birth control policy inscribed certain ideas of class and nationality, as well as economic rationale, in the representation of the research subjects, and how the demographic knowledge produced as a result of the research served to perpetuate social hierarchies that implicitly privileged the heteronormative sexual behaviors of Japanese married couples. At the same time, these policy-oriented population studies contributed to the rise of a scientific field around the Institute of Population Problems at the Ministry of Health and Welfare and the Department of Public Health Demography

---

[36] Ishii, "Statistical Visions of Humanity."
[37] Sheila Jasanoff, "The Idiom of Coproduction," in *States of Knowledge: The Co-Production of Science and the Social Order*, ed. Sheila Jasanoff (London and New York: Routledge, 2004), 2–3.
[38] Greenhalgh, *Just One Child*, 17–18.

at the Institute of Public Health, while justifying the national policy that aimed to popularize birth control practices among people for the sake of postwar reconstruction. In the case of Japan, too, the making of science, social orders, and the political regime were clearly coconstituted.

Finally, the history of science/STS helps my analysis by offering the understanding of science as essentially a human endeavor buttressed by micropolitics.[39] This view makes it easier for me to contextualize the history of population science in Japan using the framework provided by more recent work on modern governance and the state, which stresses the importance of locally grounded discourse, practice, system, and human agency. "Governing," argues historian Tom Crook, is a "matter of discourse and practice: a combination of cultural-intellectual *and* material-logistical forces," while the modern state is "rooted in ... intricate systems and the work of the myriad agents that operate and maintain them."[40] Taken together, this book emphasizes that the demographic discourse and knowledge produced for the governing of Japan's population was at times informed by fortuitous human factors that were inscribed in the everyday actions of collecting, documenting, analyzing, and storing numerical demographic data. In addition, the personal, institutional, and material conditions surrounding the population scientists sometimes brought unexpected results into their research. This individually and locally situated everyday practice shaped the contours of the population science that constituted modern governance. This perspective thus aims to interrupt the smooth narrative of the relationship between science and governance implied by governmentality studies by depicting scientific practice as a human practice that contains elements of messiness and randomness, even as it appears to be loyally fulfilling an ascribed utilitarian role for the state.

Official Administration, Bureaucrats, and the Transnational: What the Science for the Governing of Japan's Population Reveals

Narrating the story of the science for governing Japan's population, this book aims to contribute to the fields of modern Japanese history and the history of population science.[41] Below, I explain how I will achieve this

---

[39] Bruno Latour, *Reassembling the Social: An Introduction to Actor-Network-Theory* (Oxford: Oxford University Press, 2007); Bruno Latour, *Laboratory Life: The Construction of Scientific Facts* (Princeton: Princeton University Press, 1986).

[40] Tom Crook, *Governing Systems: Modernity and the Making of Public Health in England, 1830–1910* (Berkeley: California University Press, 2016), 9, 11.

[41] For the most recent works on the historical study of the formation of demography as a scientific discipline, see Heinrich Hartmann and Ellen Yutzy Glebe, *The*

objective by elaborating on three facets of the interplay between science and the state-led population management that unfolded in the Japanese context. In so doing, I incorporate keywords that highlight the critical elements that shaped the main arguments I wish to present in this book: official administration, bureaucrats, and the transnational.

First, in this book, I stress that the official administration acted as a central site where the two modes of coproduction mentioned above determined population science's relation with the Japanese state and empire. To further this line of analysis, the point of vital statistics historian Libby Schweber that the boundaries between science and administration were fluid and even endorsed the "abandonment of a priori distinctions on science on the one hand and politics, administration, and the state on the other" is useful.[42] In the case of Japan, the administrative office supporting Japanese colonial rule in Taiwan helped the technical development of population statistics in the metropole. The importance of the official administration was additionally compounded by the fact that, from the mid-1910s, population research became integrated into the official effort to solve population issues by means of policies. Similar to the population science developed in Western Europe, an official administration certainly buttressed the interplay between science and the state governance of population as a site of knowledge production and as where the state governance of population was planned and executed.

However, the Japanese case was distinctive because the role of the official administration was ascertained in a political context in which the profile of the nation-state – for which the official administration was accountable – was itself in flux a number of times in its modern history. One crucial point to note for the study of population science in Japan is that the science-state interplay emerged and was elaborated on during a specific moment in world history: When Japan had just entered international politics, and thus its status and future were yet unknown. When the official administration began to collect vital statistics in the 1870s, Japan was a novice in world politics, which was dominated by western colonial powers. Its status as an independent, sovereign nation was precarious due to the successive unequal treaties the country signed with

Body Populace Military Statistics and Demography in Europe before the First World War (Cambridge: MIT Press, 2019); Hartmann and Unger, *A World of Populations*; Karl Ittmann, Dennis D. Cordell, and Gregory Maddox, *The Demographics of Empire: The Colonial Order and the Creation of Knowledge* (Athens: Ohio University Press, 2010.

[42] Libby Schweber, *Disciplining Statistics: Demography and Vital Statistics in France and England, 1830–1885* (Durham: Duke University Press, 2006), 11.

the United States and Western European nations in the 1860s. When the Government-General of Taiwan decided to conduct population census work, Japan's authority as the only non-western colonial ruler was far from established. The Ministry of Health and Welfare authorized the state-endorsed birth control surveys and pilot projects in the late 1940s, when Japan's political profile vis-à-vis countries in the Asia Pacific region was fundamentally reconfigured. In contrast to England and France, for instance, where population science developed in relation to existing states, the science of population in Japan developed *along with* Japan as a political unit. In this context, Japan's state-making/state-running process relied heavily on the science of population. It provided tools that would facilitate the governing of populations. It stabilized knowledge about a population as a governable entity. It offered technical support to the official effort to discipline the populations in the metropole and colonies. It justified population policies that aimed to promote national productivity and colonial management. Thus, for the history of population science, the Japanese case not only illustrates the fluid boundaries between science, politics, and state administration but also shows how the science was an integral part of the process of nation- and empire-building. Population science was a constitutive force in the formation of Japan's unique position in the world as the only nonwhite, non-Christian modern nation and empire and, during the Cold War era, as an active player in constructing the "buffer zone" in East Asia. For modern Japanese history, the story of population science confirms the quintessential and continuous role that science played throughout Japan's transformations as a political entity since the 1860s – as a modern state, an empire, an occupied nation, and a postwar "reconstructed" democratized state – contributing to the normalization of the use of numerical demographic facts to govern its subjects.

The centrality of state administration for the development of population science in modern Japan also points to the second element that critically shaped the trajectory of the science-state interplay in modern Japan: The participation of bureaucrats in policy-relevant research and policymaking. Recently, historical works focusing on "technical bureaucrats" (*gijutsu kanryō*) have illustrated how they contributed directly to nation-building from the Meiji period onward by applying their technical expertise in fields such as heavy industry, railway, and mintage, which were deemed essential for the formation of a modern state.[43] These

[43] Hiroki Kashihara, *Meiji no gijutu kanryō* (Chuokoron-Shinsha, 2018); Aaron Stephen Moore, *Constructing East Asia: Technology, Ideology, and Empire in Japan's Wartime Era, 1931–1945* (Stanford: Stanford University Press, 2013); Hiromi Mizuno, *Science for the Empire: Scientific Nationalism in Modern Japan* (Stanford: Stanford General,

studies have also depicted how the group of technical bureaucrats oriented toward engineering came to constitute a powerful political force from the 1910s onward and ultimately shaped the state's science policy and colonial administration based on the vision buttressed by the mixture of nationalism and technocracy.[44] But, scientifically trained bureaucrats involved in government administration and policymaking were in fact a diverse group who went beyond the confinement of engineering.[45] Furthermore, the diversity is characterized in the ways in which their participation in these official activities constructed certain modes of delivering science and technology in Japan. To illustrate these points, this book depicts the bureaucrats as experts who engaged in modern science and medicine, and who thrived within the domain of statecraft. For this part of the analysis, population science provides a particularly effective lens precisely because of its proximity to the official administration and the vast range of subject areas it touched on. Technical and research bureaucrats mobilized for state and colonial administration for the sake of population management were indeed a diverse group. They included Mizushina Shichisaburō, a mid-ranking statistician with initial training in meteorology; Tachi Minoru, an up-and-coming social scientist with an academic background in economics; Shinozaki Nobuo, a research bureaucrat at the Institute of Population Problems who was initially trained in anthropology; and, finally, one of the most renowned medical technocrats, racial hygienists, and political advisors in wartime and postwar Japan, and the Director-General of the Institute of Public Health, Koya Yoshio. The book illustrates how these bureaucrats, each with distinctive intellectual backgrounds and ranks within the state bureaucracy, shaped different aspects of population science while engaging in the governance of Japan's population through their work as bureaucrats and scientific experts.

2009); Shoichi Oyodo, *Gijutu kanryō no seiji sankaku: Nihon no kagaku gijutu gyōsei no makuaki* (Chuokoron-sha, 1997), James R. Bartholomew, "Science, Bureaucracy and Freedom in Meiji and Taishō Japan," in Tetsuo Najita and J. Victor Koschmann eds. Conflict in Modern Japanese History (Ithaca: Cornell University Press, 1982), 295–341. Historian Mizuno Hiromi translated *gijutsu kanryō* as "technology-bureaucrats," but I have chosen to translate it as "technical bureaucrats." I believe my translation, which is broader in meaning than Mizuno's technologically focused translation, better captures the diverse backgrounds of the bureaucrats who served the state under this title. For Mizuno's translation, see Mizuno, *Science for the Empire*, 20.

[44] Moore, *Constructing East Asia*, 65–75.

[45] See Makino Kuniaki, *Senjika no keizai gakusha: Keizaigaku to sōryokusen* (Chuokoron-Shinsha, 2020); Laura E. Hein, *Reasonable Men, Powerful Words: Political Culture and Expertise in Twentieth-Century Japan* (Berkeley and London: University of California Press, 2004); Hiroyuki Takaoka, *Sōryokusen taisei to "fukushi kokka": Senjiki nihon no "shakai kaikaku" kōsō* (Iwanami Shoten, 2011).

A benefit of studying these bureaucrat-scientists is that it allows us to illustrate the diverse range of actions involved in the governing of Japan's population. Technical bureaucrats engaged in population policies and policy-relevant population studies were not just bureaucrats or scientists but often had other roles as policy advisors, public intellectuals, and activists. For this reason, they took part in a wide range of activities for the sake of population management, such as fieldwork, data collection and preservation, draft-writing, meetings, and networking. Thus, through the analysis of population bureaucrats, we can confirm that micropolitics informed the relationship between science and statecraft. State-led population governance, including policymaking, was more than a mere intellectual or political exercise, as it tends to be depicted, but took place alongside scientific activities that involved material production, circulation, paperwork, and legwork.

Another advantage of analyzing the technical bureaucrats is that it complicates a common understanding of bureaucrats as docile "servants of the state." As mentioned above, the bureaucrats participating in the science of population had multiple roles. For the most part, these multiple identities did not cause conflicts with their official duties as bureaucrats, but at times they did. In other words, precisely because of their multiple identities, the technical bureaucrats sometimes acted in ways that were not entirely aligned with official interests. Thus, the analysis of population bureaucrats, which effectively shows the elements of dissonance in the interactions between science and state governance, is another way to complicate the smooth narrative of science-statecraft interplay.

The focus on state administration and bureaucrats as state institution/actors does not necessarily mean the book privileges a domestic perspective on the subject matter. On the contrary – and this is the third point I would like to make – transnational forces molded the interplay between population science and the governing of Japan's population via nation-centered discourse and reproductive policies.[46] Because the science of population became thoroughly embedded in the transnational population control movement that was realized through fertility regulations in the middle of the twentieth century, the story of population science in

---

[46] Aiko Takeuchi-Demirci, *Contraceptive Diplomacy: Reproductive Politics and Imperial Ambitions in the United States and Japan* (Stanford: Stanford University Press, 2018); Aiko Takeuchi, "The Transnational Politics of Public Health and Population Control: The Rockefeller Foundation's Role in Japan, 1920s–1950s," Rockefeller Archive Center (RAC) Research Reports Online (2009), https://rockarch.org/publications/resrep/takeuchi.pdf.

Japan during this period effectively highlights elements of transnational medico-scientific exchanges that constituted what, on the surface, looked like categorically domestic efforts to govern the population in Japan. In turn, through the science of governing Japan's population, Japanese population scientists during the period became important constituents who shaped the transnational effort to curb the growth of the world population. Chapter 6 elaborates on the exchange between Japanese population scientists based at the Department of Public Health Demography at the Institute of Public Health and their colleagues, mostly in the United States and India, and depicts how the transnational exchange acted as a critical background for the production of knowledge about abortion and birth control that directly contributed to the domestic policy within Japan, and simultaneously to the transnational discussion on population control in Asia. With this case study, the book not only points out these transnational elements that participated in the domestic politics of population but also complicates the category of "the national" articulated in policymaking. By showing how transnational connections and vectors shaped the specific ways population science interacted with state politics in Japan, this book enriches the growing body of scholarship that contextualizes modern Japanese history within the framework of transnational history.[47]

Scope and Structure

To fulfill the abovementioned objectives, the book's main text focuses on the long period between the 1860s and the 1960s. It begins with the decade that witnessed the rise of the idea of population, which had a lasting impact on the ways in which sovereignty, society, and subjecthood were elucidated and enacted by the new generation of officials and intellectuals in Japan. The book then ends with the decade in which an even broader approach to engaging with demographic issues, with more explicit links to social welfare and international cooperation in family planning, was institutionalized within the government. This *longue durée* approach effectively illustrates how the symbiotic relationship between science making and politics, which was woven into the governing of the population, developed in tandem with the formation of a modern sovereign state in Japan. At the same time, with this scope, the book effectively problematizes the model of historical development as linear progress by showing the different social and political conditions and events that

---

[47] Sheldon Garon, "Transnational History and Japan's 'Comparative Advantage,'" *Journal of Japanese Studies* 43, no. 1 (2017): 65–92.

shaped the relationship. Along with examining various medico-scientific fields and practices, this book illustrates different ways this symbiotic relationship unfolded or waned at different points in a given historical period.

Based on this scope, the book's six chapters elaborate on either a scientific discipline developed through its engagement with the state-led population management or a topic in policymaking associated with population studies. Chapter 1 examines the development of modern population statistics c.1860s–1910s. It describes the institutionalization of population statistics in Japan, first in tandem with the making of a modern official administration in the late 1860s–80s and, from the latter half of the 1890s, alongside the colonial rule of Taiwan. I explore how the emerging cohort of individuals centering around Sugi Kōji established a scientific community, in part by taking advantage of their positions within the new government. At the same time, it depicts how these modern statisticians' position as coterminous with political authority did not automatically grant them the power to implement the scientific practices for which they lobbied. I illustrate this point by exploring their campaign to implement a national census in Japan, which, despite the authoritative positions held by statisticians, was not immediately successful: Higher-ranking officials believed the *koseki* household registration system, a survivor from the previous era and reformed in the early 1870s, adequately fulfilled this role. Their campaign only came to fruition in 1905 in the context of Japan's colonial rule over Taiwan. Gotō Shinpei, a then high-ranking officer in the Government-General of Taiwan, actively promoted a population census, deeming it a valuable tool for scientific colonial governance. Finally, I examine the activities of Mizushina Shichisaburō to describe how the scientific practice and community surrounding the census work thrived in Taiwan and how the Taiwanese experience ultimately fed back into the statistical activities in the metropole. Overall, this chapter presents a nuanced understanding of the relationship between the building of a modern sovereignty and the development of a scientific field.

Chapter 2 studies medical midwifery in the 1860s–1930s in parallel with the administrative management of vital statistics. Drawing on existing work, I depict how medical midwifery thrived as the nascent government assigned midwives a critical role in efforts to establish a reproductive surveillance system.[48] The chapter describes how the profile of midwives

---

[48] Yuki Terazawa, *Knowledge, Power, and Women's Reproductive Health in Japan, 1690–1945* (New York: Palgrave Macmillan, 2018); Shoko Ishizaki, *Kingendai nihon no kazoku keisei to shusshōjisū: Kodomo no kazu wo kimetekita mono wa nanika* (Akashi Shoten, 2015).

was significantly transformed, from regionally diverse birth attendants, often implicated in abortion and infanticide, to medically informed and licensed healthcare practitioners, defined by their role in enhancing – and simultaneously monitoring – people's everyday reproductive experiences. At the same time, this chapter goes beyond the scope of the current literature by suggesting that this transformation of midwives is a story intimately tied to public health officers' desire to collect and manage more "accurate" data about infant births and deaths, which they judged would be essential for constructing a genuinely "modern" public health system. By juxtaposing the history of the professionalization of midwives with the establishment of vital statistics in public health, this chapter shows how the burgeoning statistical rationale acted as a pivotal background for the making of medical midwifery in modern Japan.

Chapter 3 studies how an amorphous group of population experts became prominent in policymaking during the 1920s, which is when the phrase "population problem" (jinkō mondai) entered the Japanese lexicon. This catchall term was used to refer to various kinds of socioeconomic ills, many of which were deemed to require state intervention. I first describe how policy-oriented debate about the "population problem" developed in the 1920s, mostly among social scientists long familiar with the "Karl Marx versus Thomas Malthus" argument introduced from Western Europe. I then explore how the "population problem" became a policy priority in the late 1920s by examining research and policy discussions that took place in the Investigative Commission for the Population and Food Problem (Jinkō Shokuryō Mondai Chōsakai), established in 1927 as the first government organization dedicated to population issues. By scrutinizing policy deliberations within the Investigative Commission about emigration and population control, I point out that population experts, in response to the governmental endorsement of overseas migration as a solution to the "population problem," tended to value eugenic measures as well as overseas migration. I confirm that although the policy deliberation and research mobilized by the Investigative Commission did not lead directly to specific population policies, it laid a critical foundation for the institutionalization of government research on population problems and for the establishment of the Ministry of Health and Welfare, both of which were realized as Japan entered into war with China in 1937.

Chapter 4 sheds light on the population distribution under "national land planning" (kokudo keikaku), a hitherto less visible topic in modern Japan's wartime population policies. Research has thus far concentrated on eugenics and other maternal and infant health measures intended to maximize the population's potential by improving the physical and mental quality of the Japanese race.[49] Less studied, yet equally important in the minds of the contemporary policymakers and population scientists, was the balanced distribution of the population. This was deliberated in the process of creating policies for "national land planning," the wartime government's "sacred mission" to construct the "new order" in East Asia by establishing the Greater East Asia Co-Prosperity Sphere. This chapter analyzes the debates related to population distribution policies as well as policy-oriented research activities mobilized for national land planning. By focusing on the technical bureaucrat Tachi Minoru, I describe how his research reflected the political agenda of the wartime government, which primarily viewed the population as an invaluable resource to be deployed for the nation at war. I detail how the policy-oriented population research saw population in racialized and gendered terms and focused on certain demographic subjects, seeing them either as undergirding or undermining the prosperity of Japan as a nation. This chapter also illustrates the fragile nature of demographic knowledge produced for policymaking and concludes that the role of policy-oriented scientific investigation in wartime statecraft was by no means as stable as it appeared on paper.

Chapter 5 examines the birth control survey research conducted by population technocrats after World War II (WWII), c.1947–60, and analyzes how this research resonated with government efforts to manage the emerging problem of "overpopulation" via fertility regulation. Focusing on the leading population technocrat Shinozaki Nobuo, this chapter depicts how human agency participated in the at times precarious relationship between policy and practice. It also shows how the epistemological framework inscribed in the scientific knowledge produced by the survey research harmonized with the economic and political rationale that buttressed the post-WWII state's reconstruction efforts.

Chapter 6 traces the development of a field of population science that emerged from the activities at the Department of Public Health Demography, which was established in 1949 at the Institute of Public Health by Koya Yoshio, the Director-General of the Institute and leading wartime racial hygienist who became a birth control activist after the war. Drawing on existing work that locates Japanese birth control advocacy in transnational histories, I suggest that domestic efforts to discipline reproductive bodies within Japan in the 1950s, realized by population scientists such as Koya, became linked to collaborative working relationships with international colleagues in the 1960s to restrict world population

---

[49] E.g., Yutaka Fujino, *Nihon fashizumu to yūsei shisō* (Kyoto: Kamogawa Shuppan, 1998).

growth by popularizing contraceptive practices in so-called underdeveloped nations through development aid programs.[50] At the same time, going beyond the existing literature, I also depict how the transnational movement fostered inter-Asian scientific interactions between Japanese and Indian colleagues via funding support from US charitable foundations, most notably the Population Council and Clarence J. Gamble. Ultimately, this chapter portrays the Japanese state's efforts to regulate citizens' fertility as a complex practice based on the coproduction of scientific knowledge, scientific field, and social order involving multilayered interactions at local, national, regional, and transnational levels.

Finally, the concluding chapter gives a brief account of the continued interplay between population science and the governing of Japan's population from the 1960s to the present. In addition, it reflects on the Japanese science-policy nexus that became increasingly globalized in the late 1960s. In particular, it questions the autonomy of the people as a governed entity, a topic that receives limited attention in the book. I explain how the specific ways the population was imagined in relation to statecraft elided the agency of the governed population.

To fully grasp the interplay between the making of population science and the governing of the population in modern Japan, we must first comprehend how new clusters of administrative, educational, and scientific activities were organized around the modern discourse of population. The novel understanding of a "population" emerged in Japan in the nineteenth century in tandem with the transformation of Japan's polity from a feudal system based on the relationship between the shogunate and domains scattered across the country to one constitutive of the Westphalian system. The story around the development of population statistics from the 1860s onward helps us to understand this part of the history.

---

[50] Takeuchi-Demirci, *Contraceptive Diplomacy*; Homei, "The Science of Population"; Maho Toyoda, "Sengo nihon no bāsu kontorōru undō to Kurarensu Gyanburu: Dai 5 kai kokusai kazoku keikaku kaigi no kaisai wo chūshin ni," *Jendā shigaku* 6 (2010): 55–70; Miho Ogino, *"Kazoku keikaku" eno michi: Kindai nihon no seishoku wo meguru seiji* (Iwanami Shoten, 2008).

1 Population Statistics
Between Building a Modern State
and Governing Imperial Subjects

If we define population statistics as leaders' attempts to collect, record, and sort data, numerical and otherwise, about the people inhabiting their political domains, then Japan has a long history in this field. However, if we regard population statistics as the modern scientific field we know today, its history is not as long – it starts in the 1860s. The development of modern population statistics was firmly embedded in the political trans-formation of Japan from a federation of feudal domains to a centralized nation-state and empire. Population statistics was instrumental during this transformation process. The nascent Meiji government (1868–1912) used population data to understand – and if necessary mobilize – people for the sake of nation- and empire-building. This happened in part because individuals trying to establish the European-derived science of statistics in Japan took full advantage of the political conditions in their country. They promoted population statistics, arguing it was a civilizing tool that could help the government efficiently enhance Japan's national power. However, their promotion of a modern population census was not always smooth. The advocates struggled at times especially in the specific intellectual, political, and social milieu of Japan in the latter half of the nineteenth century, when the concepts of population and science were themselves in flux. In this context, a population census was conducted in colonial Taiwan, while the Japanese government dragged its feet when it came to implementing it in the metropole. This chapter depicts how population statistics as a subfield of modern statistics and the population census as a technology of governance developed in a place where Japan's effort to build a modern state met its aspiration to govern imperial subjects.

Collecting Population Statistics
in Japan: Then and Now

Morita Yūzō (1901–94), an acclaimed twentieth-century Japanese statis-tician, once claimed that population statistics in Japan "can be praised, from a statistical point of view, as [of] high value in terms of its content and periods it covers, [it is] even a global standard."[1] And population experts today note that the centralized administration in Japan compiled data to produce a population registry in the mid-seventh century and early eighth century by following legal codes they adopted from China.[2] They also point out that in the Tokugawa period (1603–1867), political rulers produced other kinds of population registers: *shūmon aratamechō* (registers of religious scrutiny) and *ninbetsu aratamechō* (person-by-person registers).[3]

This suggests there is historical continuity. However, the population data produced throughout Japan's history could not have been more different; for example, compare population registers from the Tokugawa period and today's population census. Tokugawa data recording relied on brush, paper, and ink, while today's census takers apply various electronic technologies and collaborate with scientists, such as geographers.[4] Today's population census, centrally organized by the Statistics Bureau of Japan (Sōmushō Tōkeikyoku), adopts a nationally standardized form. In contrast, the Tokugawa practice was, for the most part, neither standardized nor exhaustive.[5] Reflecting the pluralistic nature of the Tokugawa polity, in which the *bakufu* ("shogunate") governed over 250 domains, each with its own lord, the method of collating population data and the population groups that became the target of population surveys varied according to the domain. These methodological differences have also shaped what the data sets are able to show. While today's census data represent the demographics of Japan as a whole, data gleaned from the Tokugawa population registers illustrate a fragmented demographic reality.[6]

---

[1] Yūzō Morita, *Tōkei henreki shiki* (Nihon Hyoronsha, 1980); Nihon Jinkō Gakkai, ed., *Jinkō daijiten* (Baifukan, 2002), 272.

[2] Nihon Jinkō Gakkai, *Jinkō daijiten*, 272.

[3] In some domains, *ninbetsu aratamechō* and *shūmon aratamechō* were merged to form *shūmon ninbetsu (aratame)chō*. Fabian Franz Drixler, *Mabiki: Infanticide and Population Growth in Eastern Japan, 1660–1950* (Berkeley: University of California Press, 2013); Osamu Saito and Masahiro Sato, "Japan's Civil Registration Systems Before and After the Meiji Restoration," in *Registration and Recognition: Documenting the Person in World History*, eds. Keith Derek Breckenridge and Simon Szreter (Oxford: Oxford University Press, 2012), 113–35; L. L. Cornell and Akira Hayami, "The Shumon Aratame Cho: Japan's Population Registers," *Journal of Family History* 11, no. 4 (December 1986): 311–28.

[4] Takashi Abe, Shigeru Kawasaki, Atsushi Otomo et al., "Chirigaku niokeru tōkei no riyō to kongo no kadai: 'Tōkei' wo meguru kan, gaku no renkei wo mezashite," *E-Journal GEO* 6, no. 1 (2011): 81–93.

[5] Naotarō Sekiyama, *Nihon jinkōshi* (Shikai Shobo, 1942), 52–58. However, the question of universal coverage is an issue contemporary census takers are grappling with, too. Akira Ishikawa and Tsukasa Sasai, "Gyōsei kiroku ni motozuku jinkō tōkei no kenshō," *Jinkō mondai kenkyū* 64, no. 4 (December 2010): 23–40.

[6] For more, see Yuriko Yokoyama, *Meiji ishin to kinsei mibunsei no kaitai*, Dai 1-han. (Yamakawa Shuppansha, 2005).

In addition, the objectives of collecting demographic data were also different. Today, the Statistics Bureau of Japan stresses that the population census aims to provide an "informational infrastructure" supporting the activities of citizens, corporations, and the state administration.[7] Responding to its articulated goals, the census attempts, broadly speaking, to discover inhabitants' social attributes, details about their domiciles, education, work, and household information.[8] In contrast, the Tokugawa rulers recorded peoples' personal details to secure taxes or *corvée* labor.[9] They also compiled population registers because they thought the data would provide information about what they were most concerned about: social order.[10] The original objective of compiling *shūmon aratamechō* was to police people's Christian religious activities, which the Tokugawa oligarch deemed a threat to social order.[11] Tokugawa population registers recorded personal details that would immediately impact social relations, such as statuses within the household.[12]

Finally, there were distinct semantic traditions associated with the two data sets. Today, demographers define the term *jinkō* – officially used to refer to "population" and composed of the Chinese characters for "person" and "mouth" – as "a mathematical expression" for a group of people living in a given area.[13] Some, especially those focusing on economics, have used a Malthusian metaphor to explain the term as the "number of mouths to feed."[14] But, in the Tokugawa period, the character compound – also sometimes read as *ninkō* – was more commonly referred to as "gossip," or what the "mouth" of a "person" produces in social interactions. In this period, the term *minkō* seemed to refer to an idea similar to what is expressed by *jinkō* today.[15]

In addition to the abovementioned shifting meanings, the concept of "population" meant different things throughout history. Today, we tend to assume population is an aggregate of people bound by a political

---

[7] Sōmushō Tōkeikyoku, "Kokusei chōsa no kihon nikansuru Q&A (kaitō)," accessed October 17, 2019, www.stat.go.jp/data/kokusei/qa-6.html.

[8] Ibid.

[9] Drixler, *Mabiki*, 6–8.

[10] Cornell and Hayami, "The Shumon Aratame Cho."

[11] Saito and Sato, "Japan's Civil Registration Systems," 116.

[12] Sekiyama, *Nihoon jinkōshi*, 48.

[13] Jinkōgaku Kenkyūkai, ed., *Gendai jinkō jiten* (Hara Shobo, 2010), 131–32; Jinkō Daijiten Henshū Iin, *Jinkō daijiten*, 3.

[14] Masaaki Yasukawa, *Yasashii jinkōgaku kyōshitsu* (Japanese Organization for International Cooperation in Family Planning, 1978), 12.

[15] Jinkōgaku Kenkyūkai, *Gendai jinkō jiten*, 132.

entity, most commonly the nation-state.[16] Moreover, demographic studies are premised on the idea that a population exhibits patterns of societal behaviors.[17] Demographers also contend that population dynamics are directly shaped by cultural, economic, and social conditions and thus can be managed by measures that aim to improve these conditions.[18] In contrast, population in the Tokugawa period was understood as less susceptible to human intervention, though it did indicate some natural laws. For instance, in 1798, the year Thomas Robert Malthus published the canonical *An Essay on the Principle of Population*, Honda Toshiaki observed "rules from the heaven" (*tensoku*) in the patterns of fertility and, like Malthus, warned of an imbalance between the growth of the population and human subsistence.[19] However, unlike Malthus, who thought population growth could be tamed by "preventive restraints," Honda believed it was an uncontrollable phenomenon that humans could do little to counter.[20] Data predicated on these different interpretations of population tend to carry distinctive characteristics.

What made population statistics in Tokugawa and contemporary Japan so distinct? I argue that the 1860s was a watershed moment. Amid the drastic sociopolitical change Japan underwent during that decade, the recently founded government wished to gather data that showed exactly how many, and what kinds of, people lived within the territories they now reigned. Alongside this, emerging groups of intellectuals, inspired by modern statistics in Europe, tried to implement modern statistical methods within the state bureaucracy.[21]

[16] Population data published by the international organization consolidates this understanding, e.g., United Nations Department of Economic and Social Affairs, "World Population Prospects 2019," accessed May 9, 2022, https://population.un.org/wpp/Maps/.

[17] Jinkōgaku Kenkyūkai, *Gendai jinkō jiten*, 131–32.

[18] Sato and Kaneko, *Posuto jinkō tenkanki*, 21–24, 25–40.

[19] Ishii, "Statistical Visions of Humanity," 13–14.

[20] Ibid., 17–18.

[21] For the story in this chapter, I am indebted to the wealth of knowledge accumulated over the years on the history of population statistics and the census in modern Japan and Taiwan under Japanese colonial rule, e.g., Sōmushō Tōkeikyoku, "'Nihon tōkei nenkan' 120 kai no ayumi," accessed April 28, 2017, www.stat.go.jp/data/nenkan/pdf/120ayumi.pdf; Masahiro Sato, *Kokusei chōsa nihon shakai no hyakunen* (Iwanami Shoten, 2015); Masahiro Sato, *Kokusei chōsa to nihon kindai* (Iwanami Shoten, 2002); Ishii, "Statistical Visions of Humanity"; Yoshiro Matsuda, "Formation of the Census System in Japan: 1871–1945 – Development of the Statistical System in Japan Proper and Her Colonies," in *Historical Demography and Labor Markets in Prewar Japan*, ed. Michael Smitka (New York: Garland, 1998), 100–24; Takeshi Yabuuchi, *Nihon tōkei hattatsushi kenkyū* (Kyoto: Hōritsu Bunka Sha, 1994); Ryuken Ohashi, *Nihon no tōkeigaku* (Kyoto: Horitsu Bunkasha, 1965).

What follows is the story of how a small circle of intellectuals and bureaucrats that formed around Sugi Kōji (1828–1917) instigated arduous campaigns to implement a nationwide population census amidst the political changes that occurred in the 1860s. Their campaigns brought about novel understandings of population, society, and sovereignty and assigned a fundamentally new role to the act of collecting population data. In later years, the calculation of figures premised on these new conditions added new connotations to statistics produced after the 1860s.

Institutionalizing Statistics in a Burgeoning Modern State

One day in the mid-1860s, while working for the *bakufu*-sanctioned School for Western Studies as a translator of foreign materials for senior councilors, thirty-seven-year-old Sugi Kōji happened to find Dutch statistical tables from 1860 and 1861.[22] The statistics contained population figures written in decimal numbers. Sugi wondered, "how on earth could people become decimal numbers" and "felt strange about this way of studying people."[23] Yet he also "felt research of this kind would be useful for Japanese people."[24] Sugi "did not act on it then because it was a turbulent time," but later he clearly recalled that it "sowed seeds of statistical interest in me."[25]

This oft-told episode tells how Sugi, celebrated as the "father of modern statistics" in Japan, discovered the scientific field.[26] When Sugi encountered Dutch statistics, it was indeed a turbulent time in Japan. In his formative years, the arrival of the American ambassadors, led by Commodore Matthew C. Perry in 1853, and the unequal Kanagawa Treaty signed with the United States in the following year, catalyzed a succession of bloody civil wars and social unrest that lasted for more than two decades. The political turmoil profoundly impacted the lives of Sugi and his fellows, who were students of "Dutch Studies" (*rangaku*), learning western knowledge associated with Japan's relations with its only western trade partner, the Dutch East India Company.[27] While

---

[22] Taichi Sera, ed., *Sugi sensei kōenshū* (1902), 18–19, accessed May 9, 2022, https://dl.ndl.go.jp/info:ndljp/pid/898298.
[23] Ibid., 19.
[24] Sato, *Kokusei chōsa to*, 18–19.
[25] Ibid., 18.
[26] Takeshi Yabuuchi, "Nihon ni okeru minkan tōkei dantai no shōtan: 'Hyōki gakusha' to sono keifu," *Kansai daigaku keizai ronshū* 26, nos. 4–5 (January 1977): 587–88.
[27] Terrence Jackson, *Network of Knowledge: Western Science and the Tokugawa Information Revolution* (Honolulu: University of Hawai'i Press, 2016).

some collaborated with the revolutionaries and applied their knowledge to overturn the *bakufu*, others tried to secure their professional positions within the *bakufu*, which was also exploring ways to maintain its authority by utilizing western-derived scientific knowledge and technology.[28] Under the tutelage of the prominent *bakufu* retainer Katsu Awa-no-kami (aka Katsu Kaishū), Sugi was able to get closer to the latter group of Dutch Studies scholars. This became even more apparent when, in 1855, the powerful senior councilor Abe Masahiro (1819–57) employed Sugi as a retainer.[29] Later, in 1864, when the *bakufu* built the School for Western Studies, Sugi was appointed to serve as a translator and teacher for its Office of Examination of Foreign Books. However, due to his proximity to them, Sugi ultimately lost his job when the revolutionaries overthrew the *bakufu* in 1867. For a little while thereafter, he lived in Suruga (part of today's Shizuoka Prefecture) after following the former ruler, Tokugawa family, that retired there.

Although he faced a setback in his work life due to his affiliation with the now outcast oligarch, the experience Sugi gained working under the auspices of the *bakufu* paved the way for the development of statistics as a scientific field in Japan. In the mid-1860s, Sugi began to study western statistics with the lecture notes compiled by Nishi Amane (1829–97) and Tsuda Mamichi (1829–1903) who, as part of a study trip to the Netherlands with a *bakufu* stipend, learned statistics under Simon Visseling (1818–88) at Leiden University. In Suruga, under the aegis of the retired Tokugawa family, Sugi conducted a pilot population census by applying the knowledge he had obtained from the European-language books and notes. The census, published in June 1869, provided data on the population of Suruga organized by age, occupation, and marital status, and it also provided figures for population inflows and outflows.[30]

In later years, the pilot study became celebrated as "the first modern static statistical population survey" in Japan.[31] But what made the survey "modern"? To the contemporaries, the survey was modern first and foremost because the European science of statistics acting as its base

[28] For a profile of Dutch studies scholars during this period, see Tsutomu Kaneko, *Edo jinbutsu kagakushi: "Mou hitotsu no bunmei kaika" wo tazunete* (Chuokoron-Shinsha, 2005).

[29] Abe was the daimyo of the Fukuyama domain (part of today's Hiroshima Prefecture), Tokugawa *bakufu*'s chief senior councilor (*rōjū shuseki*) from 1845 to 1855, and one of the most passionate supporters of foreign diplomacy within the *bakufu* at the time.

[30] Yabuuchi, *Nihon tōkei hattatsushi kenkyū*, 149–60.

[31] Akira Hayami, "Koji Sugi and the Emergence of Modern Population Statistics in Japan: The Influence of German Statistics," *Reitaku Journal of Interdisciplinary Studies* 9, no. 2 (2001): 3; Yabuuchi, "Nihon ni okeru minkan tōkei dantai no shōtan," 589.

was a "new technology of the civilized world" and had a transformative power.[32] Sugi argued that statistics, "like the railway, telephone, and locomotive," were such technology that would "manifest power" when they were transposed to a new society and eventually modernize the society. In later years, Sugi presented another reason his census work was doubly modern: It could help Japan become a modern nation-state. In particular, he thought that a strand of statistics he called "government statistics" (*seifu sutachisuchikku*) would be particularly useful.[33] According to Sugi, "government statistics" clarified the "circumstances of [the] country's people" – including their everyday economic activities and social interactions – that had a direct impact on the "rise and fall" of a nation.[34] It would allow government officials to efficiently identify the factors in people's economic and social lives that might impede the development of new societies and industries which were currently underway as part of nation-building, thereby helping the government to proactively act on these factors and enabling Japan to "rise" as a modern nation-state.[35]

As it is clear from Sugi's description of "government statistics," when it came to the modern statistics he wished to promote in Japan, he stressed its utility for the officially endorsed social, economic, and political reforms.[36] For this reason, Sugi's idea of modern statistics garnered support from Meiji luminaries and statesmen, who were actively involved in these reforms. One of the most prominent Meiji intellectuals, Fukuzawa Yukichi (1835–1901), fervently promoted statistics, arguing it was a highly useful tool for Japan to overcome various challenges confronting the country in the process of becoming an independent nation.[37] Fukuzawa explained that Japan's independence depended on the "nation's civilization," which was composed of a "total sum of people's intellect and virtue distributed across the nation," but currently there was no way of knowing how much of a "nation's civilization" Japan possessed.[38] He then pointed out that in

---

[32] Sera, *Sugi sensei kōenshū*, 136; Sato, *Kokusei chōsa nihon shakai*, 37–38; Sato, *Kokusei chōsa to*, 23–25.

[33] Sera, *Sugi sensei kōenshū*, 143.

[34] Ibid., 158.

[35] Ibid.

[36] Sugi's utilitarian understanding of statistics was derived from German social statistics. Tadao Miyakawa, *Tōkeigaku no nihonshi: Chikoku keisei eno negai* (Tokyo Daigaku Shuppankai, 2017), 38–39.

[37] Miyakawa, *Tōkeigaku no nihonshi*, 9–15. For work that analyzes Fukuzawa's idea of "practical learning" (*jitsugaku*), see Ayumi Kaneko, "Nēshon to jitsugaku: 'Keimō' to 'gesaku' no kōten," in *Meiji, Taisho-ki no kagaku shisoshi*, ed. Osamu Kanamori (Keiso Shobo, 2017), 13–64.

[38] Yukichi Fukuzawa, *Bunmeiron no gairyaku kan no ni* (1875), 4–5, accessed May 11, 2022, https://dl.ndl.go.jp/info:ndljp/pid/993900.

Western Europe and North America, scholars used statistics to show the level of civilization effectively in concrete numbers. Fukuzawa described how statisticians recorded "how much and [how] little of the land, how many and how few of the people, the high and low of the price and wage, how many are married, how many have fallen ill, and how many die," to quantify civilization.[39] He then suggested Japanese scholars should learn Western statistics to better understand the level of the "nation's civilization" in Japan and, with that knowledge, help the country achieve the goal of national independence efficiently. Like Sugi, Fukuzawa promoted statistics because he believed it was a tool of civilization that would facilitate Japan's nation-building effort.

Within the government, Ōkuma Shigenobu (1838–1922), one of the most important statesmen of the early Meiji period, saw the benefit of statistics for statecraft.[40] While exploring ways to prove the need to conduct a land reform as a high-rank official in charge of the country's finance, Ōkuma learned about Western statistics and how the United States government even had a large office within the Department of Treasury dedicated to statistical works.[41] He then proposed that the Japanese government should set up an office for statistics. Ōkuma's idea garnered support from Itō Hirobumi (1841–1909) and Shibusawa Eiichi (1840–1931), both key figures who shaped the country's economy in the early Meiji period.[42] Consequently, in 1871, the government founded the Division of Statistics within the Ministry of Finance to facilitate official economic planning.

The increased demand for statistics coming from intellectual circles and the government gave Sugi an unprecedented opportunity to advance his career. The tangible opportunity came when delegates from the Iwakura Mission, an eighteen-month diplomatic mission in countries in Western Europe and the United States that commenced in 1871, demanded the government provide them with statistics. They were planning to use statistics to introduce Japan to these foreign powers and renegotiate the terms of the unequal treaties. This demand led the government to set up another, smaller office for statistics, the Section of Statistics within the Grand Council of State (*dajōkan*).[43] Around

---

[39] Ibid., 4–5, 10–11.
[40] Sadanori Nagayama, "Nihon no kanchō tōkei no hatten to gendai," *Nihon tōkeigakkaishi* 16, no. 1 (1986): 101–9.
[41] Miyakawa, *Tōkeigaku no nihonshi*, 18–19; Shiro Shimamura, *Nihon tōkeishi gunzō* (Nihon Tōkei Kyōkai, 2009), 31–38.
[42] Miyakawa, *Tōkeigaku no nihonshi*, 19.
[43] Shugen Takagi, "Akiyoshi Mizukuri to tōkeigaku," *Kansai daigaku keizai ronshū* 19, no. 1 (April 1969): 15.

this time, Shibusawa – who had formerly served the Tokugawa family – learned about Sugi's Suruga survey. Based on Shibusawa's recommendation, Sugi was selected to head the technical team within the section.[44] Sugi directed three staff working under him and, together, they compiled general statistics for the Iwakura Mission.[45] The section then published the first statistical almanac in 1872, which was regularly updated. In 1875, Sugi was promoted to head the entire section. These works helped Sugi restart his career, this time as a government employee under the new political regime.

While working as a government bureaucrat, Sugi also promoted statistics widely among the public. In 1876, with sixteen colleagues, Sugi founded a private research group called Society for the Statistical Science (*Hyōki Gakusha*) with the aim to "research the methods and principles of statistics and work on applying them."[46] In 1883, having failed to lobby for a state-sponsored educational institution dedicated to modern statistics, Sugi and other members of the Society for the Statistical Science established a private training and research institution, the Kyōritsu School of Statistics.[47] In 1886, the school published the *Journal of Statistics*, the field's first specialist journal in Japan.[48] Sugi was also one of the fourteen founding members of the Statistics Society (*Seihyōsha*), another statistical research community established in 1878.[49]

Throughout the 1880s, the statistical community expanded even further through the activities of Sugi and his colleagues, who created more research and training opportunities, mainly based on a private-government partnership.[50] By the early 1890s, observers felt that statistics had become established as a field of inquiry. Their feelings were confirmed when *tōkeigaku*, "a study of systematizing measurements," was adopted

[44] Miyakawa, *Tōkeigaku no nihonshi*, 19.
[45] Hayami, "Koji Sugi," 1–10; Yabuuchi, *Nihon tōkei hattatsushi kenkyū*, 3.
[46] Cited in Nihon Jinkō Gakkai, *Jinkō daijiten*, 272. *Hyōki Gakusha* could be literally translated as the "association of notation studies." For the actual members' list, see Kenta Higasa, "Sugi Kōji hakase to Meiji ishin no tōkei (7)," *Tōkeigaku zasshi*, no. 624 (June 1938): 22–34.
[47] As part of the school name, *kyōritsu* ("established in collaboration," literally translated) suggests it was a privately funded school, but many of the staff running the school were in fact civil servants such as Sugi.
[48] The original Japanese names for the journal's title changed over time, from the more phonetic *sutachisuchikku zasshi* to *tōkeigaku zasshi*.
[49] As *seihyō* was one of the words used to refer to statistics at the time, *Seihyōsha* could be translated as the "Association of Statistics."
[50] Masahiro Sato, *Teikoku nihon to tōkei chōsa: Tōchi shoki Taiwan no senmonka shūdan* (Iwanami Shoten, 2012), 184–88.

as an official translation after a heated discussion in 1889 between the statistician Imai Takeo and the military doctor and renowned writer Mori Ōgai.[51]

Yet the institutional demands of the Meiji government and statisticians' interest that emerged in the context of 1860s Japan were not the only reasons modern statistics thrived during this period. The development of population statistics as a subdiscipline of modern statistics also owed much to a fundamental shift in the meanings assigned to the relationships between population, society, and the Japanese sovereignty that emerged as Japan was turning into a modern state.

Population Statistics for the Making of a Modern State

Among the various kinds of statistics, high-rank government officials considered population statistics to be particularly significant for the Meiji government's effort to build a modern nation-state. For them, knowing roughly how many, and what kinds of, people lived and traveled within the country was a prerequisite for the smooth operation of the wide-ranging reforms they launched in the early 1870s, such as the introduction of the prefectural administrative system (1871), military conscription (1873), and land tax reform (1873). In the words of Etō Shinpei (1834–74), the first minister of justice in the nascent government, the demographic data indicating "laws of marriage, birth and death" would help the government to construct a "rich and strong" nation by "clarify[ing] the position of people."[52] Indeed, this kind of knowledge ascertaining the "position of people" – in both geographical and social terms – was particularly useful for the government because it facilitated the process of locating the group targeted for government reforms. In the early Meiji period, government officials saw the benefits of demographic data primarily because of their practical value.

Thus, from fairly early on in the reforms, various offices within the government compiled population data. First, Sugi's Section of Statistics was tasked to collect population data as part of their work compiling statistic almanacs. The police compiled population data early on as part of its independent household survey.[53] Beginning in 1871, the Ministry of

[51] Miyakawa, *Tōkeigaku no nihonshi*, 53–113.
[52] Masataka Endo, *Koseki to kokuseki no kingendaishi: Minzoku, kettō, nihonjin* (Akashi Shoten, 2013), 119–20.
[53] Masuyo Takahashi, "Meijiki wo chūshin nimita nihon no jinkō tōkei shiryō ni tsuite (keizai tōkei tokushū)," *Keizai shiryō kenkyū* 14 (June 1980): 19.

Finance's Division of Statistics compiled population statistics for a brief period. In the 1870s, the Home Ministry's Sanitary Bureau began to collect vital statistics (see Chapter 2).[54] Furthermore, some prefectural governments, such as Kumamoto, Tokyo, and Kobe, conducted population censuses in their respective administrative areas.[55] Finally, starting around 1888, the army began to conduct a population survey.[56] In the first two decades of the Meiji period, government offices independently collected population data and applied them to the reforms they were in charge of.[57]

In addition to these localized initiatives, the early Meiji government also orchestrated population work around the reform of the existing population registers, including the abovementioned *ninbetsu aratamechō* and *shūmon aratamechō*. Confronted with the need to come up with population statistics for the whole of Japan for the effective operation of various reforms but lacking the established infrastructure or manpower to conduct a comprehensive national population survey, the Meiji government pragmatically decided to use existing population registers to get a general sense of the new country's population. However, as mentioned above, the Tokugawa population registers were diverse in terms of format and content. To standardize the population registration system, the government embarked on reform as early as August 1869. On April 4, 1871, the Grand Council of State issued the Household Registration Law (*Koseki Hō*), which mandated the creation of a nationally applicable single population register called *koseki*.[58] The law was implemented on

---

[54] See Chapter 2 for details.

[55] Ibid., 20.

[56] Ibid., 18.

[57] This official exercise acted as the basis for which the biologically determined age calculated according to the newly implemented Western solar calendar became the determinant for mobilizing people for the reforms. However, implementing the new idea of age caused some tensions and confusions. See, e.g., Sayaka Chatani, "A Man at Twenty, Aged at Twenty-Five: The Conscription Exam Age in Japan," *American Historical Review* 125, no. 2 (April 2020): 427–37; Gregory M. Pflugfelder, "The Nation-State, the Age/Gender System, and the Reconstitution of Erotic Desire in Nineteenth-Century Japan," *The Journal of Asian Studies* 71, no. 4 (2012): 963–74.

[58] *Koseki* is still one of the most important official population registration systems in Japan. Karl Jacob Krogness and David Chapman define *koseki* in contemporary Japan as "fundamentally a civil registration system that records and documents individual civil status by household unit and is the definitive state mechanism for determining in individual's legal identity as Japanese." *Koseki* does not represent individuals in their own rights but locates them within a "household" (*ko*). This structure, as Krogness and Chapman argue, reflects the official understanding of civil-state relations in contemporary Japan, which pivot around the family as a social unit. Karl Jakob Krogness and David Chapman eds. *Japan's Household Registration System and Citizenship: Koseki, Identification and Documentation* (New York: Routledge, 2014), 2.

February 1, 1872. After issuing the law, the government set up the independent Bureau of Household Registration within the Grand Council of State. As early as March 1873, the bureau completed the *koseki* register, from which the government could learn roughly how many and what kinds of people lived in which parts of the country, as well as the movements of people.

In addition to pragmatic reasons, the *koseki* reform was also shaped by political concerns that cropped up as the archipelago went through drastic transformations. Specifically, the government's fervor for the *koseki* reform was fueled by government officials' anxiety over *dappansha* or *dassekisha*, people who had left their registers toward the end of the Tokugawa period.[59] *Dappansha* were a diverse migrant group with wide-ranging social and political backgrounds, and Meiji leaders striving to construct a self-contained sovereign state considered these people too unwieldy and viewed them suspiciously as a potential cause of political disturbances. Under the circumstances, high-rank officials hoped the demographic knowledge extracted from the reformed *koseki* population register would act as a form of surveillance technology. To borrow the words of David Chapman and Karl Jakob Krogness, the *koseki* reform was expected to "bring order to the previous era's disorder" and, in so doing, ensure Japan's peaceful transition from a feudal to a modern state.[60]

At the same time, the government conducted the *koseki* reform because high-rank officials believed the reform would perform another important political function in the nation-building exercise: help the government to draw the boundaries of the "Japanese."[61] After the Household Registration Law was issued, the process to implement and upgrade the *koseki* registration system coincided with the creation of a new definition of the Japanese population, which accompanied the government's effort to

---

For the *koseki* reform in early Meiji, see David Chapman, "Geographies of Self and Other: Mapping Japan through the Koseki," *The Asia-Pacific Journal: Japan Focus* 9, no. 29 (July 19, 2011), http://apjjf.org/2011/9/29/David-Chapman/3565/article .html; Kenji Mori, "The Development of the Modern Koseki," in *Japan's Household Registration System*, eds. Chapman and Krogness, 59–75; Endo, *Koseki to kokuseki*; Masataka Endo, *Kindai nihon no shokuinchi tōchi ni okeru kokuseki to koseki* (Akashi Shoten, 2010).

[59] Chapman, "Geographies of Self and Other"; Mori, "The Development of the Modern Koseki," 65–69; Endo, *Koseki to kokuseki*, 112–15.
[60] David Chapman and Karl Jakob Krogness, "The Koseki," in *Japan's Household Registration System*, eds. Chapman and Krogness, 6.
[61] David Chapman, "Managing 'Strangers' and 'Undecidables': Population Registration in Meiji Japan," in *Japan's Household Registration System*, eds. Chapman and Krogness, 96–98; Chapman, "Geographies of Self and Other"; Endo, *Koseki to kokuseki*, 23.

demarcate its sovereign boundaries. In the north, as soon as the Meiji government declared Ezochi to be the Japanese territory of Hokkaido in 1869, the *koseki* population survey was organized there and integrated the inhabitants of Hokkaido into the category of "Japanese subjects" (*nihon shinmin*).[62] After the 1871–72 *koseki* reform, the population registration system set up there also further defined the indigenous Ainu people as Japanese.[63] Similarly, in the south, the implementation of the *koseki* registration system was done in tandem with the Ryūkyū kingdom's transformation into the Okinawa Prefecture, as well as with the possession of the Ogasawara Islands, where the *koseki* also recorded the inhabitants as Japanese.[64] After the colonization of Taiwan in 1895, the modern *koseki* system also buttressed the creation of the hierarchical relationship between the Japanese colonizer, called *naichijin*, or "people of the internal land," and "external" colonial subjects.[65] The *koseki* reform facilitated Japan's transformation into a modern sovereign state by offering an administrative basis for creating the category of the Japanese population at a time when Japan's territorial border and the accompanying idea of Japanese national identity were themselves highly unstable.[66]

The compilation of population statistics was a core activity within *koseki* reform. Initially, the Ministry of Public Affairs, in charge of administering matters related to household and public infrastructure, was tasked with calculating the total population based on the Tokugawa registers. In 1873, when the Home Ministry set up the Office of Household Registration (*Naimushō Kosekiryō*), the office took over the population calculation work from the Ministry of Public Affairs. That same year, the Home Ministry proposed launching an annual population survey to perfect the population data in the first *koseki* register, and an annual nationwide population survey began on January 1, 1874.[67]

Witnessing how the population work in the *koseki* reform was firmly embedded in the nation-building exercise, Sugi once again stressed the significance of population statistics in a way that made it appear relevant to Japan's aim to become a modern state. In addition to "government

---

[62] Endo, *Koseki to kokuseki*, 149.
[63] Ibid., 151–54.
[64] Ibid., 154–59.
[65] Tessa Morris-Suzuki, "Becoming Japanese: Imperial Expansion and Identity Crises in the Early Twentieth-Century," in *Japan's Competing Modernities: Issues in Culture and Democracy*, ed. Sharon Minichiello and Gail Bernstein (Honolulu: University of Hawai'i Press, 1998), 168.
[66] David Luke Howell, *Geographies of Identity in Nineteenth-Century Japan* (Berkeley: University of California Press, 2005).
[67] However, the survey was discontinued in 1877. Endo, *Koseki to kokuseki*, 126.

statistics," Sugi also claimed that statistics in general provided the state with invaluable knowledge about "inhabitants" (*jūmin*), or "people" (*kokumin*), which he characterized as "a nation's weight and aim."[68] He stated, "if a nation embarks on an important project, it has to rely on its inhabitants. For this reason, the number and the state of people (for instance, men and women, marriage, marital relationship, births and death), as well as their occupations and business are the most needed [knowledge] for politics."[69] Sugi further argued that population statistics, showing facts about inhabitants in "big numbers," exhibited people's "living capability" (*seikatsuryoku*) and "growth and withering, rise and fall," which ultimately shaped "a country's state of affairs."[70] The role of statisticians in this context was to clarify patterns in these "big numbers" and provide materials to the political leaders so they could make an informed decision about the future of the country without the people "withering."[71] Sugi's claim represented statisticians' tireless efforts to carve out a niche for population statistics within the government's plan to establish modern sovereignty.

Significantly, through this field-building exercise, Sugi and his fellow statisticians turned out to have captured an image of the population that was distinct from the one prevalent in the previous era. Compared to Honda's characterization of population in the late eighteenth century, which stressed the population's uncontrollable nature – despite following some natural laws – the population described by statisticians in the 1870s and 1880s appeared more susceptible to human intervention *because* it followed patterns.[72] The public lecture Sugi delivered to the Tokyo Academy on November 13, 1887 attests to this. In the lecture, Sugi introduced the census data he gathered in 1879 in Kai Province (most of today's Yamanashi Prefecture). He explained how the census showed some people following "regular rules [*seisoku*]" and others who "succumbed to irregular rules [*hensoku*]."[73] When it came to marriage, for instance, 80 percent of the surveyed population followed the "regular" pattern, namely, a legal marriage with both spouses living. The other 20 percent were "irregular," referring to groups such as "wives whose husbands died" and "husbands who have divorced wives."[74] Sugi then claimed that

[68] Sera, *Sugi sensei kōenshū*, 295.
[69] Ibid., 295.
[70] Ibid., 403–4.
[71] Ibid., 403.
[72] Ishii, "Statistical Visions of Humanity," 34.
[73] Sera, *Sugi sensei kōenshū*, 168.
[74] The complete list was: "wives whose husbands died," "husbands whose wives died," "men whose personal identities [*minoue*] are unclear," "women whose personal

the larger the size of the "irregular" groups became, "the more vulgar and weak a country would become, and it would naturally fall."[75] However, instead of leaving it there, as Honda would probably have done, Sugi recommended that a sign of irregularity, once detected, "should be questioned and researched," and if the irregularity was judged as "negative" (i.e., detrimental to the country), the government should consider "setting up legislations to prevent it from corrupting" the country.[76]

Sugi's lecture, which characterized population as manipulable, clearly indicated the emergence of a new population discourse. However, the significance of Sugi's lecture did not end here. It showed just how congruent the new population discourse was with the government's nation-building effort, which it carried out by modifying people's everyday practices. Sugi's theory of population thus opened up a discursive space in which "population" was transformed into a governable entity, and this justified the government's involvement in people's everyday lives under the name of nation-building.

However, this new population discourse did not emerge out of a vacuum. Behind the statisticians' claims, there was an idea they had just become familiar with: "facts" (*jijitsu*) expressed in large numbers could account for the internal workings of a society. Indeed, statisticians in the 1870s and 1880s were excited that statistics permitted them to see what they called "human society" (*jinrui shakai*) in a novel way.[77] In particular, they found it fascinating that they could extract a society's "natural laws" (*tenpō* or *ten'nen no hōsoku*), or what Sugi sometimes called "reasons" (*dōri*), through an analysis of the numerical "facts."[78] As Sugi's student Kure Ayatoshi also explained, they were able to do this because these numbers were not just random, they revealed regular patterns in human phenomena.[79] Sugi then claimed the pursuit of these numerical "facts" through observation was "the academic style today" and no longer something to be looked down on, as past academics had on "ideological grounds."[80] Sugi and his statistician colleagues were convinced statistics was cutting-edge precisely because of their trust in numerical facts and the empiricism engrained in their pursuit of knowledge.

---

identities are unclear," "wives who have divorced from husbands," "husbands who have divorced from wives," "men whose whereabouts are uncertain," and "women whose whereabouts are uncertain." Ibid.

[75] Ibid.
[76] Ibid., 166.
[77] Ayatoshi Kure, *Riron tōkeigaku, jissai tōkeigaku* (Senshū Gakkō, 1890), 7.
[78] Sera, *Sugi sensei kōenshū*, 147–48.
[79] Kure, *Riron tōkeigaku, jissai tōkeigaku*, 7.
[80] Sera, *Sugi sensei kōenshū*, 147–48.

Sugi's characterization of the "academic style today" was not so far-fetched; within a broader community of social commentators and thinkers, statistical rationality was becoming vogue for the theorization of society. As suggested above, Fukuzawa Yukichi agreed that the numerical facts of people as a collective would display patterns that shaped social phenomena.[81] Inspired by Henry Thomas Buckle's *History of Civilization in England*, Fukuzawa maintained that what determined a society was not the power coming out of a person's individualized free will but natural laws inscribed in the details of people's everyday lives that, if analyzed en masse, exhibited statistical regularities.[82] Fukuzawa thus concluded that statistics, aiming to identify these regularities, was an effective tool for understanding society, and this was behind his endorsement of statistics.[83] As a frequent visitor to gatherings reserved for Meiji luminaries such as Fukuzawa, Sugi had ample opportunity to exchange ideas about the benefits of his science in the intellectual community.[84]

Capitalizing on the intellectual milieu that was largely in favor of statistics, statisticians tried to persuade the government to conduct a scientifically informed, nationwide population census. However, in the early years, the Meiji government dragged its feet regarding the census, although it otherwise fervently endorsed modern science for the sake of nation-building. The story surrounding the census indicates that the relationship between the making of scientific population statistics and the making of a nation-state was more complex than it appeared on the surface.

Statisticians' Struggles to Implement a Nationwide Population Census

Once they embarked on the campaign to popularize statistics in Japan, the biggest ambition of Sugi and his subordinates at the Section of Statistics was to conduct a nationwide population census. As mentioned above, Sugi conducted the population census in Suruga in the late 1860s. However, he was not entirely satisfied with it because it was small in scale, covering only three regions in the domain of Suruga; he thought

---

[81] Ishii, "Statistical Visions of Humanity," 29–32.
[82] Ibid., 29–30.
[83] Ibid., 29.
[84] Sugi was indeed a constituent of the community of Meiji intellectuals. He was a founding member of the *Meiroku zasshi* magazine, which attests to his central position within the society of Meiji luminaries. Sōmushō Tōkeikyoku, "Nihon kindai tōkei no so 'Sugi Koji,'" accessed May 7, 2017, www.stat.go.jp/library/shiryo/sugi.htm.

he could hardly claim to have collected numbers large enough to analyze "natural laws." Wishing to conduct a population census of an old domain, Sugi submitted a proposal to the government in March 1873. The proposal was approved in 1879. On December 31, 1879, Sugi's section conducted the aforementioned demographic survey in Kai Province using techniques derived from the Eighth International Statistics Congress that was held in Saint Petersburg in 1872.[85] The team published the results in 1882 under the title *The Survey of the de facto Population of Kai Province*.[86] The staff at the section hoped the survey would act as a prequel to a full-fledged, nationwide census.

At a glance, this episode seems to imply that Sugi and the government worked well together to gradually implement a nationwide population census over the 1870s and 1880s. In reality, however, their collaboration was not so smooth. To start with, despite Sugi's incessant appeals, it took the government nearly six years to approve the Kai Province population census project. Furthermore, while Sugi wanted the government to support a comprehensive nationwide survey following the Kai Province project, the government only agreed to fund the project and nothing more.[87] Consequently, the census study Sugi's team conducted in the late 1870s was confined to the former *bakufu* domain, covering a mere 110,000 households. Clearly, Sugi and high-rank government officials did not see eye to eye on the population census.

The government's lukewarm attitude to the population census seems at odds with the story presented so far. If the high-rank officials recognized the value of population statistics for nation-building, why did the government hesitate to support the national census promoted by Sugi? There were chiefly two interlinked reasons for the government's response: The institutional constraint and the different opinions about the role of statistics in the governance of Japan's population. The first was derived from the shifting internal structure within the government, which ultimately acted unfavorably toward Sugi. In the 1870s, the government, eager to set up the basic infrastructures of a modern state, expanded the administrative office, but it did so too quickly and haphazardly. Thus, when inflation hit Japan in the early 1880s, the government adopted retrenchment policies and, as part of the policies, downsized statistical bureaus.

---

[85] Sato, *Teikoku nihon*, 184.
[86] "Kai no kuni genzai ninbetsu shirabe [1882]," in *Sōrifu tōkeikyoku hyakunenshi shiryō shūsei*, vol. 2, ed. Sōrifu Tōkeikyoku, (Sōrifu tōkeikyoku, 1976), 161. See also Hayami, "Koji Sugi," 3–5.
[87] Kōji Sugi, "Yamanashi-ken genzai ninbetu shirabe shogen [August 1882,]" in *Sōrifu tōkeikyoku hyakunenshi*, ed. Sōrifu Tōkeikyoku, 162–64.

The restructuring came as a blow to Sugi and his Section of Statistics. From the onset, the Section of Statistics was the weakest of the three statistical offices, lacking in specialisms the other two (the Ministry of Finance's Division of Statistics and the Home Ministry's Sanitary Bureau) had.[88] Furthermore, Ōkuma Shigenobu, who institutionalized statistical work within the Ministry of Finance in the first place, had lost the political battle within the government office and, consequently, in 1880 was demoted to head the Grand Council of State's Department of Accounting, which included Sugi's Section of Statistics.[89] Under the circumstances, the Section of Statistics became exposed to scrutiny when the government decided to reorganize the administrative offices.

After the restructuring, the Section of Statistics ceased to exist, and Sugi was dismissed from the government office.[90] Sugi's dismissal had a lasting impact on the statisticians' campaign to introduce a nationwide population census. With Sugi's departure, bureaucrat-statisticians lobbying for the national census within the government administration lost a leader and came to possess even less power than before. Consequently, their campaign became less effective, and a population census remained a low priority within the government. The specific way the government office developed in the first two decades, and the accompanying internal political struggle, prevented Sugi and his subordinates from effectively persuading the government to set up a national population census system in Japan.

The gap that existed between statisticians and political leaders over the interpretation of official population statistics was another reason why the government was initially reluctant to take up the national census.[91] From the statisticians' point of view, official population statistics should be informed by the principles of statistical science and accurately reflect the demographic reality of the entire population. Based on this understanding, they stressed that only a nationwide census deserved the title of official population statistics, because it aimed to collect data from the whole population through the scientifically rigorous method of fieldwork. Only data compiled in this way would allow population statisticians to conduct an efficient and sophisticated analysis of probability in the patterns of social behavior, and only the natural laws identified in this analysis could present an accurate picture of the current societal

---

[88] Matsuda, "Formation of the Census System in Japan," 49–50.
[89] Shimamura, *Nihon tōkeishi gunzō*, 34.
[90] Sōmushō Tōkeikyoku, "'Nihon tōkei nenkan.'"
[91] Sato, *Kokusei chōsa nihon shakai*, 35–37.

situation, which the government should use to come up with effective strategies for running the country.[92]

Behind this assertion was statisticians' dissatisfaction over the government's use of the *koseki* to calculate population figures. They claimed the current population statistics were incomplete and full of errors because the *koseki* register, compiled by "untrained, low-rank town and village officers," was prone to produce duplicates and omissions of data.[93] Statisticians further argued that the current *koseki* registration system relied on people's goodwill to notify the local authorities about their personal details, and this was causing additional errors in the statistics.[94] Thus, they claimed, a nationwide census conducted by trained fieldworkers should replace the *koseki* register as a source of official population statistics, precisely because the former would generate more accurate and comprehensive knowledge about the Japanese population. This perspective, which no doubt was infused by statisticians' desire to carve out a niche for their science within the government, also undergirded the statisticians' campaign to promote a nationwide population census in Japan.

Unfortunately for the proponents of the census, top government officials did not share this sentiment. As far as the government was concerned, the population data thus far collated through the *koseki* reform, though far from perfect, were sufficient for government reforms.[95] Furthermore, high-rank government officials wished to invest in improving the existing system rather than building an entirely new infrastructure for the census.[96] Based on the calculation of cost against effect, the government decided to prioritize the existing system over a census and, in 1886, conducted another *koseki* reform, which mandated people register births, deaths, and their whereabouts.[97] High-rank government officials did not think the census would add value. This perspective was behind the government's, at best tepid, response to the call for a nationwide census.

Population statisticians were not at all satisfied with the government's decision.[98] Throughout the 1880s, they continued to insist that the government take up a population census. Following in the footsteps of Sugi and Fukuzawa from a decade earlier, proponents of the population census

---

[92] See Kure Ayatoshi, *Jissai tōkeigaku* (Senshū Gakkō, 1895), 79–88.
[93] Sera, *Sugi sensei kōenshū*, 51–52.
[94] Ibid., 48–50.
[95] Ibid., 47.
[96] For the internal politics taking place for the legal reform surrounding the *koseki* during the period, see Endo, *Koseki to kokuseki*, 125–32.
[97] Ibid., 127–30.
[98] For more details on why statisticians mistrusted the *koseki*, see Sato, *Kokusei chōsa to*, 48–52.

stressed that the nationwide census was a symbol of modernity and national power that Japan should be equipped with if the country wished to be seen as a "civilized country" (*bunmeikoku*) by the rest of the world.[99] In 1886, the Tokyo Statistics Association (formerly *Seihyōsha*) submitted the "Proposal for the Population Census" to the newly formed Cabinet Bureau of Statistics (CBS),[100] arguing:

> There is no civilized country with an organized government that does not conduct a population census, ... it is an urgent task of the government to clarify people's power [*minryoku*], specifically, the physical strength, level of knowledge, popular custom, economy and industry of the governed people to which the government serves. This is the reason why [we argue that the government should] adopt population census derived from western countries and consider it to be one of the important national ... projects.[101]

After reading the proposal, some politicians expressed agreement with the statisticians' argument. However, the majority were of the same opinion as the government officials, preferring to adhere to *koseki* reform.[102] Politicians, too, saw little benefit in the government investing in the population census.

However, by the 1890s, the tide had changed. This time, more and more political leaders began to consider endorsing the census. A direct trigger came from outside Japan, from the International Statistical Institute (ISI).[103] During the 1890s, the ISI was planning to compile a worldwide census to commemorate the year 1900. As part of the campaign, it made an inquiry to the Japanese government in 1895, asking if Japan would be interested in participating in the project.

Japanese political leaders in 1895 had plenty of reasons to answer positively to the request from the ISI, especially since the request came at a time when Japan's international position was still precarious. In 1894, Japan took an important step toward revising the unequal treaties by signing the Anglo-Japanese Commercial Treaty, which would end British extraterritorial rights and partially restore Japan's rights of tariff autonomy. Despite the diplomatic success, a long-drawn-out process awaited the country before it could regain complete tariff autonomy. Furthermore, government leaders were to feel bitter about political

[99] Sato, *Kokusei chōsa nihon shakai*, 37–38; Sato, *Kokusei chōsa to*, 23–25.

[100] The CBS was founded in 1885 as a successor organization to the Bureau of Statistics within the Grand Council of State, along with the Grand Council of State's replacement with the cabinet that same year. Upon establishment, it also took charge of the existing *koseki* registration work.

[101] Sōrifu Tōkeikyoku, ed., *Sōrifu tōkeikyoku hyakunenshi shiryō shūsei*, vol. 2 (Sōrifu tōkeikyoku, 1976), 193.

[102] Ishii, "Statistical Visions of Humanity," 87–88.

[103] W. Winkler, *A History of the International Statistical Institute 1885–1960* (Oxford: Blackwell, 1962).

interference from Russia, France, and Germany following the Sino-Japanese War (1894–95), as a result of which Japan, despite being the victorious nation, was forced to relinquish the Liaodong Peninsula, the greatest prize the country had won in the war. In a world order dominated by western imperial powers, political leaders poignantly felt Japan would have to show the world what the country was capable of, even if it was emerging belatedly and as the only non-western imperial power.

Under these circumstances, political leaders understood the ISI's call as a matter of national pride. They considered it a sign that the international community recognized Japan's achievements thus far.[104] Thus, the government acted on the request immediately. As soon as he received the message from the ISI, Itō Hirobumi, now prime minister, forwarded it to the head of the CBS, with a note that the matter required an urgent response.[105] Itō also urged the government to form a census executive committee in the House of Peers and House of Representatives. At the same time, more and more politicians came to endorse the argument statisticians had been presenting for a long time: Japan, as the "civilized nation of the East," ought to conduct a population census.[106] As a result, in 1902, the government issued the Population Census Law. The external pressure acted as a catalyst for the implementation of a national population census within Japan's governing body.

Nevertheless, even after the law was enacted, the path toward materializing the census was not straightforward. The original 1902 law stipulated that the first national census would be taken on October 1, 1905 and would cover Hokkaido, Okinawa, South Sakhalin, and Taiwan in addition to the Japanese archipelago. However, the plan had to be shelved in the wake of the Russo-Japanese War (1904–5).[107] In the end, it was only in 1920 that the first national census took place in Japan.[108]

In contrast, in colonial Taiwan, the census was taken in 1905 as originally planned according to the 1902 law. The call for the first census in Taiwan was enabled by the political structure and discourse of scientific colonialism specific to the Japanese rule there.

---

[104] Takahashi Jirō, "Meiji jūninenmatsu Kai no kuni genzai ninbetsu shirabe tenmatsu," August 1905, in Sōrifu Tōkeikyoku ed., *Sōrifu tōkeikyoku hyakunenshi*, 189.

[105] Sato, *Kokusei chōsa to*, 26. For the correspondence between the ISI and the Japanese government on this subject, see Sōrifu Tōkeikyoku, *Sōrifu tōkeikyoku hyakunenshi*, vol. 2, 201–6.

[106] Sato, *Kokusei chōsa to*, 26–27.

[107] Sōrifu Tōkeikyoku, *Sōrifu tōkeikyoku hyakunenshi*, 972–78.

[108] For the actual process of implementing the 1920 census, see Sato, *Kokusei chōsa nihon shakai*, 43–64; Sato, *Kokusei chōsa to*, 53–67, 103–39.

Population Census in Taiwan: Scientific Colonialism in Action

In stark contrast to Japan, in Taiwan, which became Japan's first official colony in 1895 as a result of Japan's victory in the first Sino-Japanese War, the colonial government enthusiastically took up the population census.[109] On October 1, 1905, the Government-General of Taiwan (GGT) organized the first population census, the Temporary Taiwan Household Investigation.[110] Thereafter, under the aegis of the GGT, the population census was conducted every five years, on October 1, until the end of Japanese colonial rule. In addition, the GGT compiled and published vital statistics based on the population census. In Taiwan, census-taking quickly developed into routine work within the colonial administration.

Indeed, in colonial Taiwan, statistics in general enjoyed a high status, being "at the heart of colonial statecraft from the beginning."[111] Already in 1897, only two years after Japan occupied the land, the GGT compiled statistics on matters relating to public health.[112] However, more fully-fledged statistical surveys bloomed soon after Gotō Shinpei (1857–1929) arrived in Taiwan in 1898 as the civilian governor directly

---

[109] Pei-Hsin Lin, "The Unfolding and Significance of the Temporary Taiwan Household Investigation in Japanese Taiwan (1905–1915)" [in Chinese], *Cheng Kung Journal of Historical Studies* 45 (December 2013): 87–128; Ishii, "Statistical Visions of Humanity," 73–107; Sato, *Teikoku nihon*; Mau-Shan Shi et al., "A Conversion of Population Statistics of Taiwan at the Sub-Provincial Layer: 1897–1943" [in Chinese], *Renkouxuekan*, 2010, 157–202, https://doi.org/10.6191/jps.2010.4; Jen-to Yao, "The Japanese Colonial State and Its Form of Knowledge in Taiwan," in *Taiwan Under Japanese Colonial Rule, 1895–1945: History, Culture, Memory*, eds. Binghui Liao and Dewei Wang (New York: Columbia University Press, 2006), 37–61; Kurihara, Jun, "The National Census, Family Registrations and the Extraordinary Taiwan Census of 1905" [in Japanese], *Tokyo joshi daigaku hikaku bunka kenkyūsho kiyō* 65, 2004, 33–77. In contrast, in Korea under Japanese rule (1910–45), it was not until 1925 that even a simplified general census was taken. Following the census law in the metropole, the general population census was scheduled for 1920, but it was cancelled due to the disruptions of the March First Movement and the independent movement that ensued in 1919. For a fresh take on gender and the workings of the household registry in colonial Korea, see Sungyun Lim, *Rules of the House: Family Law and Domestic Disputes in Colonial Korea* (Berkeley: University of California Press, 2018).

[110] Yao, "The Japanese Colonial State," 53.

[111] Ibid., 42.

[112] Michael Shiyung Liu, *Prescribing Colonization: The Role of Medical Practices and Policies in Japan-Ruled Taiwan, 1895–1945* (Ann Arbor, Michigan: Association for Asian Studies, 2009); Chin Hsien-Yu, "Colonial Medical Police and Postcolonial Medical Surveillance Systems in Taiwan, 1895–1950s," *Osiris* 13, no. 1 (January 1998): 326–38.

accountable to the governor-general, Kodama Gentarō (1852–1906).[113] The GGT under Kodama and Gotō quickly sponsored a succession of statistical surveys.[114] In September 1898, it established the Temporary Land Survey Group to prepare for a Taiwan-wide land survey. In 1901, it further set up the Temporary Group for the Investigation of Traditional Customs in Taiwan, which aimed to study the legal structures, kinship, rituals, and customs of Taiwan.[115] Finally, in 1905, the aforementioned population survey was conducted under the aegis of the GGT.

Gotō's enthusiasm for the statistical surveys, including the population census, was informed by the specific understanding of Japan's colonies and colonial development that he had nurtured over the years as a medically trained bureaucrat who became interested in German colonial policy.[116] Through his experience as a high-rank officer serving the Home Ministry's Sanitary Bureau, he came to hold a unique view on colonies, which he expressed through the term "biological principle" (*seibutsugaku no gensoku*).[117] Using this term, Gotō likened a colony to a human body. Just as eyes and arms have prefixed functions for the human body, existing cultural practices and social organizations fulfill certain predetermined roles for the colony. Based on this idea, Gotō maintained that colonial rule was best done when it made use of existing local customs and structures. As a person would not get rid of body parts and replace them with prostheses each time they became ill, a colonial government should not wipe out existing local customs or impose entirely new

---

[113] There are a countless number of biographies of Gotō but for a more recent and comprehensive one, see the volumes authored by Yusuke Tsurumi, *Seiden Gotō Shinpei* (Fujiwara Shoten, 2004).

[114] Akihiro Nomura, "Shokuminchi ni okeru kindaiteki tōchi ni kansuru shakaigaku: Gotō Shinpei no Taiwan tōchi wo megutte," *Kyoto shakaigaku nenpō* 7 (1999): 1–24.

[115] Makito Saya, *Minzokugaku, Taiwan, kokusai renmei: Yanagita Kunio to Nitobe Inazō* (Kodansha, 2015); Katsumi Nakao, "Taihoku teikoku daigaku bunsei gakubu no dozoku jinshugaku kyōshitsu ni okeru fīrudo wāku," in *Teikoku nihon to shokuinchi daigaku*, eds. Sakai Tetsuya and Matsuda Toshihiko (Yumani Shobō, 2014), 221–50; Timothy Y. Tsu, "Japanese Colonialism and the Investigation of Taiwanese 'Old Customs,'" in *Anthropology and Colonialism in Asia and Oceania*, eds. Jan Van Bremen and Akitoshi Shimizu (London: Routledge, 1999), 197–218.

[116] Haruyama Meitetsu, "Meiji kenpō taisei to Taiwan tōchi," in Shinobu Oe et al., eds., *Iwanami kōza kindai nihon to shokuminchi 4 tōchi to shihai no ronri* (Iwanami Shoten, 1993), 47–48; Mark R. Peattie, "Japanese Attitudes toward Colonialism, 1895–1945," in *The Japanese Wartime Empire, 1931–1945*, eds. Peter Duus, Ramon H. Myers, and Mark R. Peattie (Princeton: Princeton University Press, 1996), 80–127.

[117] Yao, "The Japanese Colonial State," 45–47.

systems in order to maintain the health of a colony.[118] This bodily metaphor buttressed Gotō's attitude toward Japan's colonial rule.[119]

Gotō then applied the "biological principle" to develop a theory of "scientific colonialism," which he claimed was a systematic and research-driven approach to colonial development that Japan should incorporate.[120] Gotō contended that scientific investigations into the preexisting cultural, social, political, and environmental conditions would yield in-depth knowledge of the colony, and the knowledge gleaned from such organized studies would ultimately lead to more efficient colonial management based on the effective application of the local systems that were working well. Through this theory, Gotō endorsed colonial studies as a scientific field and campaigned for the creation of the chair of Colonial Studies at the University of Tokyo, which came to fruition in 1908.

This was the backdrop against which Gotō pressed for statistical surveys as soon as he arrived in Taiwan. With absolute trust in large numbers, Gotō was convinced that statistical knowledge about land, customs, and population would display natural laws governing the colonial society, and he judged that the statistical surveys providing such knowledge would be essential for materializing the vision of colonial governance expressed in the "biological principle." Specifically in Taiwan, statistical knowledge would build foundations for policymaking, which would then be used to cultivate the island as Japan's "model colony."[121] In Gotō's terms, statistical surveys, including the population census, were one critical item ensuring his grand experiment with "scientific colonialism."

The 1905 population census displayed many features of the scientific investigations conducted under the banner of "scientific colonialism." Most conspicuously, the preparation for the census was based on a top-down command system, with the GGT at the top. After issuing the Population Census Law in the metropole in 1902, the GGT set up

---

[118] Yet, the support for local autonomy and respect for local cultures implied in the "biological principles" did not automatically lead to peaceful governance. In fact, the "biological principle" justified an armed control of the factions the GGT deemed to be the ills of a colonial society. As Gotō testified, under his administration a total of 11,950 "native bandits" (dohi) were executed between 1898 and 1902. Shin'ichi Kitaoka, Gotō Shinpei (Chuokoron-sha, 1988), 44.

[119] Huiyu Cai, Taiwan in Japan's Empire-Building an Institutional Approach to Colonial Engineering (New York: Routledge, 2009).

[120] For "scientific colonialism," see Ming-cheng Miriam Lo, Doctors within Borders: Profession, Ethnicity, and Modernity in Colonial Taiwan (Berkeley: University of California Press, 2002), 35–39; Peattie, "Japanese Attitudes," 83–86.

[121] Masahiro Sato, "Chōsa tōkei no keifu: Shokuminchi niokeru tōkei chōsa shisutemu," in "Teikoku" nihon no gakuchi dai 6 kan kenkyū chiiki toshite no ajia, ed. Akira Suehiro (Iwanami Shoten, 2006), 191.

the Temporary Taiwan Population Survey Group (TTPSG), with Gotō assuming its directorship. Between 1902 and 1905, the TTPSG created the top-down population census network by liaising with local authorities, while also keeping in close communication with the administrative offices in Tokyo, most notably the CBS. In August 1904, the TTPSG organized a pilot survey in Taoyuan as part of a training course. On January 1, 1905, leading up to the actual survey, the TTPSG conducted a local survey in Taipei.

However, while at the top of the command chain, the GGT administrators did not work alone. In fact, the police played a pivotal role in the execution process.[122] For the census, the GGT appointed the Commander in Chief of Police Inspectors Ōshima Kumaji (1865–1918), vice-director for the TTPSG. Together with Ōshima, Gotō decided the census fieldwork would be arranged through the police network and policemen would be mobilized as fieldworkers for the census.[123] The police, thus, provided the bulk of resources for the census work.

The reasons for the involvement of the police in the census work were multifaceted. First, Gotō, once a student of the *Medizinische Polizei* system of Prussian Germany, trusted the police when it came to medical and public health administration and saw them as a positive and productive force for colonial governance.[124] Second, the police had the network appropriate for the census work. Since the 1901 bureaucratic reform, the police had an island-wide network evolved from the existing mutual surveillance structure called *baojia*, and it was hoped that this *baojia* system, embedded within the administration of local affairs, would facilitate the fieldwork.[125] Finally, the police were already conducting a household survey as part of the attempt to implement the *koseki* registration system in Taiwan, and the GGT concluded it would be more efficient to

---

[122] Ishii, "Statistical Visions of Humanity," 78–85.

[123] Sato, *Teikoku nihon*, 228–29.

[124] Ishii, "Statistical Visions of Humanity," 81–82. Historians of medicine have described the *Medizinische Polizei* system as a state system in Germany that aimed to ensure the health of citizens by deploying various forms of policing, e.g., the policing of physical environments, dangerous materials, elements of everyday lives deemed hazardous, and practices and individuals deemed "nuisances." For works describing how the originally German *Medizinische Polizei* system was tied to Japanese colonial rule in Taiwan, see Liu, *Prescribing Colonization*; Hsien-Yu, "Colonial Medical Police." 326–38.

[125] Hui-Yu Ts'ai, "Shaping Administration in Colonial Taiwan, 1895–1945," in *Taiwan Under Japanese Colonial Rule, 1895–1945: History, Culture, Memory*, eds. Binghui Liao and Dewei Wang (New York: Columbia University Press, 2006), 99–104; Ching-Chih Chen, "The Japanese Adaptation of the Pao-Chia System in Taiwan, 1895–1945," *The Journal of Asian Studies* 34, no. 2 (February 1975): 391–416.

take advantage of the existing practice instead of making a completely new system from scratch.[126] Based on these factors, Gotō thought police involvement would make census work more efficient. In turn, police participation in census work indicated another important aspect of colonial governance. In contrast to the ideal presented by Gotō, which alluded to local autonomy, in reality, the scientific knowledge buttressing the colonial rule under the Kodama-Gotō administration was premised on the modern system of control and surveillance that encroached on people's everyday lives.

The census illustrates another aspect of the scientific colonial rule of Taiwan that is linked to the point above: Japanese staff dominated the census's organization. The overwhelming majority of local fieldworkers and supervisors recruited for the census work were Japanese; the ratio of Japanese to local Taiwanese officers was nine to one.[127] One reason for this outcome was the suspicious attitude toward the native administrators that some high-rank Japanese officers harbored. For instance, Ōshima noted that "low-rank administrative personnel" in the police organization could "obstruct our effort by spreading groundless rumors."[128] Some Japanese officials believed it would be inappropriate to involve these local officers because of the sensitive nature of the census data. They pointed out that the colonial government might use these data for intelligence purposes, thus the data should be kept among the Japanese and not shared with the local officers.[129]

This kind of condescending attitude toward the local populations, which shaped the 1905 census, was also a characteristic of "scientific colonialism." "Scientific colonialism," as Mark R. Peattie once explained, also referred to "a way of looking at differences in political capacity between ruler and ruled," wherein the superiority of the ruler was implied.[130] This "way of seeing" directly shaped the process of organizing the population census. The census organizers insisted on Japanese control over the investigation because they were convinced of the natural superiority of the Japanese in the political *and* scientific management of the colony. The domination of the Japanese staff in the population surveys was therefore indicative of the Japanese attempt to assert their authoritative position based on the line they had drawn between

[126] Endo, *Kinai Nihon no*, 137–38.
[127] Sato, *Teikoku nihon*, 244.
[128] Cited in Sato, *Kokuzei chōsa to*, 74.
[129] Ibid., 74–75.
[130] Peattie, "Japanese Attitudes," 85.

the Japanese as the colonial investigator and the colonial subjects as the object of investigation.[131]

However, this clear-cut hierarchical positioning between the ruler and the ruled did not always apply in the actual fieldwork.[132] First, while the census's primary target group was the local populations, the census also aimed to comprehensively investigate the total population *in* Taiwan. The census therefore ended up collecting data from Japanese expats, making the Japanese colonizers as exposed to the demographic survey as the colonial subjects. On the one hand, this practice threatened the binary of Japanese colonizer-investigator versus the colonized Taiwanese under investigation. On the other, by covering the whole population living on the island, the census was actually assisting the GGT to govern colonial Taiwan more effectively than otherwise. Evidence shows that Japanese expats in the early years of Japan's colonial rule in Taiwan were as diverse as the local populations, ranging from socially respected high-rank bureaucrats and businesspersons to hustlers, human traffickers, and others thriving on the margins of the emerging nation-states.[133] In this context, the census, uncovering the details of the whereabouts of Japanese expats of all walks of life as much as those of the local populations, would work as a technology of surveillance that would buttress the efficient colonial governance, even though it might have eroded the framework that expressed the idealized relationship between the colonizer and colonized.

The actual work involved in collecting the census data was another site where the dichotomous framework was overridden. In day-to-day fieldwork, Taiwanese fieldworkers contributed as much as the Japanese, mostly acting as interpreters for the Japanese fieldworkers. For the 1905 census, 1,431 out of a total of 4,369 badges – handed out to the fieldworkers to indicate their participation in the census project – were reserved for local interpreters.[134] These interpreters played a pivotal role in the success of the 1905 census. They facilitated the census-taking process by liaising between the Japanese fieldworkers and the local research subjects who shared neither linguistic traditions nor attitudes to census-taking. Evidence also suggests they might have brought success to the

---

[131] This was done in parallel with various attempts to "Japanize" the people in Taiwan. Leo T. S. Ching, *Becoming "Japanese": Colonial Taiwan and the Politics of Identity Formation* (Berkeley: University of California Press, 2001).

[132] Toru Sakano, "Joron 'teikoku nihon' 'posto teikoku' jidai no fīrudowāku wo toinaosu," in *Teikoku wo shiraberu*, ed. Toru Sakano (Keiso Shobo, 2016), 4.

[133] David Richard Ambaras, *Japan's Imperial Underworlds: Intimate Encounters at the Borders of Empire*, Asian Connections (Cambridge: Cambridge University Press, 2018).

[134] Sato, *Teikoku nihon*, 224–25.

census indirectly by enhancing local populations' levels of understanding about the census.[135] Local fieldworkers were certainly indispensable cogs in the operation by playing effective mediator roles.[136]

In part thanks to the smooth collaboration between the Japanese and local fieldworkers, the 1905 census went relatively well. Yet, this was not to say the fieldwork was without complications. While the Japanese plan to conduct a population census came to be understood among the upper echelons of local communities, it seemed to cause some rumors and confusion among many population groups, and this disrupted the census survey to some extent. For instance, many local people panicked because they believed the Japanese government was conducting the census to impose tax or conscription duties on the natives.[137] Others feared they would lose their nationality or a position in the household register if they were not registered in their hometowns. For this reason, many returned home immediately before the time designated for the census, which ended up causing heavy traffic and overcrowded public transportation.[138] Some women who had abandoned the custom of foot-binding bound their feet specifically for the occasion, because they thought they would lose the right to bind feet if they were registered as having "natural feet."[139] Others pretended they were blind, hoping they would receive government subsidies.[140] The census work was carried out while negotiating these chaotic situations it had caused.

The result of the 1905 population census was over 450 pages of tabulations published by the TTPSG in 1908. An impressive array of figures described diverse aspects of the demographic and living conditions of the colonial subjects, of the Japanese expats recorded as *naichijin*, and of other "foreigners" (*gaikokujin*) in Taiwan at the time.[141] It covered topics as diverse as population, race, physicality, age, kinship, occupation (adults and children), language, education, disability, foot-binding practice, the place of origin of the Japanese living in Taiwan, the details of travel to Taiwan experienced by the Japanese, foreign nationals, the

---

[135] Ibid., 223–26.
[136] In addition to the interpreters, the leaders of the local *baojia* system also played a pivotal role, even though they were not officially registered as local collaborators. For details, see Ibid., 230.
[137] Lin, "The Unfolding and Significance," 107.
[138] Ibid., 108.
[139] Ibid., 109.
[140] Ibid.
[141] Rinji Taiwan Kokō Chōsa Bu, "Meiji sanjū hachinen rinji taiwan kokō chōsa kekkahyō," National Archives of Japan Digital Archive (1908), accessed August 18, 2019, www.digital.archives.go.jp/das/image/F1000000000000061277.

use of opium, household, and finally, type of domicile.[142] As Jen-to Yao once argued, through the census, everything became unambiguous, and nobody in the colony had "the privilege of remaining anonymous."[143] Furthermore, the knowledge, by presenting colonial reality in numbers, transformed the population "from estimation and imagination to calculation and classification."[144] The numerical facts in the census offered a foundation for the colonial governance by displaying the lives of Taiwan's inhabitants as a mathematically categorizable, clear-cut entity, and in so doing, turned the people living in Taiwan into a more governable population.

The making of the 1905 census thus exhibits how statistics became quickly integrated into the colonial administration in Taiwan, in part due to Gotō's fervor for scientific colonialism. At the same time, it also shows how the census work in colonial Taiwan paved the way for the development of a statistical community there. Furthermore, the work involved in the preparations for the census also shaped statistical practice in the metropole, suggesting that the science surrounding population statistics developed in tandem with the construction of the Japanese Empire through coordination between the metropole and the colony.[145]

The science of Population Statistics in Taiwan and "Japan Proper"

Since Gotō firmly believed in "scientific" colonial management, he considered the statistical science that Sugi and his colleagues had been promoting in Japan since the 1860s should be the pivot for the census work in Taiwan, too. Thus, to prepare for the 1905 census, Gotō invited Mizushina Shichisaburō from Japan to act as the GGT technical bureaucrat specializing in statistics.

---

[142] Su-chuan Chan, "Identification and Transformation of Plains Aborigines, 1895–1960: Based on the 'Racial' Classification of Household System and Census" [in Chinese], *Taiwanshi yanjiu* [Taiwan historical research] 12, no. 2 (December 2005): 134–37; Akira Tomita, "1905-nen rinji Taiwan kokō chōsa ga kataru Taiwan shakai: Shuzoku, gengo, kyōiku wo chūshin ni," *Nihon taiwangakkaihō* 5 (May 2003): 87–106.

[143] Yao, "The Japanese Colonial State," 56.

[144] Ibid., 54.

[145] Sato, *Teikoku nihon*; Sato, "Chōsa tōkei no keifu," 179–204. Further research is required to uncover how multivalent politics in Japan's colonies – including not only Taiwan but also Korea and other "informal" colonies – shaped population statistics in the metropole and the colonies.

Mizushina arrived at statistics in the 1880s after starting his career in meteorology.[146] In 1883, he entered the Kyōritsu School of Statistics to improve the statistical side of his meteorological work. After graduating, Mizushima initially returned to meteorological work, first at the Nemuro Weather Station and then at the Hokkaido Prefectural Government. However, in the 1890s, Mizushina set up prefectural-level statistics training courses with colleagues he had met at the Kyōritsu School of Statistics. Eventually, in 1899, the Ministry of the Navy appointed him to work on general statistical administration and education. In August 1903, probably through his connection with Nitobe Inazō (1862–1933), who acted as his supervisor while in Hokkaido and moved to Taiwan thereafter, Mizushina was appointed to establish the statistical administration in Taiwan.[147] He stayed in the position until the 1915 census was conducted.[148]

Mizushina was the general manager of the 1905 census. He worked directly under Gotō and Ōshima in the TTPSG. He laid the groundwork for the fieldwork, and once the fieldwork was completed, he oversaw the compilation of raw data and the publication of the aforementioned report. In December 1903, he toured Taiwan and the Pescadores Islands (Penghudao) to investigate the current state of the infrastructure for the census fieldwork.[149] Based on the trip, he made recommendations to the GGT in February 1904. These recommendations were detailed, but they mostly aimed to explain the locally variable cultural practices to the Japanese fieldworkers. For instance, he called for the documentation of reference tables that listed local traditional Chinese medicine names for diseases and presented them alongside the corresponding modern medicine names.[150] Mizushina's legwork eventually helped to generate the relatively smooth operation of the 1905 census work.

In addition, to aid the census work, Mizushina ran the training sessions under the aegis of the GGT. Based on the Regulation for GGT Statistics Training issued in September 1903, Mizushina organized the

---

[146] For Mizushina's biography, see Yoshinori Ishimura and Sakura Ishimura, "'Mizushina Shichisaburō' nōto (oboegaki): Hokkaido ni okeru sangaku kōsō kansoku, kishō kansoku gyōsei oyobi tōkei kyōiku, Taiwan kokō chōsa wo chūshin toshite," *Takushoku daigaku ronshū* 2, no. 3 (July 1994): 143–95.

[147] Tiejun Wang, "Kindai nihon bunkan kanryō seido nonakano Taiwan sōtokufu kanryō," *Chūkyō hōgaku* 45, no. 1–2 (2010): 117. Nitobe was another important individual lobbying to make Japanese population science an international science that was simultaneously embedded in Japan's colonial rule.

[148] Ishimura and Ishimura, "'Mizushina Shichisaburō' nōto," 184–85.

[149] Sato, "Chōsa tōkei no keifu," 192–93.

[150] Ibid., 192–93.

training course between October 20 and November 24, targeting senior officers in the local authorities who would be involved in the administration of fieldwork. In the following year, he ran the same course, this time aimed at officers working for the Ministry of Arms.[151] From February 1904 onward, the GGT mandated that police superintendents and prison guards be taught about the population census in their respective training courses, which Mizushina was involved in organizing. Between 1903 and 1911, with Mizushina's help, the GGT organized a total of six training courses.

Mizushina's training work seemed to catalyze the local-level training courses targeting the fieldworkers. Between 1903 and 1919, a total of fifty-five training courses of this kind were organized in the *chō/ting*, *gaishō/jiezhuang*, and other, smaller administrative units. Most of the attendees of these local-level courses were policemen and other low-rank administrators. Through the training, they learned both about the theories of statistics and the techniques in actual practice. Though they were expected to study statistics in general, given the imminence of the census work, the population census occupied a large part of the learned content. Including those in the GGT training sessions, a total of 1,968 mid- to low-rank civil servants were taught statistics in Taiwan between 1903 and 1919, and they supported the first and second (1915) population census surveys.

Through the training activity, a community of statisticians was quickly formed in colonial Taiwan. Mizushina also played a central role in the development of this community. It was Mizushina who launched the Taiwan Statistical Association that was attached to the GGT training courses. Mizushina also safeguarded the association's *Journal of the Taiwan Statistical Association* as its editor-in-chief, and the journal published between 1,000 and 2,000 copies of each issue.[152] Finally, Mizushina ensured that the GGT training sessions fostered a sense of community through events such as a social gathering held after completing the courses.[153] Through these activities, Mizushina contributed to the development of a community around statistical work.

Thus, the scientific field of population statistics developed in Taiwan was certainly situated in a colonial context. Furthermore, among the Japanese based in Taiwan, there was a high hope that the knowledge

---

[151] By 1905, a total of 540 personnel had taken part in the training sessions organized by the GGT. Sato, *Kokusei chōsa to*, 72.

[152] Masuyo Takahashi, "'Taiwan tōkei kyōkai zasshi' sōmokuji kaidai" (Hitotsubashi University Research Unit for Statistical Analysis in Social Sciences, May 2005).

[153] Sato, *Teikoku nihon*, 21–23.

the community of statisticians had produced would genuinely facilitate the governing of populations in Taiwan. Attesting to the shared sense of optimism, the highest echelon of Taiwan's Japanese expat communities, as well as generally renowned Japanese figures in business and state bureaucracy, including Nitobe Inazō, Yagyū Kazuyoshi, Kinoshita Shinsaburō, and Nakamura Tetsuji, took time to attend the grand launch ceremony of the Taiwan Statistical Association.[154] In Taiwan, the growth of population statistics certainly hinged on the local infrastructure supporting the island's transformation into a Japanese colony.

At the same time, census work in Taiwan had a symbiotic relationship with the development of the science of population statistics of "Japan Proper." Having overseen the second population census in Taiwan, in 1919, Mizushina was transferred to Tokyo, this time appointed to serve the CBS as a commissioned technical bureaucrat. At the CBS, he joined forces with the team charged with organizing the first census in Japan, which, as mentioned above, was eventually conducted in 1920. Once back in Tokyo, Mizushina was also elected to serve as a councilor for the Tokyo Statistics Association. Finally, once the first census work was done, he left the CBS in 1924 to teach statistics at the Tokyo-based Takushoku University, which maintained strong ties with the colonial government in Taiwan.[155] Arguably, the early years of the colonization of Taiwan offered a critical background to what statistician Matsuda Yoshio once called a "great leap toward the modernization of the statistical survey system" in Japan between the 1890s and 1920s.[156] In part, via the network centering around Mizushina, the expertise in building population statistics that was nurtured in colonial Taiwan circulated between metropole and colony, making the metropolitan and colonial contexts more intertwined than before.

Conclusion

As a subdiscipline of modern statistics, population statistics owes much to the dramatic sociopolitical transformation Japan witnessed from the 1860s onward. It emerged along with Meiji statesmen's efforts to build a modern nation-state and empire, which necessitated numerical

---

[154] Sato, *Teikoku nihon*, 10. Yagyū was the president of the Bank of Taiwan, Kinoshita the chief editor of the *Taiwan Nichinichi Shinpō* newspaper, and Nakamura the executive director of Taiwan Minpōsha.
[155] Ishimura and Ishimura, "'Mizushina Shichisaburō' nōto," 184–85.
[156] Sato, *Teikoku nihon*; Kurihara, "'Taiwan sōtokufu kōbunruisan'"; Lin, "The Unfolding and Significance."

knowledge that would be readily applicable to their efforts. At the same time, population statistics was actively promoted in the burgeoning intellectual scenes where bureaucrat-statisticians like Sugi and a new generation of thinkers caricatured European statistics as a civilizing tool that could potentially enhance Japan's national power. In this context, modern statistics presented a concept of "population" that resonated with Japan's nation- and empire-building efforts.

However, the story of the population census also confirms how complex the relationship between science and politics was. Despite their coterminous position with political power, Sugi and his colleagues failed to persuade government officials to conduct a scientific nationwide population census within Japan. Instead, the population census thrived in colonial Taiwan, where high-rank officials such as Gotō believed in its usefulness as a tool of governance. The census-taking expertise accumulated in Taiwan then helped to implement the census in the metropole via bureaucrat-statisticians such as Mizushina. The trajectory of the population census was not determined merely by the statisticians' position in relation to the political organization but was also contingent upon the practice's perceived utility for the demands of the political exigencies.

Vital statistics was another kind of population statistics that intimately interacted with national politics. Similar to the population census, vital statistics evolved into a technology of statecraft over the Meiji period. Along with this, the profile of midwives as witnesses to births and deaths in childbirth labor changed greatly, from a group subject to state surveillance to healthcare professionals ensuring the smooth operation of the state health administration. The following chapter describes how medical midwifery developed in tandem with vital statistics.

2   Medical Midwifery and Vital Statistics
For the Health of Japan's Population

The Essentials for Midwives and Nurses, published in the provincial city of Gifu (approx. 135 kilometers northeast of Kyoto) in April 1902, was a rather plain booklet.[1] The booklet taught local midwives about the latest state legislation regulating their professional conduct. It opened with the Midwives' Ordinance, the most important state regulation for midwives, issued as an imperial edict in 1899. In the middle of the booklet, twelve (out of the booklet's eighty) pages were dedicated to the new ministerial ordinance issued in 1910 by the Home Ministry, which was in charge of the central medical and public health administration. It instructed them on how to fill out a death certificate.[2] Just by looking at the booklet, midwives could tell exactly what the state expected from them during their everyday work.

The booklet, despite its sober appearance, tells us a lot about the state of midwifery in Japan at the turn of the twentieth century.[3] In particular, it shows the extent to which medical midwifery had become a state matter by this period. As the booklet indicates, the state provided the law defining midwives' expertise and also set up regulations dictating their

---

[1] Katsumu Katayama, Sanba Kangofu Hikkei (Gifu-shi, Japan, 1902), accessed June 1, 2022, https://ndlonline.ndl.go.jp/#!/detail/R300000001-I000000475584-00.

[2] Ibid., 16–27.

[3] Academic works on the history of modern midwifery in Japan abound. Those published in the last ten years include: Manami Abe, "Meijiki no Osaka niokeru sanba seido no hensen," Nihon ishigaku zasshi 65, no. 1 (2019): 3–18; Terazawa, Knowledge, Power, and Women's Reproductive Health; Chiaki Shirai, ed., Umisodate to josan no rekishi: Kindaika no 200-nen wo furikaeru (Igaku Shoin, 2016); Naoko Kimura, Shussan to seishoku womeguru kōbō: Sanba, josanpu dantai to sankai no 100-nen (Otsuki Shoten, 2013); Aya Homei, "Midwife and Public Health Nurse Tatsuyo Amari and a State-Endorsed Birth Control Campaign in 1950s Japan," Nursing History Review 24, no. 1 (January 2016): 41–64; Aya Homei, "Midwives and the Medical Marketplace in Modern Japan," Japanese Studies (Australia) 32, no. 2 (2012): 275–93. For an example of the canonical works, see Mugiko Nishikawa, Aru kindai sanba no monogatari: Noto, Takeshima Mii no katari yori (Toyama: Katsura shobo, 1997); Michiko Obayashi, Josanpu no sengo (Keiso Shobo, 1989).

everyday work. The state had an overpowering presence in the work lives of medical midwives, even in a provincial city like Gifu.

Indeed, the state was at the center of the creation of medical midwifery in Japan. In the 1870s, the nascent government proactively worked to replace "old midwives" – those perceived as outdated, vernacular, and superstitious granny midwives – with "modern midwives," licensed midwives familiar with the principles of medicine and hygiene derived from Western Europe. The government's enthusiasm for midwifery reform partly came from the consensus that modern medicine and public health – the areas midwifery was immediately associated with in the process of nation-building – were a critical foundation for making Japan a civilized modern state and empire.[4] To a great extent, due to the assigned role of medicine and public health within nation-building, the state was heavily involved in introducing medical midwifery into Japan and turning it into an auxiliary field of state-sanctioned modern medicine integral to the state public health system.

The elephant in the room in this narrative, which I argue was a defining factor in justifying state involvement in midwifery reform, was the specific concept of population that emerged in the process of constructing a modern state health administration. Similar to the idea of population presented by Sugi and his fellow statisticians, the notion of population that prevailed in the state health administration was a dynamic force that directly shaped "national power" (*kokuryoku*). However, compared to Sugi's conceptualization of population, this discourse stressed its corporeal aspect; population as an aggregate of biological bodies that reproduce, grow, fall ill, age, and perish. In the process of nation-building, this formulation of population made midwifery a concern of the state.[5] Midwives were associated with birth and death in childbirth, which were among the most important events for the population as a biological entity.

---

[4] Hoi-eun Kim, *Doctors of Empire: Medical and Cultural Encounters between Imperial Germany and Meiji Japan* (Toronto: University of Toronto Press, 2016); Masahira Anesaki, "History of Public Health in Modern Japan: The Road to Becoming the Healthiest Nation in the World," in *Public Health in Asia and the Pacific: Historical and Comparative Perspectives*, eds. Milton James Lewis and Kerrie L. Macpherson (London: Routledge, 2011), 55–58; Susan L. Burns, "Constructing the National Body: Public Health and the Body in Nineteenth-Century Japan," in *Nation Work: Asian Elites and National Identities*, eds. Timothy Brook and André Schmid (Ann Arbor: University of Michigan Press, 2000), 17–49.

[5] Ishizaki, *Kingendai nihon no*, 13–14, 70–77; Yuki Fujime, *Sei no rekishigaku: Kōshō seido, dataizai taisei kara baishun bōshihō, yūsei hogohō taisei e* (Fuji Shuppan, 1997), 117–18.

This chapter reappraises the history of medical midwifery and statecraft in modern Japan with this idea of population in mind. Specifically, it locates the development of medical midwifery within the making of a state health system predicated on this idea of population. Thus, I tell the story of medical midwifery alongside the government's endeavor to compile vital statistics.[6] Vital statistics – the collection, classification, recording, and preservation of the numerical facts about people's life events, such as birth, marriage, and death – is a great lens through which to see how the notion of a corporeal population buttressed the state health system, as well as negotiations for establishing the relationship between midwives and the modern state. In contrast to the population census, the government had already set up vital statistics in the state health administration in the 1870s.[7] This happened precisely because high-rank health officials quickly adopted the idea that the sum of people's bodily experiences represented the nation's health and wealth, and vital statistics, which quantified these experiences, was an effective tool for visualizing the actual state of the nation in tangible numbers. This idea of corporeal population also exhorted the government to take the lead in midwifery reform. The government acted, expecting that reformed midwives would consolidate the nation's power by improving women's bodily experiences during pregnancy and at birth through the application of modern, medical, and hygienic birth attendance methods. In the 1880s, the government once again reached out to midwives, this time including them in its effort to improve vital statistics.[8] Consequently, midwives became even more firmly entrenched in a state health system that aimed to promote the health of Japan's population.

Within vital statistics, death, especially infant mortality, was where medical midwifery and state health politics crossed paths the most.[9] The

---

[6] A common method in historical demography is to corroborate a hypothesis by comparing historical phenomena against a statistical trend of the time. Kyoko Miyamoto, "Meiji-ki kara no josanpu shoku no hatten to nyūji shibō no kanren: Shimane-ken no baai," *Shakai igaku kenkyū* 31, no. 2 (2014): 93–105; Osamu Saito, "Senzen nihon ni okeru nyūji shibō mondai to aiikuson jigyō," *Shakai keizaishigaku* 73, no. 6 (March 2008): 611–33. My interest, however, lies in the ways in which the professionalization of midwifery became integral to the process of building a modern statistical infrastructure within public health.

[7] The official effort to compile vital statistics commenced after Nagayo Sensai, introduced in this chapter, returned from the trip with the Iwakura Mission (see Chapter 1). Many thanks to Dr. Reiko Hayashi for this invaluable comment.

[8] Unfortunately, sources by midwives that reflect this aspect of their activities are hard to come by. This chapter therefore tries to compensate by consulting various other sources.

[9] Historical demographers generally agree that official mortality figures, until the issuing of the Graveyard and Burial Regulation Law in 1884, which will be discussed

politicization of infant death from abortion and infanticide – practices midwives had been implicated in for a long time – was another crucial reason the recently formed government instigated midwifery reform in the 1860s.[10] In the 1910s, government officials stressed medical midwives' role in protecting maternal and infant health, as official statisticians singled out child mortality as a cause of Japan's compromised "national power." Between these decades, during the process of midwifery reform, the government involved midwives in its efforts to improve mortality figures in vital statistics. The government officials identified midwives as a suitable group for producing more accurate data on deaths in pregnancy and childbirth. But, they were equally anxious that some "unreformed" midwives might still betray the government by illicitly *causing* infant death through their involvement in now illegal abortion and infanticide. To overcome this tension, the government did what was described in *The Essentials for Midwives and Nurse*: It mandated that midwives notify the government of every death in childbirth and regulated their professional conduct. These state actions represented an official strategy to place midwives within its reach at a time when midwives' allegiances to the state were tenuous. Through these state actions, the government hoped midwives would turn into a body of healthcare practitioners who wholeheartedly embraced their assigned roles and would facilitate the state's efforts to strengthen national power through active population management.[11] A significant result of this was that the state became even more present in the professional lives of midwives.

Midwives diligently responded to the government's demands; however, this compliance should not be read uncritically as a gesture of loyalty to the state. Behind it were ongoing struggles with obstetrician-gynecologists,

---

later, were not statistically authentic due to the inconsistent methods used to collect data. I respect this point, but I am more interested in the numerical representation of child mortality alongside the textual representation. For this reason, I will examine pre-1884 statistics in the same vein as post-1884 figures. Kazunori Murakoshi, "Meiji, Taisho, Showa zenki ni okeru shizan tōkei no shinraisei," *Jinkōgaku kenkyū*, no. 49 (June 2013): 1–16; Osamu Saito, "Jinkō tenkan izen no nihon ni okeru mortality: Patān to henka," *Keizai kenkyū* 43, no. 3 (July 1992): 248–67; Masato Takase, "1890nen–1920nen no wagakuni no jinkō dōtai to jinkō seitai," *Jinkōgaku kenkyū*, no. 14 (May 1991): 21–34.

[10] Shoko Ishizaki, "Meijiki no shussan wo meguru kokka seisaku," *Rekishi hyōron*, no. 600 (April 2000): 39–53; Shoko Ishizaki, "Kindai nihon no sanji chōsetsu to kokka seisaku," *Sōgō joseishi*, no. 15 (1998): 15–32.

[11] For recent works describing how these national-level attempts were translated into practice on the regional level, see, e.g., Kyoko Miyamoto, "Shimane-ken ni okeru kindai sanba seido unyō ni kansuru kenkyū," *Shakai bunka ronshū* 11 (March 2015): 37–54; Kahoru Sasaki, "Meiji-ki niokeru Gunma-ken no sanba yōsei no hajimari," *Gunma kenritsu kenmin kenkō kagaku daigaku kiyō* 4 (March 2009): 1–11.

whose professional domains often overlapped with those of midwives. From the 1920s on, as state public health and health activism collaborated to tackle the problem of infant mortality, midwives asserted their professional raison d'être even further. As this chapter shows through the case of Osaka, many actors used the narrative of infant death and the nation's health, as well as their privileged position within the state health system, to advance their cause.

Administering the Number of Deaths for the Meiji State

When Sugi began lobbying for a national population census, his colleagues in the Ministry of Finance Division of Household Registration (*Ōkurashō Kosekiryō*), which was in charge of the *koseki*, were diligently compiling "details such as the birth, death, entry, and exit of the members of a household, as well as the numbers in each," the kind of information that comprised vital statistics in later years.[12] In January 1873, the Home Ministry took over the task after it established the Division of Household Registration.[13] From 1875 on, the ministry had another office that collected vital statistics, with the foundation of the Sanitary Bureau (*Naimushō Eiseikyoku*) in that year, which consolidated the medical and public health administration.[14] Over the next few decades, this ministry acted as *the* government office in charge of vital statistics, until the responsibility moved to the Cabinet Bureau of Statistics in 1898.[15]

Behind the Home Ministry's engagement with vital statistics was the quickly forming consensus that people's health and physical constitutions were not just an individual matter but directly determined the nation's power; therefore, the government should invest in medicine and public

---

[12] Sōrifu Tōkeikyoku, *Sōrifu tōkeikyoku hyakunenshi*, 2: 9.

[13] Sadanori Nagayama, "Nihon no kanchō tōkei," 102. For the actual statistics, see Naimushō, ed., *Kokusei chōsa izen nihon jinkō tōkei shūsei 1 (Meiji 5-nen – 18-nen)*, vol. 1 (Tōyō Shorin, 1992).

[14] For recent and representative works on public health administration in the Meiji period, see Kazutaka Kojima, *Nagayo Sensai to naimushō no eisei gyōsei* (Keio Gijuku Daigaku Shuppankai, 2021); Yoko Yokota, *Gijutsu karamita nihon eisei gyōsei shi* (Kyoto: Kōyō Shobō, 2011); Anesaki, "History of Public Health in Modern Japan"; Hidehiko Kasahara and Kazutaka Kojima, *Meijiki iryō, eisei gyōsei no kenkyū: Nagayo Sensai kara Gotō Shinpei e* (Kyoto: Mineruva Shobō, 2011); Burns, "Constructing the National Body"; Hidehiko Kasahara, *Nihon no iryō gyōsei* (Keiō gijuku daigaku shuppankai, 1999); Shiro Oguri, *Chihō eisei gyōsei no sōsetsu katei* (Iryō Tosho Shuppansha, 1981).

[15] Takahashi, "Meijiki wo chūshin nimita," 20–21.

health in order to construct a strong nation.[16] Office statistician Kure Ayatoshi's brother, renowned psychiatrist Shūzō (1865–1932), once said that an individual's "sickness and health, robustness and weakness, are related to … the prosperity and decline of a nation," therefore, the fate of the new Japanese nation was now "in doctors' hands."[17] Arguments such as Kure's confirmed an official scheme already underway to implement European-derived modern medicine and public health in Japan. At the same time, it exhorted the government to adopt vital statistics. In this context, the high-rank health officials understood vital statistics as a highly useful device that could effectively guide the government to maneuver through the potentially tumultuous process of building a nationwide public health and medical system. By presenting the patterns of people's life events and bodily experiences in numbers and in an aggregate form, vital statistics helped the government identify personal factors that could lead to the "decline of a nation" and come up with countermeasures for the sake of the nation's "prosperity." The government, informed by this type of logic, assigned the vital statistical work to the Home Ministry, the government office in charge of public health and medical affairs.

While compiling vital statistics, the Home Ministry Sanitary Bureau privileged the death figure in its official publications.[18] The Sanitary Bureau's interest in death was initially driven by acute infectious diseases. Of those, cholera epidemics left the most profound demographic, social, and political impact.[19] According to the *Statistical Yearbook of Imperial Japan*, the devastating 1879 epidemic caused 105,789 deaths in that year alone.[20] The dramatic effect of the epidemics incited fear among people and the fact that they coincided with Japan opening up diplomatic relations shaped the public image of cholera as a monstrous foreign disease.[21] The epidemics quickly affected politics, too. Cumbersome negotiations over quarantining in the face of extraterritoriality gave

---

[16] Miyakawa, *Tōkeigaku no nihonshi*, 115–33.

[17] Shuzo Kure, "Keizai oyobi tōkei to igaku shakai," *Keizai oyobi tōkei*, no. 3 (March 1889): 128.

[18] Regarding statistics on birth, the number of births was added to the official spreadsheet in 1877, and it was not until 1905 that the crude birth rate began to be published in official vital statistics. Reiko Hayashi, "Perception and Response to the Population Dynamics – on Fertility (Pre-war Period)" [In Japanese], *Jinkō mondai kenkyū* 73, no. 4 (December 2017): 271.

[19] Shunichi Yamamoto, *Nihon korera shi* (Tokyo Daigaku Shuppankai, 1982).

[20] Naimushō, "Dainihon teikoku naimushō tōkei hōkoku dai 1-kai" (1886), 46–47.

[21] Miri Nakamura, *The Monstrous Bodies: The Rise of the Uncanny in Modern Japan* (Cambridge: Harvard University Asia Center, 2015), 13–20; Yoshiro Ono and Isao Somiya, "Meijiki nihon no kōshū eisei nikansuru jōhō kankyō," *Papers of the Research Meeting on the Civil Engineering History in Japan* 4 (1984): 41–48.

rise to the argument that Western powers were undermining the independence of Japan as a burgeoning nation-state.[22] Under these circumstances, cholera epidemics set the tone for the medical administration in the first years of its existence. Under its first director, Nagayo Sensai (1838–1902), the Sanitary Bureau orchestrated quarantine, isolation, and disinfection initiatives with the help of local sanitary health officers, doctors, police, and religious institutions.[23] The central government became keen to know about mortality and morbidity patterns, in addition to information about the disease's topographical profile.[24]

Against this backdrop, the government poured energy into collecting mortality figures. *Isei*, the first state medical policy issued in 1874, stipulated doctors should report "the name of the disease, the days in which the patient suffered from the disease, and the cause of death within three days of the death of the patient."[25] In February 1876, the Home Ministry issued an edict that mandated all prefectural authorities should fulfill the reporting duty stipulated in *Isei* and ordered the Sanitary Bureau to administer the mortality data sent by the prefectural offices.[26] Finally, in 1884, the Grand Council of State and Home Ministry jointly issued the Graveyard and Burial Regulation Law in 1884. Article 8 of the law made it compulsory for local authorities to report burials to the home minister. This law opened up another route for the central administrative office to obtain mortality figures.[27]

Consequently, death figures came to dominate vital statistics published in the *Report of the Sanitary Bureau* series in the 1870s and 1880s.

---

[22] Mark Harrison, "Health, Sovereignty and Imperialism: The Royal Navy and Infectious Disease in Japan's Treaty Ports," *Social Science Diliman* 14, no. 2 (2018): 49–75; Harald Fuess, "Informal Imperialism and the 1879 'Hesperia' Incident: Containing Cholera and Challenging Extraterritoriality in Japan," *Japan Review*, no. 27 (2014): 103–40; Tomoo Ichikawa, "Kindai nihon no kaikōchi niokeru densenbyō ryūkō to gaikokujin kyoryūchi: 1879-nen 'Kanagawa-ken chihō eiseikai niyoru korera taisaku," *Shigaku zasshi*, no. 117 (June 2008): 1–38.

[23] In addition to the works cited so far, for the mobilization of religion against cholera epidemics, see William D. Johnston, "Buddhism Contra Cholera: How the Meiji State Recruited Religion against Epidemic Disease," in *Science, Technology, and Medicine in the Modern Japanese Empire*, eds. David G. Wittner and Philip C. Brown (London: Routledge, 2016), 62–78.

[24] William Johnston, "Cholera and the Environment in Nineteenth-Century Japan," *Cross-Currents: East Asian History and Culture Review* 8, no. 1 (2019): 105–38.

[25] For the *Isei*, see Kasahara, *Nihon no iryō gyōsei*, 1–26.

[26] Naimushō Eiseikyoku, "Eiseikyoku hōkoku" (July 1877), 6.

[27] Sōrifu Tōkeikyoku, *Sōrifu tōkeikyoku hyakunenshi*, 2:28. The mortality and morbidity data collected as a result of the law have been used for analysis in historical demography. See, e.g., Murakoshi, "Meiji, Taisho, Showa zenki"; Hiroshi Iki, "Meiji, taisho-ki no maisō kyokashō ni miru yamai to shibō nenrei," *Nihon ishigaku zasshi* 45, no. 2 (1999): 246–47.

The first vital statistics introduced in *The First Report of the Sanitary Bureau*, published in 1877, was a death table that presented the number of deaths in Tokyo, Kyoto, and Osaka between July and December 1875 according to disease categories.[28] *The Third Report of the Sanitary Bureau*, published in November 1877, had a large statistical table dedicated to cholera in every prefecture, including Hokkaido, as well as among army soldiers, navy personnel, and those on the Mitsubishi ships.[29] Due to the Sanitary Bureau's prioritization of acute infectious diseases, the report ended up emphasizing death figures above all other vital statistics.

While this trend continued, in the 1880s, the Sanitary Bureau introduced a different kind of mortality: death from childbirth. *The Sixth Report of the Sanitary Bureau*, published in July 1880, had figures for "stillbirth" (*shizan*) for the first time, which were presented along with numbers for "live childbirth" (*seisan*) and "marriage" (*kekkon*) (Figure 2.1).[30] The report also showed the stillbirth figure next to the total population figure. Finally, it presented the ratios of stillbirths per 100 births, of the total population per stillbirth, of stillbirths per 100 births among married couples, and finally of married couples per stillbirth (Figure 2.2).[31] After this issue, stillbirth figures became a staple in the section on vital statistics until 1886, when the Sanitary Bureau ceased to be responsible for vital statistics.[32]

This trend in the *Report of the Sanitary Bureau* series coincided with the Home Ministry's effort to improve the existing administrative infrastructure to facilitate the collection of stillbirth figures.[33] In June 1883, it issued a ministerial notification to prefectures, informing them that it had set up separate forms for tabulating childbirth, marriage, and death figures, and mandating the prefectures to send these tables every month, beginning in July of that year.[34] For the birth table, it instructed

[28] Naimushō Eiseikyoku, "Eiseikyoku hōkoku," 16–17. See also Kazuo Takehara, "Meiji shoki no eisei seisaku kōsō: 'Naimushō eiseikyoku zasshi' wo chūshin ni," *Nihon ishigaku zasshi* 55, no. 4 (2009): 509–20.

[29] See table 3 inserted in Naimushō Eiseikyoku, "Eiseikyoku hōkoku dai 3-ji nenpō" (November 1877).

[30] Naimushō Eiseikyoku, "Eiseikyoku nenpō dai 6-ji" (July 1880), 16–17. Initially, the definition or terminology referring to stillbirth that appeared in the statistics was not standardized. For instance, the *Isei* used the term *ryūzan*, today translated as "abortion" or "miscarriage," for stillbirths. As previously mentioned, the situation changed when the Sanitary Bureau gave a definition in 1883, which was applied to the bylaw of the Graveyard and Burial Regulation Law. Murakoshi, "Meiji, Taisho, Showa zenki," 2–3.

[31] Naimushō Eiseikyoku, "Eiseikyoku nenpō dai 6-ji," 23–25.

[32] The last Sanitary Bureau report that presented stillbirth figures was Naimushō Eiseikyoku, "Eiseikyoku nenpō Meiji 17-nen 7-gatsu – Meiji 20-nen 12-gatsu" (n.d., c.1887), 1–8.

[33] Naimushō Kosekikyoku, "Nihon zenkoku kokōhyō Meiji 10 nen, 11 nen," in *Kokusei chōsa izen*, ed. Naimushō, 1–3.

[34] Sōrifu Tōkeikyoku, *Sōrifu tōkeikyoku hyakunenshi*, 2:26.

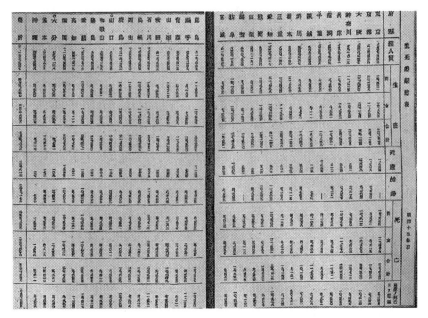

Figure 2.1 The number of births and deaths in each prefecture in 1880.
Source: Naimushō Eiseikyoku, "Eiseikyoku nenpō dai 6-ji" (July 1880), 16–17. From the National Diet Library Digital Collections (https://dl.ndl.go.jp/).

prefectures to specify whether or not each birth was living or dead, defining stillbirth as the birth of a dead fetus after the fourth month of pregnancy. Furthermore, the aforementioned Graveyard and Burial Regulation Law of 1884 endorsed the submission of a death certificate if a dead fetus in the fourth month of pregnancy or later was buried or cremated.[35] These government regulations enabled the production of stillbirth figures in the report.

Why did the Sanitary Bureau only start to systematically report on stillbirth figures in 1880, not in 1875, when it was established? In fact, prior to 1880, the Sanitary Bureau had acknowledged "the statistics for childbirth … is an urgent matter for the administration of public health" and had even begun entering childbirth figures in the statistical spreadsheet in 1877. If the Sanitary Bureau was so keen to collect the "statistics

35 Murakoshi, "Meiji, Taisho, Showa zenki," 3.

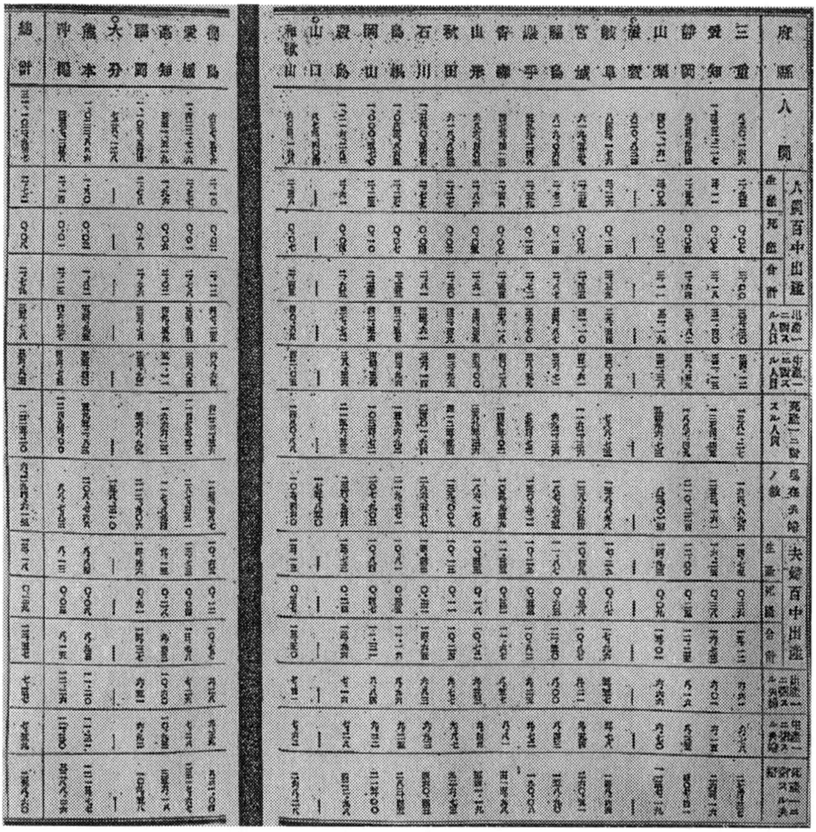

Figure 2.2 The number of live births and stillbirths, 1880.
Source: Naimushō Eiseikyoku, "Eiseikyoku nenpō dai 6-ji" (July 1880), 20–21. From the National Diet Library Digital Collections (https://dl.ndl.go.jp/).

for childbirth," why did it wait until 1880 to create an independent category for recording stillbirth figures?

While there is little conclusive evidence for answering these specific questions, the timing of the first mention of the stillbirth figure in *The Sixth Report of the Sanitary Bureau* (1880) is suggestive, especially considering abortion became illegal that year.[36] The visibility of

---

[36] Susan L. Burns, "Gender in the Arena of the Courts: The Prosecution of Abortion and Infanticide in Early Meiji Japan," in *Gender and Law in the Japanese Imperium*, eds. Susan L. Burns and Barbara J. Brooks (Honolulu: University of Hawai'i Press, 2014), 81–108; Shigenori Iwata, *"Inochi" wo meguru kindaishi: Datai kara jinkō ninshin*

stillbirth figures in the report, I argue, was connected to the process of criminalizing abortion, since the boundaries between stillbirth and abortion were often fuzzy. Specifically, it embodied the official effort to cultivate a discursive space that would facilitate state control over abortion – and more generally, reproductive bodies – under the name of public health.

Though it was only in 1880 that abortion became illegal, political oligarchs had been interested in controlling abortion and infanticide even before the Meiji period.[37] From the late 1860s onward, Meiji statesmen's aspirations to establish a civilized state compelled the nascent government to turn its attention to the practice of abortion and infanticide.[38] In October 1868, the Grand Council of State issued an edict that banned midwives from selling abortifacients and practicing infanticide. In 1869, the new Kochi governor proclaimed that abortion and infanticide would be banned, and soon after, the prefectural governments of Iwate, Hita, Kisarazu, Kagoshima, Wakamatsu, and Aomori followed suit.[39] On December 27, 1870, the central government issued the Outline of the New Criminal Code (*Shinritsu Kōryō*), modeled on the Chinese Ming and Qing codes, which stipulated that a man would be sentenced to third-degree exile if he committed adultery, conspired for abortion with his pregnant partner, and the partner died as a result of it.[40] Further, the Amended Criminal Regulations (*Kaitei Ritsurei*), issued on June 13, 1873, detailed other conditions under which abortion would become subject to punishment. Finally, modeled on the French law, in 1880 abortion became illegal under the new Criminal Code. The Criminal Code, which went into effect in 1882, stipulated that the pregnant woman, anyone conspiring abortion with her, and any doctors, midwives, or pharmacists who practiced abortion would face criminal charges.[41]

chūzetsu e (Yoshikawa Kobunsha, 2009); Hidemi Kanazu and Marjan Boogert, "The Criminalization of Abortion in Meiji Japan," *U.S.-Japan Women's Journal*, no. 24 (2003): 37–42; Shoko Ishizaki, "Nihon no dataizai no seiritsu," *Rekishi hyōron*, no. 571 (November 1997): 53–70; Fujime, *Sei no rekishigaku.*

[37] See Eiko Saeki, "Abortion, Infanticide, and a Return to the Gods: Politics of Pregnancy in Early Modern Japan," in *Transcending Borders*, eds. Shannon Stettner et al. (New York: Palgrave Macmillan, 2017), 19–33; Drixler, *Mabiki*; Motoko Ota, *Kodakara to kogaeshi: Kinsei nōson to kazoku seikatsu to kosodate* (Fujiwara Shoten, 2007); Mikako Sawayama, *Sei to seishoku no kinsei* (Keiso Shobo, 2005); Taku Shinmura, *Shussan to seishokukan no rekishi* (Hosei Daigaku Shuppankyoku, 1996).

[38] For the debate over the reasons why the Meiji government criminalized abortion and infanticide, see Burns, "Gender in the Arena of the Courts," 85; Drixler, *Mabiki*, 199; Ishizaki, "Nihon no dataizai no seiritsu"; Fujime, *Sei no rekishigaku.*

[39] Burns, "Gender in the Arena of the Courts," 85.

[40] Kanazu and Boogert, "The Criminalization of Abortion," 37.

[41] Ibid., 37–45.

The Sanitary Bureau began publishing stillbirth figures at the time as the government was preparing to implement the law. Stillbirth was associated with abortion and infanticide, and the report showing the stillbirth figures symbolized the state's attempt to regulate these now illicit acts. However, the way the report supported the attempt was subtle. Instead of offering the data on stillbirths for a punitive purpose, for example, the report facilitated the state's abortion control effort by providing an epistemological ground for such control. First, by mentioning stillbirth, the report transformed it into a public matter, specifically a matter of public health. Second, by assigning an independent category to stillbirth in the vital statistic chart, it broadcast the view that death from childbirth was a national fact, just like other demographic phenomena. Finally, by presenting stillbirth in numbers, the report portrayed it as following regular patterns, thus suggesting the state could analyze and predict it. In other words, the report projected the idea that stillbirth was a nationwide, statistical phenomenon that could be, and needed to be, managed by the state public health authorities. By portraying stillbirth this way, the report laid a rhetorical foundation justifying state intervention in stillbirth/abortion via public health. The visibility of stillbirth statistics in the report, therefore, represented public health officials' heightened interest in creating an apparatus that would support the state effort to regulate reproductive bodies – at the time when abortion, a form of death in childbirth, came under state jurisdiction.

The process of making infant mortality visible in the Sanitary Bureau's official report coincided not only with the criminalization of abortions but also with the state regulation of midwives. Over the Meiji period, midwifery developed into an officially recognized medical field and a socially respected profession for women, in part due to its position vis-à-vis the newly formed nation-state.

Medical Midwifery: Specialists in "Normal" Birth and Advocating Public Health

In the 1870s, the vital statistical figures calculated from the *koseki* register were incomplete, and this was a serious headache for official statisticians. The Household Registration Law mostly relied on voluntary notification, and without an effective system of communication in place, people tended to take lax attitudes toward reporting deaths and births to the government office. As a result, vital statistics hardly captured the demographic reality of the entire population. One way to tackle this issue was to employ individuals within the local community as informants. In this context, midwives, along with doctors, were identified as particularly suitable for the task.

Yet, during this period, midwives were regarded as in need of official control rather than as appropriate for this informant task. To start with, a dominant popular image of midwives was as pernicious practitioners of abortion and infanticide.[42] Thus, the Meiji government tried to control midwives' practices in the aforementioned edict issued in October 1868. Furthermore, the 1880 Criminal Code stated that midwives would receive a degree of punishment one higher than the pregnant woman committing abortion, carrying a prison sentence of two months to two years as well as a fine of between two and twenty yen.[43] Government officials subjugated midwives to state control because of their popular image, which put them in close proximity to the shady business of infant death.

While the aforementioned image persisted, starting in the 1870s, a competing perception gradually prevailed within the state administration, which portrayed midwives as trained healthcare professionals.[44] This image came with the new government's effort to reform medicine, modeled primarily on the traditions of Prussian Germany.[45] In 1873, the provincial Gunma Prefecture defined midwifery as an officially licensed occupation in its Outline of the Rules of Medical Administration (*Imu gaisoku*). In 1874, the *Isei* included midwifery in the list of eleven major medical fields to go through government reforms. It defined a midwife as a person forty years old or over who must be familiar with the general anatomy, physiology, and pathology of women and children. The midwife must have a license, which would be granted after demonstrating at least ten normal births and two difficult births in front of obstetricians.[46] In 1899, the Home Ministry issued the Midwives' Ordinance as an imperial edict, which was followed by the Legislation for Midwives' Examination and the Legislation for the Licensing of Midwives. The Midwives' Ordinance defined midwifery as a profession reserved for women. It also

---

[42] See, for instance, the front cover of Drixler, *Mabiki*. For the textual representation, see the works of Hidemi Kanazu and Eiko Saeki; Eiko Saeki, "Abortion, Infanticide, and a Return to the Gods"; Hidemi Kanazu, "Edo sankasho ni mirareru seishoku-ron: 'Umu Shintai' towa dareno Shintai ka," *Nihon shisōshi kenkyūkai kaihō*, no. 20 (2003): 152–64.

[43] Kanazu and Boogert, "The Criminalization of Abortion in Meiji Japan," 44.

[44] Shirai, *Umisodate to josan no rekishi*; Homei, Aya. "Birth Attendants in Meiji Japan: The Rise of the Biomedical Birth Model and a New Division of Labour," *Social History of Medicine* 19, no. 3 (2006): 407–24; Terazawa, "The State, Midwives, and Reproductive Surveillance"; Brigitte Steger, "From Impurity to Hygiene: The Role of Midwives in the Modernisation of Japan," *Japan Forum* 2 (1994): 175–87.

[45] Keiko Ogawa, "Seiyō kindai igaku no dōnyū to sanba no yōsei," in *Umisodate to josan no rekishi*, ed. Shirai (Igaku Shoin, 2016), 26–46.

[46] The rule was originally applied only in Osaka and Tokyo. In other areas, prefectural authorities set up their own education and licensing schemes following *Isei*.

lowered the minimum age of eligibility to twenty years old and mandated a midwife to complete at least a year's academic training and pass the nationwide licensing examination.[47] The government regulations issued throughout the Meiji period were intended to generate female healthcare professionals who could replace the aforementioned granny midwives who were complicit with abortion and infanticide.

The government was not the sole player in the construction of medical midwifery. The new generation of doctors forming the modern field of "obstetrics-gynecology" (sanfujinka) also aided in turning midwifery into a medical subdivision.[48] In the 1880s, as obstetrics-gynecology was being established as a medical discipline within universities, obstetric specialists began to engage in midwifery education.[49] In 1880, Sakurai Ikujirō opened a private midwifery training school, Kōkyōjuku, in Tokyo. In April 1890, Hamada Gen'tatsu (1854–1915), the second Japanese professor of obstetrics-gynecology at the University of Tokyo, established a midwifery training school affiliated with his Section of Obstetrics-Gynecology at the University of Tokyo.[50] In Osaka, Ogata Masakiyo (1864–1919), the most renowned obstetric specialist in the city at the time, set up the Ogata Midwifery Training School in October 1892 in his family-owned Ogata Hospital.[51] In subsequent years, the disciples of these first-generation obstetrician-gynecologists built midwifery schools in provincial prefectures such as Yamagata, Niigata, and Miyagi.[52] After the Midwives' Ordinance, there was a boom in midwifery schools across the nation, by both private benefactors and local authorities. By the early 1910s, there was at least one midwifery training school in each prefecture.[53]

For the obstetrician-gynecologists, training midwives was a strategy to establish their position in the crowded market of childbirth medicine. Despite practicing government-approved orthodox medicine, the status of obstetrician-gynecologists in the 1880s was not stable. First, obstetricians and gynecologists trained under the old regime were still practicing,

[47] Shirai, Umisodate to josan no rekishi, 24.
[48] Masakiyo Ogata, Nihon sanka gakushi (Kyoto: Maruzen, 1919), 1164–65.
[49] Prior to these doctors, local authorities – especially in cities – engaged in midwifery training following Isei. Keiko Ogawa, "Seiyō kindai igaku no dōnyū," 27–29; Kimura, Shussan to seishoku, 19–29.
[50] For Hamada, see Riichiro Saeki, "Hamada Gen'tatsu sensei no omoide banashi Hamada Gen'tatsu sensei no nijukkaiki wo shinobite," Sanka to fujinka 2, no. 2 (1934): 63–69.
[51] Ogata, Nihon sanka gakushi, 1328–29.
[52] Ogawa, "Seiyō kindai igaku no dōnyū," 26–37.
[53] Kiyoko Okamoto, "Josanpu katsudō no rekishiteki igi: Meiji jidai wo chūshin ni," in Nippon no josanpu Showa no shigoto, ed. Reborn Henshūbu (Reborn, 2009), 182–84.

although the government had been trying to disqualify their practices through regulations that privileged German-derived medicine.[54] Second, during this period, the number of female doctors trained under the new regime was on the rise.[55] Many of them specialized in areas of medicine linked to women's health, so their existence was threatening to (male) obstetrician-gynecologists. Finally, under the government's protection, more and more midwives were trained in modern medicine, and some seemed to practice medicine just like the obstetrician-gynecologists. Under these circumstances, obstetrician-gynecologists propagated a German model based on the gendered division of labor in their midwifery training: Female midwives were specialists in low-tech "normal" birth and male obstetrician-gynecologists specialized in "abnormal" birth requiring surgical procedures.[56] Furthermore, in the 1890s, they lobbied for the official implementation of the gendered division of labor; they succeeded when the Midwives' Ordinance of 1899 was issued. Male obstetrician-gynecologists thought this model would allow them to cultivate their own niche from which they could compete against their rivals. In particular, it was an effective way to bring their closest rivals, medically au fait licensed midwives, under their control. This was the rationale behind the male obstetrician-gynecologists' involvement in midwifery training.

In part, due to the efforts of the government and obstetrician-gynecologists, the number of certified midwives specializing in "normal" birth rose over the course of the Meiji period. In 1878, there were only 12,007 certified midwives, but within a decade, the number grew to 30,862. After 1899, and until the end of the Meiji period, the number was greatly reduced (25,000–30,000) due to the restructuring of the licensing scheme and the categorization of different groups of midwives.[57] In 1913, the number of midwives licensed under the 1899 ordinance surpassed those certified under the old regime for the first time. At least in

---

[54] Yuko Misaki, "Jūrai kaigyō joi nitsuite no ichi kōsatsu," *Nihon ishigaku zasshi* 65, no. 3 (September 2019): 301–13.

[55] Hiro Fujimoto, "Women, Missionaries, and Medical Professions: The History of Overseas Female Students in Meiji Japan," *Japan Forum* 32, no. 2 (2020): 185–208; Ellen Nakamura, "Ogino Ginko's Vision: 'The Past and Future of Women Doctors in Japan' (1893)," *U.S.-Japan Women's Journal*, no. 34 (2008): 3–18.

[56] Kimura, *Shussan to seishoku*, 19–42; Homei, "Birth Attendants in Meiji Japan." A similar type of struggle took place in Prussian Germany, the place where Japanese obstetrician-gynecologists learned about medical midwifery. Lynne Anne Fallwell, *Modern German Midwifery, 1885–1960* (London: Routledge, 2015).

[57] The new licensing scheme introduced three categories of midwives. The first was the "passing the examination" category (*shiken kyūdai*), referring to midwives who passed the midwifery exam after a year's training at a formal school or under a midwife or

numerical terms, the effort to generate midwives who specialized in "normal" births seemed to have succeeded by the mid-1910s.

From the government's point of view, this development represented a shift in midwives' positions vis-à-vis the state. At the beginning of the Meiji period, midwives were subject to state control because of their association with abortion and infanticide. As a new generation of midwives went through the reform and became integrated into state-endorsed medicine and public health, government officials came to trust them more. They now expected these midwives to partake in the government's efforts to reform people's reproductive practices. At the same time, through teaching, obstetrician-gynecologists instilled a sense of nationalism in their student midwives. The obstetrician-gynecologists calculated that midwives would help strengthen the imperial state by promoting hygienic childbirth.[58] The reform, therefore, intended to transform midwives into loyal agents of the state.

While the effect of the midwifery reform varied across different classes and regions, on the whole, the midwives licensed from the 1890s on diligently internalized the role ascribed to them. First, they tried to implement new cultures of childbirth that were informed by the state-sanctioned modern medicine and hygiene inculcated in them by their teachers. For instance, applying western germ theory, midwife Morita Mariko from Hiroshima washed her hands in a saponated cresol solution before internal examinations to avoid puerperal fever.[59] Second, responding to the obstetrician-gynecologists' call for a clearer division of labor, midwives publicly confirmed their specialism in "normal" births. In *Josan no shiori* (*Midwives' Leaflet*), the midwifery journal launched by Ogata, midwives who contributed clinical case reports time and again stressed that they attended childbirth labor only in so far as it was "normal" and called in medical doctors as soon as they detected signs of abnormality.[60] Thus, midwifery reform succeeded not only in numbers but also in practice.

Yet, these midwives never blindly followed the government regulations or the obstetrician-gynecologists' teachings; many did so to improve

---

obstetrician. The second was the "locally limited practice" category, in which midwives in areas experiencing a shortage of midwives were given a limited five-year license based on their career record. The final category was the "existing midwives," who had already been licensed either by the Home Ministry or by prefectural governments under the scheme implemented by the 1874 medical regulation.

[58] Terazawa, *Knowledge, Power, and Women's Reproductive Health*, 138–43.

[59] Makiko Morita, "Zenchi taiban no ichijikken," *Josan no shiori*, no. 40 (September 1899): 226–27. Also see Terazawa, *Knowledge, Power, and Women's Reproductive Health*, 143–57.

[60] Homei, "Birth Attendants in Meiji Japan."

their otherwise precarious status in the local birth culture. Although modern midwives were sanctioned by state authority and armed with cutting-edge knowledge and techniques in medical childbirth, in many communities, people hardly recognized these qualities because they preferred to adhere to the existing birth customs. In many places, this meant hiring existing birth attendants in their neighborhood instead of qualified midwives.[61] To tackle this situation, new midwives stressed their unique attribute as experts in "normal" childbirth as well as medical professionals able to *recognize* "abnormal" births.[62] This position enabled midwives to establish their status within the local community. By asserting this position, they were, on the one hand, able to show doctors that they were conforming to their assigned role. On the other, under the circumstances in which many villages lacked doctors, midwives could sell themselves as the only available medically trained practitioners and thereby carve out a niche in the local birth culture that the existing birth attendants, who lacked medical knowledge, were unable to enter.

Another strategy modern midwives took to consolidate their status was to actively distance themselves from abortion. Beginning around the late 1890s, some midwives tirelessly produced case reports to expose the wrongdoings of the "old midwives" (*kyūsanba*) and how their abhorrent illegal practices caused suffering to the families that received their care.[63] On the one hand, this tactic could be risky for modern midwives. In many places, people were still practicing abortion even after they were made illegal.[64] Under these circumstances, this attitude could alienate midwives from their local communities. On the other, the same tactic could work in their favor. By adopting this tactic, medical midwives could create another niche in local birth culture: a local watchperson ensuring, on behalf of the state, that people would not engage in abortion. In other words, midwives denounced the practice of abortion primarily to survive in this competitive environment, but in so doing, they ended up attaching themselves to the government's effort to lay a nationwide reproductive surveillance system. Consequently, midwives became even more entwined in the state's effort to control reproductive bodies.

---

[61] Aya Homei, "Sanba and Their Clients: Midwives and the Medicalization of Childbirth in Japan," in *New Directions in History of Nursing: International Perspectives*, eds. Barbara Mortimer and Susan McGann (London: Routledge, 2005), 68–85.

[62] Kimura, *Shussan to seishoku*, 59–63.

[63] Homei, "Sanba and Their Clients."

[64] Iwata, *"Inochi" wo meguru kindaishi*, 2.

It was against this backdrop that the government included midwives in its effort to improve statistics on infant mortality. While reforming midwives, the government assigned them the task of officially notifying the state of any stillbirths. However, the official process for doing this was gradual. The *Isei* of 1874 allowed midwives to record "birth or death, male or female, and the date of birth" as well as "any incidence of spontaneous abortion or stillbirth occurring in the three months of pregnancy and later" and submit the birth or death certificate to the respective medical office. But it also set conditions: Midwives were able to undertake these tasks only in case of emergency and in the absence of obstetric doctors.[65] Later, Article 11 of the 1884 Graveyard and Burial Regulation Law's bylaw stipulated that persons dealing with the burial of dead fetuses of four months or older would have to seek a certificate from doctors or midwives prior to the burial and that these medical practitioners must report the stillbirth if they were asked to produce a certificate. The bylaw was not compulsory, and the decision to entrust midwives with this task was made on the prefectural level; however, with the Midwives' Ordinance, the midwives' notification duty became compulsory. The ordinance stated that midwives must certify every stillbirth they witnessed, and the Home Ministry made an official template to this effect.[66] This template appeared in *The Essentials for Midwives and Female Nurses* – introduced at the beginning of this chapter.

The notification of stillbirth on the local level was an important first step for compiling infant mortality data on the national level, and these legislations indicate how the government gradually came to trust midwives as data collectors. From a statistical point of view, what was particularly significant about these legislations, in particular the bylaw of the Graveyard and Burial Regulation Law, was that they laid a foundation for improving statistical accuracy by standardizing the notification procedure. Prior to the bylaw, every prefecture adopted its own mechanisms for reporting deaths, and this was causing errors in vital statistics at the central level. The bylaw was a tactic to minimize statistical errors by streamlining the collection method. The fact that the government included midwives in the effort to improve statistics suggests policymakers thought that a sufficient number of midwives were reformed and could carry out this important task for official statistics.

The new procedure for infant death notifications involving midwives seemed to improve official vital statistics. The demographers Takase

[65] Sōrifu Tōkeikyoku, *Sōrifu tōkeikyoku hyakunenshi*, 2: 93.
[66] Murakoshi, "Meiji, Taisho, Showa zenki," 3–4.

Makoto and Murakoshi Kazunori pointed out that the infant mortality rate in Japan became more accurate beginning in the late 1890s. Takase attributes this to the Graveyard and Burial Regulation Law, while Murakoshi went further and suggests that the reporting duty assigned to midwives under the Midwives' Ordinance, in addition to the bylaw, might have contributed to the changing profile of the data.[67] These studies indicate that midwives, in particular after the issuing of the Midwives' Ordinance, internalized their professional duty as ascribed by the state and diligently submitted death certificates to their local authorities when they witnessed deaths in childbirth.

The official understanding of death in childbirth – or infant death, more generally – changed in the early twentieth century as Japan went through an epidemiological transition. Health officials began to perceive the infant as a self-contained, age-specific population group and infant death as a demographic phenomenon that had a significant impact on Japan's economic and political future. In this context, medical midwifery was also mobilized for maternal and infant health.

Problematizing the Infant as a Population Group

After inheriting vital statistical work from the Home Ministry in 1898, from 1899 onward, the Cabinet Bureau of Statistics (CBS) published official vital statistics annually, as well as, from 1906 on, statistics on the cause of death. In the 1910s, these data clarified that the morbidity and mortality rates from acute infectious diseases had significantly dropped at the turn of the century, while the morbidity rate of chronic infections, most conspicuously tuberculosis, venereal diseases, and cancer, remained high.

Patterns in mortality and morbidity changed the contours of public health administration significantly. Until the turn of the twentieth century, the question of how to counter acute infectious diseases dominated policy discussions within the Home Ministry. In the 1910s, it began to proactively explore measures for raising the general standard of health and hygiene, since it deemed that many of the emerging epidemiological challenges stemmed from everyday health and hygiene practices.[68] As a tangible first step, the Home Ministry launched the Health and

[67] Murakoshi, "Meiji, Taisho, Showa zenki," 1–16; Takase, "1890nen–1920nen no wagakuni."
[68] Kōseishō Gojūnenshi Henshū Iinkai, ed., *Kōseishō gojūnenshi* (Kōseishō Mondai Kenkyūkai and Chūō Hōki Shuppan, May 1988).

Hygiene Survey Group (*Hoken Eisei Chōsakai*, HHSG) on June 27, 1916 to investigate the state of health and hygiene across the country. Under the supervision of the home minister, thirty-four members, consisting of academics, members of the half-government, half-private Central Hygiene Association, and high-rank officials from the Home Ministry, Metropolitan Police Department, and army were tasked with investigating and making official recommendations on eight topics related to health and hygiene practices.[69]

From the perspective of population history, the HHSG is highly important because it highlights that, by this period, the mortality trend in vital statistics had come to occupy a special position within the state public health administration due to its perceived significance for Japan as a nation-state. The preamble of the first HHSG report, published in 1917, was about Japan's high mortality rates compared to the "civilized" nations of France, England, and Germany.[70] The report claimed that this trend represented a "national scandal," and the government should tackle the problem to shield the "nation's fortune and power."[71] The report mirrored the burgeoning understanding within the government that high death rates symbolized Japan's lack of "fortune and power" and its internationally crumbling status.[72]

Among many other mortality categories, the HHSG report singled out the high mortality rate among children under the age of five, who were referred to as *nyūji* (infant) and *yōji* (small children), as particularly problematic.[73] Echoing high-rank health officials' anxieties about Japan's inferior health vis-à-vis "civilized" countries in Europe, the report explained how the phenomenon of high child mortality was disturbing precisely because the reverse was the case in Europe. In Germany, for instance, the figure had recently decreased from 250–300 to less than 160 per 1,000 births. In contrast, in Japan, the mortality rate among children under one year old increased from 110 per 1,000 births in 1888 to over 160 per 1,000 births in recent years. The mortality rate of children over the age of one was so high that it could "not be compared to any other civilized nations."[74] The report stated that "if we do

---

[69] Hoken Eisei Chōsakai, "Hoken eisei chōsakai dai ikkai hōkokusho" (Naimushō Hoken Eisei Chōsakai, April 1917).
[70] Hoken Eisei Chōsakai, "Hoken eisei chōsakai dai ikkai," 1–4.
[71] Ibid., 3.
[72] Ibid. This understanding came to buttress more routinized statistical work during the period: Kenichi Ohmi, "Dainijisekaitaisen izen no wagakuni niokeru jinkō dōtai tōkei," *Nihon kōshū eisei zasshi* 51, no. 6 (2004): 452–60.
[73] Another was the mortality of young men, which I will touch on in Chapter 4.
[74] Hoken Eisei Chōsakai, "Hoken eisei chōsakai dai ikkai," 3.

not explore and investigate its cause and do not set appropriate measures against infant mortality, we will ... fail to establish a solid foundation for the health of young men," another important population
group the report characterized as the "nucleus of our nation [who] ...
shoulder the burden of national security.. and the driving force for [the
nation's] industry."[75] The report reflected the widely held view within
the government office that the infant, along with adult men, was an
independent demographic subject. As infants eventually become adults
and play a pivotal role in the nation's economy and military capability, the demographic behavior of the infant as a population group was
a state matter. This was why the report expressed concerns about high
child mortality.

However, the idea of the infant comprising an independent category
in official vital statistics was not always self-evident. It gradually formed
over the course of the Meiji period, along with the conceptualization of
age categories for mortality figures in the statistics.[76] A critical moment
came in the late 1890s, when the CBS took over official vital statistics.
The first vital statistics published by the CBS in 1899 had a table showing the number and rate of deaths categorized by sex and age. Compared to the earlier tables presented by the Sanitary Bureau, the age
range was more fine-tuned and included the age range of zero to five
years old. From then on, the infant, as with other age-specific population categories, was a standard part of the statistical tables showing
mortality figures.[77]

This trend corresponded with the burgeoning interest among official statisticians for a thorough death table. At the turn of the twentieth century, the CBS employed statistician Yano Kōta to compile a
death table.[78] Following his mission, Yano created death tables that
had mortality figures for each year of age, starting from zero.[79] A table

[75] Ibid., 2–3.
[76] Sōrifu Tōkeikyoku, *Sōrifu tōkeikyoku hyakunenshi*, 2:13. However, in 1880, the
population table made by the Home Ministry Division of Household Registration
introduced the classification of living people by age. Naimushō Eiseikyoku, "Meiji
13-nen 1-gatsu 1-nichi shirabe nihon zenkoku jinkōhyō," in *Kokusei chōsa izen*, ed.
Naimushō.
[77] Naikaku tōkeikyoku, "Nihon teikoku jinkō tōkei Meiji 31-nen" (March 1901), https://
dl.ndl.go.jp/info:ndljp/pid/805976.
[78] Sōrifu Tōkeikyoku, *Sōrifu tōkeikyoku hyakunenshi*, 2: 992. The original term for
"death table" is *shibōhyō*. It correlates with today's "life table" *(seimeihyō)*. For the
etymology of *shibōhyō* in Japanese, see Kiichi Yamaguchi et al., *Seimeihyō kenkyū*
(Tokyo: Kokon Shoin, 1995), 3–6. I would like to thank Professor Ryuzaburo Sato
for giving me advice on this.
[79] Kōta Yano, "Nihonjin no seimei ni kansuru kenkyū," in *Sōrifu tōkeikyoku hyakunenshi*, Sōrifu Tōkeikyoku, 339.

in one of his publications had a detailed description of infant mortality. It even showed mortality figures for the neonate, down to days (zero, five, ten, fifteen days) and months (one, two, three, six, twelve months) in the first year after birth, in addition to the figures for each year of age.[80]

While Yano was working on a death table, a consensus was forming among the CBS statisticians that infant mortality was a noticeable demographic phenomenon and thus should be regarded as a critical factor in the composition of general mortality. Kure Ayatoshi, now serving in the CBS, was among the first to seek a link between infant and general mortality.[81] While "observing stillbirths and other child deaths across the country" for the CBS's first annual statistics published in 1899, he also "calculated a percentage of childhood death to total death ... in order to study what kind of relationships there are between child mortality and general mortality [ippan shibō]."[82] Following Kure, in 1904, Aihara Shigemasa (1846–1914), another prominent CBS statistician published "Child Mortality in Japan."[83] In the paper, Aihara introduced the results of vital statistics in 1899 and 1900, pointing out that 64.6 percent and 68.4 percent of the children between the ages of zero and five, in 1899 and 1900 respectively, died before their first birthday.[84] By the mid-1900s, official statisticians had seen the correlation between child mortality and the trend in general mortality.

However, it was only in the mid-1910s that official statisticians began to characterize infant mortality explicitly as a cause for the rising mortality rate of the Japanese population. The senior official statistician Nikaidō Yasunori (1865–1925) played a pivotal role in the popularization of this view within the government.[85] While compiling vital statistics at the CBS, Nikaidō observed the Japanese population exhibited some disturbing signs, compared to the demographic trend in Western Europe. In Europe, particularly in England and Germany, fertility rates were decreasing in recent years. However, mortality was

[80] Ibid., 350–51.
[81] Ayatoshi Kure, "Meiji 32-nen nihon teikoku jinkō dōtai tōkei gaikyō," in Sōrifu tōkeikyoku hyakunenshi, ed. Sōrifu Tōkeikyoku, 384–97.
[82] Ibid., 389.
[83] Shigemasa Aihara, "Nihon ni okeru shōni no shibō," Tōkei shūshi, no. 284 (November 1904): 568–71; Shigemasa Aihara, "Nihon ni okeru shōni no shibō (dai 284 gō no tsuzuki)," Tōkei shūshi, no. 289 (April 1905): 151–54. For Aihara's biography, see Toshiyasu Kawai, "Aa Aihara Shigemasa kun ikeri," Tōkei shūshi, no. 339 (July 1914): 252.
[84] Aihara, "Nihon ni okeru shōni no shibō," 568–69.
[85] For Nikaidō, see Ai Chuman, "Hoken eisei chōsakai hossoku eno michi: Nyūji shibōritsu modai no shiten kara," Rekishigaku kenkyū, no. 788 (2004): 16–26.

also in decline, so the population overall was still expanding. In turn, in Japan, while fertility rates were still high, mortality rates were even higher. Even more disturbing, what he called "civilization" seemed to lower marriage and fertility rates, but not mortality, in Japan. In demographic terms, this phenomenon heralded a doomsday picture: a contracting population caused by declining fertility and rising mortality. Looking at the current demographic trend, Nikaidō judged that Japan was lagging behind Western Europe by half a century.[86] He concluded that the increasing mortality rate in particular was a "serious problem [to which] the Japanese hygiene [administration] must pay the utmost attention."[87]

For this reason, sometime in 1913–14, Nikaidō investigated child mortality in Japan and found three unique features.[88] First, it went against the general trend in Western Europe, where infant mortality was in sharp decline in recent years.[89] Second, in Japan, the mortality rate among children between one and two years old was rising the most, while in other countries, the rate was typically the highest among babies before their first birthdays.[90] Third, the most common cause of death among children between one and two years old differed from the trend in Europe; in Europe, it was typically respiratory disease, while in Japan it was gastrointestinal disease.[91] From these observations, Nikaidō concluded that the high infant mortality in Japan was due to the nutrition disorder children experienced after they were weaned off mother's milk, and that the "changing societal structure" that compelled women to engage in the waged work, coupled with the "uneducated people" who used artificial formula incorrectly, were causing the nutrition disorder.[92]

---

[86] Yasunori Nikaidō, "Honpōjin no seishi ni kansuru tōkeiteki hihan no gaiyō," *Tōkei shūshi*, no. 413 (July 1915): 337.

[87] Ibid., 340.

[88] Yasunori Nikaidō, "Honpō shōni shibō no sūsei," *Nihon gakkō eisei* 2, no. 8 (1914): 567–68; Yasunori Nikaidō, "Honpō shōni shibō no tokuchō (ichi)," *Tōkei shūshi*, no. 404 (1914): 473–80; Yasunori Nikaidō, "Honpō shōni shibō no tokuchō (ni)," *Tōkei shūshi*, no. 411 (May 1915): 237–45; Yasunori Nikaidō, "Honpō shōni shibō no tokuchō (san)," *Tōkei shūshi*, no. 412 (June 1915): 289–300.

[89] Nikaidō, "Honpō shōni shibō no tokuchō (ichi)."

[90] Nikaidō, "Honpō shōni shibō no tokuchō (ni)."

[91] Ibid.

[92] Nikaidō, "Honpō shōni shibō no tokuchō (ni)," 442. Nikaidō's argument resonates with the narrative stressing the superiority of mother's milk over formula, which emerged during this period as women's reproductive role vis-à-vis the state was being naturalized. Izumi Nakayama, "Moral Responsibility for Nutritional Milk: Motherhood and Breastfeeding in Modern Japan," in *Moral Foods* eds. Angela Ki Che Leung et al. (Honolulu: University of Hawai'i Press, 2020), 66–88.

Nikaidō's studies mobilized the government. In the 1910s, children's health also surged as a subject of debate within the government after the influential navy doctor, Takagi Kanehiro (1849–1920), argued that the nation's physical capability had been "lowered" in recent years due to the compromised ability of mothers to care for their children.[93] Responding to Takagi's warning, on February 5, 1915, the Sanitary Bureau invited Nikaidō to provide statistical evidence that could verify Takagi's claim.[94] On May 7, Nikaidō submitted a report to the Sanitary Bureau. The Bureau immediately forwarded it to the prime minister, with a note urging the government to organize research on the rising mortality rate.[95] Thereafter, the Home Ministry secured a government budget for the research, which was used to launch the HHSG. At the launch meeting, Home Minister Ichiki Tokurō (1867–1944) publicly acknowledged that one of the main objectives of the HHSG was to identify the reasons for the rising mortality rate, with a special focus on the high infant mortality.

The HHSG report, which presented the problem of infant mortality in numbers, paved the way for social movements and policies promoting maternal and infant health from the 1920s onward. As healthcare professionals linked to maternal and infant health, midwives enthusiastically took part in the movements. For these midwives, participation in the movements was a strategy to secure their position within the broader arena of infant health, where the state politics and health activism coalesced. Under the circumstances, the statistical rationale buttressed midwives' struggles.

Midwives and the Discourse of Infant Care

The HHSG was significant not only because it consolidated the official narrative that infant mortality was damaging the nation's health but also because the narrative borne out of it catalyzed a number of initiatives promoting maternal and infant health. The discussion of infant mortality within the HHSG paved a way for the establishment of the Bureau of Social Affairs (*Shakaikyoku*) in 1920, which listed maternal and infant health as a priority area for its child protection administration.[96] Social policy intellectuals submitted a proposal requesting government

---

[93] Chuman, "Hoken eisei chōsakai," 21.
[94] Ibid., 18.
[95] Ibid.
[96] For the social policy debate on infant and maternal protection during this period, see Naho Sugita, *"Yūsei," "yūkyō" to shakai seisaku: Jinkō mondai no nihonteki tenkai* (Kyoto: Horitsu Bunka Sha, 2013), 179–80; Naho Sugita, *Jinkō, kazoku, seimei to shakai seisaku: Nihon no keiken* (Kyoto: Hōritsu Bunka Sha, 2010), 86–107.

subsidies for building maternal and childcare consultation clinics.[97] At the same time, this period's thriving feminist, labor, and socialist movements demanded state subsidies for childbirth and childrearing for the "protection of motherhood" (*bosei hogo*).[98] Finally, pediatricians and department stores jumped at the opportunity created by the burgeoning discourse of child protection.[99] They authored prescriptive literature teaching childcare techniques and organized exhibitions on the theme of hygiene in childbirth and childcare, which primarily targeted middle-class consumers.[100] The discourse of infant mortality broadcast by the HHSG resonated with the rising public consciousness of child and motherhood protection, and triggered cooperation between social and official movements to promote maternal and infant health during the 1920s.

In large cities, bureaucrats and reformers in Osaka were among those taking up the discourse of infant mortality most actively in order to implement social work for poor working mothers and their babies.[101] Osaka, a long-standing merchant city that quickly became industrialized in the early Meiji period, attracted teenage girls from impoverished neighboring villages who were looking for opportunities to work as factory workers or maids.[102] Well into the 1910s, their living conditions were harsh, far from being conducive to raising healthy babies.[103] Working long hours was common, and these girls were often assigned to night shifts. Even if they became pregnant, many could not afford nutritious meals due to their low wages. After they gave birth, they had to return to work immediately so

---

[97] Chuman, "Hoken eisei chōsakai," 26.

[98] Vera C. Mackie, *Feminism in Modern Japan: Citizenship, Embodiment, and Sexuality* (Cambridge: Cambridge University Press, 2003); Barbara Molony, "Equality Versus Difference: The Japanese Debate over 'Motherhood Protection', 1915–50," in *Japanese Women Working*, ed. Janet Hunter (London and New York: Routledge, 1993), 123–48; Hiroko Tomida, "The Controversy over the Protection of Motherhood and its impact upon the Japanese Women's Movement," *European Journal of East Asian Studies* 3, no. 2 (2004): 243–71.

[99] Mikako Sawayama, *Kindai kazoku to kosodate* (Yoshikawa Kobunkan, 2013), 128–55.

[100] Mark A. Jones, *Children as Treasures: Childhood and the Middle Class in Early Twentieth Century Japan* (Cambridge, MA and London: Harvard University Press, 2010); Louise Young, "Marketing the Modern: Department Stores, Consumer Culture, and the New Middle Class in Interwar Japan," *International Labor and Working-Class History* 55 (April 1999): 52–70.

[101] Below, I rely on the description in Higami Emiko's impressive work on the subject. Emiko Higami, *Kindai Osaka no nyūji shibō to shakai jigyō* (Osaka: Osaka Daigaku Shuppankai, 2016).

[102] Jeffrey E. Hanes, *The City as Subject: Seki Hajime and the Reinvention of Modern Osaka*, Twentieth-Century Japan (Berkeley: University of California Press, 2002); James L. McClain and Osamu Wakita, eds., in *Osaka: the Merchant's Capital of Early Modern Japan* (Ithaca: Cornell University Press, 1999).

[103] Higami, *Kindai Osaka no nyūji shibō*, 5–6.

they could earn a living. This situation did not allow the young mothers to nurse their babies, and in many cases, the babies received less than standard formula milk.[104] These circumstances surrounding poor mothers and their young children were directly reflected in the child mortality trend. From 1913 on, the city's infant mortality rate increased significantly, to as high as 238.6 per 1,000 births in 1919. Thus, when the HHSG was launched in 1916, the infant mortality rate in Osaka was alarmingly high.[105]

Responding to the demographic trend, in the early 1910s, local governments, reformers, philanthropists, and volunteers organized social work activities with the specific aim of improving maternal and infant health, especially in the areas where people with the lowest socioeconomic status lived. In 1911, the president of Osaka Mainichi Newspaper Publishing Company, Yamamoto Hikoichi, authorized the launch of the Osaka Mainichi Newspaper Charity Group, which dispatched mobile clinics for people who could not afford medical care.[106] In July 1919, the Osaka Municipal Government also set up the Osaka City Child Consultation Station (Osaka-shi Jidō Sōdanjo) and offered a wide range of services, such as infant and childcare guidance, medical consultation, diagnosis of disabled children, and consultation for a child's education.[107] These activities in the 1910s led to the rapid growth of maternal and infant welfare schemes in the following decade.

An important organization running such schemes in the 1920s was the half-private, half-public Osaka Infant Protection Society (Osaka Nyūyōji Hogo Kyōkai, hereafter OIPS).[108] The OIPS, launched in July 1927 with the mayor of Osaka at the helm, was based on the collaboration between the Osaka Municipal Government Social Section's supervisor, Kawakami Kan'ichi (1888–1961), and Okubo Naomutsu, Medical Director of the Department of Pediatrics at the Osaka branch of the Japan Red Cross.[109] The OIPS not only worked with medical professionals but also with the commercial sector, most notably with the department stores Mitsukoshi, Takashimaya, Matsuzakaya, and Sogō to set up free, temporary infant consultation clinics.[110]

Significantly, when activists and local authorities promoted maternal and infant protection work, they used a statistical rationale and

[104] Ibid.
[105] For more detail, see Higami, Kindai Osaka no nyūji shibō, 80–83.
[106] Ibid., 94.
[107] Ibid., 141–44.
[108] Ibid., 178–81.
[109] Ibid., 179.
[110] Ibid., 180.

the perspective of the nation's health as if they were a prerequisite for advancing their cause. *The Guideline for Infant Protection* published by the OIPS opened with the sentence: "It is one of the most serious national problems in recent years: how to decrease infant mortality rates. In particular, in cities like Osaka where the infant mortality rate is high ... this problem should not be neglected for even one day."[111] The statement was followed by statistical tables showing infant mortality rates in Japan, and then in Osaka. Similar to official publications of this kind, the first table showed the Japanese infant mortality rate as compared to the "civilized nations," in this case British Empire, the United States, Germany, France, Italy, Austria, Holland, and New Zealand (Figure 2.3). It showed that the mortality rate in Japan in 1905–24 hovered between 15.6 and 18.9 deaths per 100 births, while in other countries the rates showed a downward trend – one-digit numbers in most cases. This table was followed by two others (see e.g., Figure 2.4) that indicated the rates in Osaka were by far higher than the national average.[112] Together, these tables made it clear that infant death in Osaka was not a local incidence but a national affair that, like general mortality, had ramifications for Japan's self-identity as a member of the "civilized" nations. Social reformers used this narrative of infant mortality and nationhood to justify their cause, which clearly shows how much faith people grew to have in vital statistics and statistical reasoning.

Midwives were a major player in maternal and infant healthcare schemes organized by the aforementioned social work organizations.[113] In 1914, the Osaka Mainichi Newspaper Charity Group employed five midwives to run a free birth attendance scheme.[114] In the 1920s, the group collaborated with the OIPS to expand the scheme, and in 1921, it increased the number of commissioned midwives to seven. In 1923, it stationed a commissioned midwife in every designated district – forty in total – and employed Inoue Matsuyo as a special home visitor. In 1925, there were fifty commissioned midwives in the city, which expanded to

---

[111] Naomutsu Okubo and Yoshitoshi Misugi, *Nyūyōji hogo shishin* (Osaka: Osaka Nyūyōji Hogo Kyōkai, 1928), 1.

[112] Ibid., 1–3.

[113] For more details about the campaign to offer free midwifery services in Osaka and Tokyo during this period, see Terazawa, *Knowledge, Power, and Women's Reproductive Health*, 228–34.

[114] Mayumi Wada, "Osaka mainichi shinbun jizendan no nyūyōji hogo katsudō to katei eno shien: Muryō josan jigyō to hoiku gakuen no sōsetsu wo chūshin ni," *Himeji daigaku kyōiku gakubu kiyō*, no. 11 (2018): 171–74; Osaka Mainichi Shinbun Jizendan, *Osaka mainichi shinbun jizendan nijūnen-shi* (Osaka: Osaka Mainichi Shinbun Jizendan, 1931), 178–79.

各國の乳児死亡率（社會事業講座生江氏）

國別 ＼ 年次	英威	米國	獨逸	佛國	伊國	墺國	和蘭	ニュージーランド	日本
一九〇五 一九一〇（明治三十八年 明治三十九年）	一一·七	—	一七·四	一二·七	一五·三	二〇·二	一二·四	六·九	一五·七
一九一五（大正四年）	一一·〇	一一·〇	一五·四	一四·一	一四·六	二三·一	八·七	五·〇	一六·〇
一九一六（大正五年）	九·一	一〇·一	一三·六	一三·二	一八·七	二二·〇	八·三	五·〇	一七·〇
一九一七（大正六年）	九·六	九·四	一五·〇	一三·五	二一·一	一八·六	八·九	四·八	一七·三
一九一八（大正七年）	九·七	一〇·一	一五·八	一四·一	—	一九·三	九·三	四·八	一八·九
一九一九（大正八年）	八·九	八·七	一四·四	一二·二	—	一五·六	五·〇	四·五	一七·一
一九二〇（大正九年）	八·〇	八·六	一三·一	九·八	—	一五·七	七·三	五·〇	一六·六
一九二一（大正十年）	八·三	七·六	一四·一	一一·八	—	一五·七	七·六	四·七	一六·八
一九二二（大正十一年）	七·七	七·六	一三·四	一二·四	—	—	六·七	四·一	一六·六
一九二三（大正十二年）	六·九	七·七	—	—	八·八	—	六·七	四·三	一六·三
一九二四（大正十三年）	七·五	七·二	一〇·八	—	八·五	—	五·一	四·〇	一五·六

Figure 2.3 Comparison of infant mortality rates: British Empire, United States, Germany, France, Italy, Austria, Holland, New Zealand, Japan, 1905–24

Reproduced from Naomutsu Okubo and Yoshitoshi Misugi, *Nyūyōji hogo shishin* (Osaka: Osaka Nyūyōji Hogo Kyōkai, 1928), 30.

全國及六大都市乳児死亡率（人口動態統計）

年次	東京	大阪	京都	神戸	名古屋	横濱	市區	全國
大正三年（一九一四年）	一五.四	二三.五	二二.三	一九.九	一五.六	一六.九	一七.六	一五.九
大正四年（一九一五年）	一七.八	二五.八	二〇.八	二二.一	一四.七	一九.九	一八.七	一五.七
大正五年（一九一六年）	一七.二	二三.八	二〇.八	一九.三	一八.六	一九.五	一九.〇	一七.一
大正六年（一九一七年）	一七.七	二五.四	二〇.一	二二.五	一六.四	一九.九	一九.三	一七.三
大正七年（一九一八年）	一七.九	二五.七	二五.〇	一九.五	二〇.六	二〇.五	二〇.六	一八.九
大正八年（一九一九年）	一六.六	二二.五	二二.四	二二.二	一五.四	一八.〇	一八.〇	一七.〇
大正九年（一九二〇年）	一五.九	二三.一	二〇.九	一九.七	一七.〇	一八.三	一七.六	一六.六
大正十年（一九二一年）	一五.五	二三.一	二三.五	一九.八	二一.〇	一六.三	一八.四	一六.八
大正十一年（一九二二年）	一五.二	二三.八	一八.六	二三.六	一七.七	一六.七	一七.七	一六.六
大正十二年（一九二三年）	一七.五	二二.三	二一.三	二〇.五	二〇.四	一九.九	一九.三	一六.七
大正十三年（一九二四年）	—	一九.七	一六.四	二〇.三	一七.五	一六.六	一六.二	一五.六

Figure 2.4 Comparison of infant mortality rates: Tokyo, Osaka, Kyoto, Kobe, Nagoya, Yokohama, 1905–24
Reproduced from Naomutsu Okubo and Yoshitoshi Misugi, *Nyūyōji hogo shishin* (Osaka: Osaka Nyūyōji Hogo Kyōkai, 1928), 31

100 by the end of the decade. By the mid-1930s, 110 commissioned midwives had been registered to work under the scheme.[115]

The commissioned midwife played a key role in the scheme's day-to-day maternity and neonatal care work, which was provided primarily for less affluent households. They did prenatal check-ups, attended childbirth labor, bathed and disinfected the baby for the first week after childbirth, and did the home visit for postnatal care. In addition, the commissioned midwife helped to arrange for formula feeding if requested. On behalf of the mother, they ordered a bottle and arranged for her to be able to purchase cow milk at a wholesale price.[116] Finally, the midwife administered eye drops at the time of birth and during home visits and checked for any deformity or dislocation of bones, etc. so babies could be treated early.[117]

However, midwives were not passively co-opted into the social work scheme. In fact, starting in the 1920s, local midwifery organizations actively participated in the booming social work initiatives. For instance, just after a year of its existence, the Osaka City Midwives' Association (OCMA, est. May 31, 1920) decided to issue 600 free birth attendant vouchers, which were distributed to "the proletariats" via the Social and Hygiene Sections of the Osaka Municipal and Prefectural Governments.[118] Between 1927 and June 1929, midwives affiliated with the Awabori, Honjō, and Imamiya birth clinics funded by the Osaka Municipal Government attended an average of 87, 124, and 56 unpaid childbirths, respectively.[119] Finally, at an emergency meeting among the senior councilors on January 14, 1930, the OCMA decided to establish its own "social birth clinics" (*shakaiteki san'in*).[120] The midwives as a collective took up their assigned role with fervor.

Why did the Osaka midwives take up this social work with fervor?[121] In the 1920s and early 1930s, midwives had plenty of reasons. The most crucial was the struggle to expand their area of expertise vis-à-vis

---

[115] Osaka Mainichi Shinbun Jizendan, *Osaka mainichi shinbun jizendan nijūnen-shi*, 179–81.

[116] Wada, "Osaka mainichi shinbun jizendan."

[117] Osaka Mainichi Shinbun Shakai Jigyōdan, *Osaka no sanba wa kataru taisetsu na osan no hanashi* (Osaka: Osaka Mainichi Shinbun Shakai Jigyōdan, 1936), 121.

[118] Hidetora Aoki, *Osaka-shi sanba dantaishi* (Osaka: Osaka-shi Sanbakai, 1935), 174.

[119] Ibid., 238.

[120] Ibid., 238–42.

[121] It must be stressed that not all midwives characterized their cause in relation to the state. Some were urged by the sense of a cause, similar to how the famous midwife Shibahara Urako from Onomichi became an advocate of birth control activism as part of the proletariat liberation movement unfolding in Osaka. For Shibaura, see Fujime, *Sei no rekishigaku*, 125–29, 136–39, 247–52.

obstetrician-gynecologists. From the 1910s onward, midwives' interests shifted from establishing their professional domain by excluding their then closest rivals – the old generation of experienced but "unlicensed" midwives – to becoming the equals of obstetrician-gynecologists.[122] In the 1920s, midwives acted on their interests, requesting the government to amend the Health Insurance Law (est. 1922), which privileged obstetrician-gynecologists over midwives as insured childbirth attendants. To counter the government's argument in support of its partnership with obstetrician-gynecologists – it was administratively easier because they belonged to a nationwide organization (i.e., the Japan Medical Association) – in April 1927, locally-based midwives' groups (including the Osaka Midwives League formed in 1925) established the nationwide Greater Japan Midwives Association. Through the nationwide organization, midwives would be able to sign a contract with the state, and they would be insured, exactly like the obstetrician-gynecologists. Furthermore, in 1931, through politicians, midwives submitted a "Midwives Law" bill in order to become as competitive as the obstetrician-gynecologists. If passed, the bill would have mandated midwives to raise the standard of their medical education and to form a Midwives Association through an imperial edict. It would also reserve for midwives the sole right to practice "normal" births. The political campaign did not materialize in the end, but it clearly demonstrated that midwives tried to establish their professional territory by borrowing state authority, just as the obstetrician-gynecologists did in the 1890s.[123]

In turn, obstetrician-gynecologists were not passive observers of the midwives' moves. Obstetrician-gynecologists opposed the proposal to raise the educational standard for midwives, arguing it would lead to a shortage of midwives in rural areas.[124] On a more discursive level, obstetrician-gynecologists tried to protect their vested interests by creating another model of the gendered division of labor that would not threaten their position. One model they explored in the 1930s, as the government was strengthening the maternal and infant health provisions in preparation for war, was to let midwives be the experts in "motherhood protection," while obstetrician-gynecologists took over the domain of medical midwifery altogether.[125]

---

[122] Kimura, *Shussan to seishoku*, 75.
[123] Harue Oide suggests that these moves jumpstarted the process to institutionalize childbirth. Harue Oide, "Byōin shussan no seiritsu to kasoku: Seijōsan wo meguru kōbō to sanshihō seitei wo undō wo chūshin toshite," *Ningenkankeigaku kenkyū*, no. 7 (2006): 25–39.
[124] Kimura, *Shussan to seishoku*, 129.
[125] Ibid., 138–39.

Against this backdrop, midwives in Osaka participated in social work and enthusiastically affirmed their roles within it. In so doing, midwives expressed devotion to the imperial state, as midwife Watanabe Tomoe did when she talked of her work:

The reason why I chose this profession was because I, as a child of the Emperor, desperately wished to contribute to ... the nation. In this sense, to engage with the birth attendance work ... is a truly responsible work, especially in times of emergency such as today, because the infant is the foundation for the rise or fall of the nation in the future. For this reason, whenever I attend childbirth labor, I go to the [woman] imagining as if I was running toward a war front, wishing that the baby, as a future national subject, would be born with both the baby and mother intact.[126]

How should we read Osaka midwives' enthusiastic enactment of their identities as "children of the Emperor" and their assigned roles in maternal and infant healthcare, especially in a context where their local and national representatives were struggling to fend off pressure from the obstetrician-gynecologists, who were trying to confine them to the domain of "motherhood protection"? I argue that the specific way midwives portrayed their role in infant and maternal care work embodied their strategy to further secure their professional domain. Through constant negotiations with obstetrician-gynecologists to demarcate professional boundaries, midwives turned their opponent's demands to their own advantage. While, in the minds of obstetrician-gynecologists, it might have been their professional strategy to reduce the midwives' field of expertise to "motherhood protection," midwives saw this as an *additional* opportunity they could exploit to expand their area of expertise. Furthermore, in order to maximize the benefits of this opportunity, midwives mobilized the oft-used narrative about their service to the imperial state. They stressed how their work, ensuring the healthy growth of babies, was directly contributing to the prosperity of the nation-state and empire by ensuring a constant supply of future workers and soldiers.[127] This logic, in addition to the sense of professional duty, buttressed the Osaka midwives' participation in infant and maternal care work and the characterization of their work as a national and imperial mission.

Significantly, when expressing their devotion to the imperial state, these midwives also incorporated the argument of high infant mortality. According to midwife Kishida Tome:

---

[126] Tomoe Watanabe, "Josanpu de nakereba ajienai shokugyōjō no taiken ninshiki nit-suite," in *Osaka no sanba wa kataru*, 55.
[127] Kimura, *Shussan to seishoku*, 109–74.

It is a shame that Japan's mortality rate is still embarrassingly high, although it seems it has been decreasing somewhat with the development and popularization of hygiene knowledge. Japan, a world's first-class nation. Our Japan is aggressively striding forward, to the land, to the sea, and to the sky. But, when we hear we are falling far behind European and American nations when it comes to infant issues, we, as Japanese women, and as someone engaging in the field, cannot condone it.[128]

Kishida's claim indicates the extent to which statistical rationale had spread in society: A rank-and-file midwife felt comfortable using it to describe her cause. It also vividly illustrates how much authority was conferred upon vital statistics as a rhetorical tool to uphold the existence of Japan as a nation-state and empire.

Conclusion

Medical midwifery and vital statistics, today regarded as coterminous yet separate fields, were once intimately intertwined, as they were both formed in modern Japan. In the 1870s, the development of both fields was guided by the nascent government's interest in managing aspects of the corporeal population for nation-building. Medical midwifery was promoted by the government, in part to reduce the number of deaths during pregnancy and childbirth. Alongside this, the Meiji government readily implemented vital statistics as part of state bureaucracy, because the government saw death as a pressing national issue and vital statistics as an effective tool for visualizing the actual state of the nation's health in numbers. From the 1880s onward, the two fields crossed paths in the government's decision to involve midwives in the official effort to improve the quality of vital statistics. From the 1910s onward, when infant death became singled out as a critical factor in the health of the nation, vital statistics generated relevant data that mobilized medical midwifery for official and public actions to improve maternal and infant health. In turn, midwives capitalized on the power of statistical rationale to advance their professional position.

Ultimately, this story, constructed by weaving together the histories of official vital statistics and midwifery, highlights the centrality of the state's population-governing exercises for the formation of medical midwifery as a modern healthcare profession. It points out that the idea of a corporeal population and the statistical rationale that were behind the government's quick adoption of vital statistics were indispensable

---

[128] Tome Kishida, "Sanba dē ni saishite," in *Osaka no sanba wa kataru*, 93.

for the establishment of the link between midwifery and statecraft. Yet, in the context of Japan in the last quarter of the nineteenth century, when state actors consciously abolished old customs and implemented new knowledge and practices for the construction of a modern state, this had to be explored while the key concepts for the link – individual bodies, population, health, medicine, and even the modern state – were still being formed. This meant that negotiations of different kinds, with various foci, were required, and in the case of medical midwifery, they manifested in, for instance, making the infant a statistical subject or the struggles between obstetrician-gynecologists and midwives.

In the 1920s, as administrative work on vital statistics became routinized, the population itself became regarded as a source of concern. The rising discourse of Japan's "population problem" mirrored anxieties that prevailed in the government office and among the burgeoning communities of population experts over a number of social issues that were emerging as Japan was confronted with new political challenges.

3 Policy Experts
Tackling Japan's "Population Problems"

The 1920s was another turning point in population history in modern Japan. This decade witnessed the emergence of another discourse, summed up in the expression "population problem," which claimed the population *itself* – its size, quality, and mobility – was a liability for economic, social, and political stability. Over the decade, public narratives around the "population problem" spurred the government toward establishing policies dedicated to the problem. To fulfill this goal, the government appointed a broad range of population scientists within policymaking and assigned them to collaborate with bureaucrats and deliberate on policies based on their expert knowledge. While the collaboration during the 1920s failed to lead to a national population policy, it exhibited a novel way of coproduction between population science and the governing of Japan's population. The "population problems" imperative compelled the government to work together with population experts in order to manage Japan's population through policies, while in turn, through policymaking, state bureaucracy became a central site for the population experts to generate and apply their expert knowledge.

The Emergence of the "Population Problem"

The term "population problem" (*jinkō mondai*), now normalized in contemporary Japanese, was foreign to many people at the turn of the twentieth century.[1] However, starting in the 1910s, the expression began to circulate with the rise of popular media. The popular magazine *Central Review* (*Chūō Kōron*) introduced the term as early as 1913 in an article authored by Fukuhara Ryōjirō (1868–1932), a Ministry of Education bureaucrat.[2] "Population problem" was also the theme

---

[1] See Kure, *Riron tōkeigaku*. Nobuo Haruna, *Jinkō, shigen, ryōdo: Kindai nihon no gaikō shisō to kokusai seijigaku* (Chikura Shobo, 2015), 29–66.

[2] Ryōjirō Fukuhara, "Yō wa jinkō mondai no kaiketsu ni ari," *Chūō kōron* 28, no. 8 (July 1913): 54–55.

of the third major conference of the Japan Sociological Institute (est. 1913) held in 1915.[3]

However, it was only in the 1920s that the term became genuinely widespread. In 1924, the popular women's magazine *Housewives' Friend* (*Shufu no Tomo*) defined "population problem" as a current affairs topic and claimed an educated "good wife, [and] wise mothers" ought to know about it.[4] The term also appeared in *Diamond* (*Daiyamondo*), a business magazine, where the now mature statesman Gotō Shinpei reflected on the subject.[5] Finally, it was a core subject in the booklet aimed at "proletariats" that was published by the Industrial Labor Research Institute, a front organization of the Japanese Communist Party.[6] By the end of the decade, "population problem" had become a common expression in mass-circulated print media.

The presence of the term during this period was catalyzed by the increased availability of statistical knowledge on the Japanese population.[7] The government had been publishing vital statistics annually since 1899 (see Chapter 2). Furthermore, in 1920, it organized the first population census of "Japan Proper" (*naichi*) (see Chapter 1).[8] Looking at these population statistics in the first half of the 1920s, experts noted a sign of population growth. According to the 1920 census, the population of Japan – excluding the populations of Karafuto, Taiwan, Korea, the Guangdong Territory, and Micronesia – was 55,963,053, with a total of 11,122,120 households, making the average household 4.89 persons.[9] Between 1921 and 1925, the population kept expanding at a rate between 10.8 and 14.7 per 1,000 persons, which amounted to a natural

[3] Haruna, *Jinkō, shigen, ryōdo*, 122. The Japan Sociological Institute, though short-lived (dissolved in 1922), was a historically significant academic organization. It was founded by Takebe Tongo (1871–1945), known as the founder of sociology in Japan, and attracted leading social scientists during the period. On the institute's role in the debate on population, see ibid., 121–24.

[4] Soho Tokutomi, "Jinkō mondai to ryōsai kenbo shugi no kyōyō sono yon Taishō fujin no kyōyō jūni kō," *Shufu no tomo* 8, no. 4 (April 1924): 4–11.

[5] Shinpei Goto, "Jinkō mondai kaiketsusaku," *Daiyamondo keizai zasshi* 14, no. 17 (June 1, 1926): 19.

[6] Sangyō Rōdō Chōsajo, *Musansha seiji hikkei* (Dōjinsha shoten, 1926).

[7] Ryōsaburō Minami, *Jinkōron hattenshi: Nihon ni okeru saikin jūnen no sōgyōseki* (Sanseido, 1936).

[8] For a discussion of how *naichi* became a legal category in the process of building the Japanese imperial order, see Toyomi Asano, "Kokusai chitsujo to teikoku chitsujo womeguru nihon teikoku saihen no kōzō: Kyōtsūhō no rippō katei to hōteki kūkan no saiteigi," in *Shokuminchi teikoku nihon no hōteki tenkai*, ed. Toshihiko Matsuda and Toyomi Asano (Shinzansha Shuppan, 2004), 61–136.

[9] Sōmushō Tōkeikyoku, "Tōkei de miru anotoki to ima no. 3: Dai 1-kai kokuzei chōsaji (taishō 9-nen) to ima," October 1, 2014, www.stat.go.jp/info/anotoki/pdf/census.pdf, accessed 22 July 2019.

increase of 673,000 to 913,000 annually. In 1926, more than one million people – 1,004,000, precisely – were added to the existing population for the first time since the Meiji government began to compile population statistics in 1872. The year also witnessed the population figure going over 60 million for the first time, reaching 60,741,000.[10]

Witnessing this demographic trend, journalists and scholars warned that Japan was currently facing a serious problem in regard to population growth.[11] In his seminal article "Some Reflections on Overpopulation," published in *Chūō Kōron* in July 1927, Yanaihara Tadao, a professor at Tokyo Imperial University and an authority on colonial policy, wrote:

When it was announced that the natural increase of the population in 1924 was 743,000 and in 1925 it was 875,000, people simply were shocked at how large the sheer size of it was, and we heard much noise about the population problem.... Now, hearing this year's population increase is one million, do people not feel as if the fire is engulfing them so fast that they cannot do anything but gaze at it?[12]

Many social commentators and policy intellectuals shared Yanaihara's sentiments, defining population growth as a current "population problem."

The idea of population expansion creating issues for the collective was not new. In the early Meiji period, noted intellectuals expressed their concerns over the sudden visibility of an expanding group of struggling samurais after the government dismantled the social structure that had long privileged them. They characterized samurais as politically volatile, "surplus people" and suggested that the government should come up with appropriate measures to tame them for peaceful nation-building.[13] Likewise, in the mid-1880s, when the fiscal policies under Finance Minister Matsukata Masayoshi caused an economic depression, population expansionists characterized impoverished farmers as "surplus people."[14] In the Meiji period, public intellectuals and officials problematized politically and economically burdensome groups by mobilizing the language of population expansion.[15]

---

[10] Sōmushō Tōkeikyoku, "2-1 danjobetsu jinkō, jinkō zōgen oyobi jinkō mitsudo (Meiji 5-nen-heisei 21-nen)," n.d., www.stat.go.jp/data/chouki/02.htm, accessed 22 July 2019.
[11] Minami, *Jinkōron hattenshi*, 1.
[12] Tadao Yanaihara, "Jiron toshiteno jinkō mondai," *Chūō kōron* 42, no. 7 (July 1, 1927): 34–35. For more discussion on Yanaihara's notion of the population problem, see, e.g., Hiroyuki Shiode, *Ekkyōsha no seijishi* (Nagoya: Nagoya Daigaku Shuppankai, 2015), 162–65; Ryoko Nakano, "Uncovering Shokumin: Yanaihara Tadao's Concept of Global Civil Society," *Social Science Japan Journal* 9, no. 2 (2006): 187–202.
[13] Lu, *The Making of Japanese Settler Colonialism*, 40–41.
[14] Ibid., 58–63.
[15] Ibid., 67; Lee, "Problematizing Population," 139–48.

In the 1920s, an additional perspective found a voice in the debate over population growth. This time, the focus was placed on the effect of overpopulation on the economy, specifically on people's living conditions. Reminiscing about Japan's population problems in the 1920s, in 1937 the population policy researcher Ishii Ryōichi identified events that brought misery into people's economic lives.[16] One was the economic recession after WWI. It was exacerbated by the Washington Naval Treaty in 1922, which caused unemployment in the shipbuilding sector due to the limitations imposed on the construction of capital ships in Japan. The following year, the Great Kanto Earthquake hit the capital Tokyo and its vicinity and cost the government billions of yen. Population growth in this context was seen to place pressure on the nation's economy, causing further unemployment, lowering the living standard, and eventually leading to "difficulty in living" (seikatsunan) – a buzzword within the socialist and labor activism that was burgeoning during this period in response to the economic depression.[17] In addition, Ishii thought the United States Immigration Act of 1924, which limited Japanese migration, indirectly worsened people's living conditions in Japan. When the act was issued, Japan lost an important outlet for its excess population. Consequently, the little island country became even more crowded, which eventually led to the erosion of already limited natural resources. As Ishii saw it, population growth in the 1920s was a problem related to "economic life," which was made worse by Japan's disadvantaged position in world politics.[18]

This way of problematizing population growth triggered an academic debate on the nature of "population problems" among social scientists, especially those with backgrounds in economics. It was sparked by the provocative article, "Give Birth, Multiply," presented on June 4, 1926 by sociologist Takata Yasuma (1883–1972).[19] Going against the mainstream opinion that problematized population growth, Takata claimed that the real population problem was the population contraction due to fertility decline. In stark contrast to the 1910s, the birth rates declined

---

[16] Ryoichi Ishii, *Population Problems and Economic Life in Japan* (Chicago: University of Chicago Press, 1937).

[17] A critical background to this phenomenon is the emergence of a moralistic discourse of poverty, which stigmatized the poor as undeserving. Elise K. Tipton, "Defining the Poor in Early Twentieth-Century Japan," *Japan Forum* 20, no. 3 (October 2008): 361–82.

[18] Ishii, *Population Problems and Economic Life.*

[19] Yasuma Takata, *Jinkō to binbō* (Nihon Hyoronsha, 1927), 90–96. Takata was a disciple of Yoneda Shōtarō (1873–1945), who is credited with introducing sociology into Japan, along with Takebe Tongo. For more on Takata's theory on population, see Sugita, *Jinkō, kazoku, seimei,* 22–32.

over the 1920s, from 36.2 births per 1,000 population in 1920 to 34.4 in 1928. Takata pointed out how this trend was caused by the growing popularity of artificial birth control among the urban intellectual class. He argued that the population stagnation resulting from the fertility decline among this class was "the most dangerous" for the Japanese as "a weak and disadvantaged racial group" in world politics.[20] Based on this logic, in the article, Takata urged married couples to make more babies.

As Takata's statement indicates, his concern was fueled less by the argument linking population growth with poverty and more by the geopolitical perspective, which viewed a large population size as an asset that would help raise Japan's global position.[21] Takata admitted that the recent economic recession had certainly led to "difficulty in living" among the people in Japan but simultaneously criticized this argument as being based on an introverted perspective that narrowly focused on domestic situations. Takata then urged his colleagues to adopt a broader view and consider Japan's growing population in relation to its racial and geopolitical struggles within world politics, which were dominated by white, western colonial powers. He explained, "population is indeed a source of every kind of power for a race [*minzoku*]," therefore, "the larger a population is, the more the race will thrive and its power expand in every possible direction."[22] Takata believed Japan could take advantage of its currently growing population to alter its disadvantaged position as the colored colonial power. He claimed, "population increase" was "the only weapon a colored race has to fight against white people." He even argued that "the colored race must be prepared to face self-destruction [*jimetsu*]" if it lost the power to grow its population.[23] This logic of racial struggle compelled Takata to warn about the danger of declining fertility.

Marxist economist Kawakami Hajime (1879–1946) at the University of Kyoto immediately refuted Takata's argument.[24] As Kawakami saw it, population surplus was a problem inherent to capitalism. Kawakami argued that under capitalism, the proportion of invariant capital – the capital allocated to equipment and raw materials – would increase with

---

[20] Takata, *Jinkō to binbō*, 90–96.
[21] Haruna, *Jinkō, shigen, ryōdo*, 116–28, 158–64.
[22] Takata, *Jinkō to binbō*, 90–96.
[23] Ibid. The implication with eugenics will be discussed subsequently.
[24] Nihon Jinkō Gakkai, *Jinkō daijiten*, 274; Eric G. Dinmore, "A Small Island Nation Poor in Resources: Natural and Human Resource Anxieties in Trans-World War II Japan" (PhD diss., Princeton University, 2006), 32–33; Koji Yoshino, "Yutakana shakai no binbōron: Takata Yasuma to Kawakami Hajime," *Hitotsubashi kenkyū* 30, no. 3 (October 2005): 35–52.

the mechanization and the increasing scale of production, whereas the proportion of variable capital allotted to wages would decline. Under these circumstances, even if the amount of capital as a whole increased, it would not catch up with the increase in the number of workers, thus eventually generating surplus population and poverty (i.e., "difficulty in living") among the waged population. For Kawakami, the population problem was by nature a problem connected to the capitalist economy, and overturning capitalism would be the only way to solve it.[25]

Kawakami's response led to a heated population debate that last for nearly a decade – one involving prominent social scientists such as Nasu Shiroshi, Shinomiya Kyōji, Yoshida Hideo, Sakisaka Itsurō, and Ōuchi Hyōe.[26] Reminiscing about the decade-long debate, in 1936 the leading economist and population theorist Minami Ryōsaburō concluded that the debate ultimately failed to establish any concentrated efforts, though it did jumpstart a number of research projects.[27] In turn, postwar social scientists, starting with Ichihara Ryōhei, portrayed the debate as manifesting a deep-seated, inherently unresolvable conflict between the distinctive population theories of Karl Marx and Thomas Malthus.[28] From a historical point of view, the academic population debate was significant, because it illuminated the growing understanding that population was a dynamic social factor directly impacting the living conditions of the masses and, therefore, "population problems" were invariably synonymous with "social problems" (*shakai mondai*), another catchword in the public, intellectual, and policy debates during this period.

The public discourse and debate around "population problems" urged government officials to explore policies to tackle wide-ranging social issues associated with population phenomena. The policymaking process showed how the official agenda was shaped by the broader sociopolitical circumstances surrounding Japan at the time.

The Official Response to the Population Problem

Among the many population problems, the government perceived the "food problem" (*shokuryō mondai*), heightened by population expansion, as the most urgent.[29] In particular, after the inflation of the cost

---

[25] Hajime Kawakami, *Jinkō mondai hihan* (Sobunkaku, 1927), 41.
[26] Yasuyuki Nakanishi, "Takata Yasuma no jinkō riron to shakaigaku," *Keizai ronsō* 140, no. 5–6 (1987): 59–63.
[27] Minami, *Jinkōron hattenshi*, 18–19.
[28] Nakanishi, "Takata Yasuma no jinkō riron," 60–61.
[29] Dinmore, "A Small Island Nation Poor in Resources," 26.

of rice caused mass protests in 1918, high-rank government officials became concerned that food shortage would not only promote "difficulty in living" but also stimulate political unrest.[30] To tame the protests, government officials set price controls on rice in the metropole and encouraged the cultivation of Japanese-style rice in colonial Taiwan and Korea. However, this policy led to a sudden rise in the amount of less-expensive rice from Taiwan and Korea in the metropole, and consequently, the government met with harsh criticism.[31] Social commentators explained that the policy of importing colonial rice exposed the government's failure to raise domestic agricultural productivity enough to feed the "population of Japan Proper" (*naichi jinkō*).[32] Under these circumstances, policymakers were presented with the thorny question of how to feed the expanding Japanese population using the country's limited resource capacity.

Against this backdrop, the government, under the premiership of Tanaka Giichi (1927–29), established the Investigative Commission for Population and Food Problems (*Jinkō Shokuryō Mondai Chōsakai*, hereafter IC-PFP) on July 7, 1927. Headed by Tanaka and located directly under the cabinet, the IC-PFP was a major government initiative responsible for research and policy deliberation over population and food supply issues. Under Tanaka, Home Minister Suzuki Kisaburō (1867–1940) and Minister of Agriculture Yamamoto Teijirō (1870–1937) became the vice-chairmen. Reflecting the joint vice-chairmanship, the government identified two enquires for the IC-PFP to address: one on population and the other on food issues. The IC-PFP Secretariat met six times between July 17 and October 5, 1927 to deliberate on official inquiries and set the agenda, and it set up two sections (i.e., the Population Section and the Food Section) as specialist deliberation committees.

As is apparent from its name and organizational structure, the IC-PFP's understanding of population problems was through the lens of the Malthusian dilemma (i.e., the imbalance between population growth and the pace of food production). At the same time, the conflicting view, such as Takata's that saw population growth as a symbol of expanding national

---

[30] The inflation of the cost of rice was originally caused by its reduced availability in Japan due to it being diverted to Japanese soldiers in Siberia who were participating in the Allied forces' intervention in the Bolshevik Revolution (Siberian Intervention, 1918–22).

[31] Dinmore, "A Small Island Nation Poor in Resources," 25. For a different interpretation, see Penelope Francks, "Rice for the Masses: Food Policy and the Adoption of Imperial Self-Sufficiency in Early Twentieth-Century Japan," *Japan Forum* 15, no. 1 (January 2003): 125–46.

[32] Dinmore, "A Small Island Nation Poor in Resources," 25.

power, also influenced the official characterization of population issues. This was succinctly summarized in the speech Tanaka delivered at the IC-PFP First General Assembly:

Our imperial state's population is growing year by year. [Population growth] should be truly commendable because it … not only enhances the brilliance of our race but also enriches the national territory. However, Japan is a small islands nation and poor in natural resources. Furthermore, our industrial economy is still not fully developed. As the population density intensifies and as the demand for food increases, the supply and demand in labor is bound to become imbalanced, and this tends to ferment unrest in people's living conditions. Thus, it is truly important and urgent to establish measures to improve situations surrounding the growing population of the Japanese Empire and the enrichment of the food supply and to consider economic and societal methods with which to solve the problem.[33]

As expressed in the speech, the official stance was that the population increase was not ipso facto problematic and, in fact, should be celebrated as a factor promoting the "brilliance of our race." However, it became a problem when put into the specific context of Japan: Poor resource availability, coupled with the compromised state of the nation's industry, meant that the country had a limited ability to absorb surplus population. Under these circumstances, continuous population growth would weaken the nation's capacity to feed its people, saturate the labor market, and eventually lower people's living standards. Tanaka's speech implied that the IC-PFP's agenda in relation to the population problem was in fact more expansive than the simple Malthusian predicament and dovetailed with issues related to the national economy and industry, labor, and resources.

Another significant component of Tanaka's speech was his explanation of issues related to the national economy and Japanese territory through the intensification of population density. This caricature of population problems was not at all original to Tanaka; on the contrary, it was currently in vogue within the international community of population experts. The idea of population density was at the heart of the World Population Conference held in Geneva in the summer of 1927.[34] At the conference, participants highlighted issues related to sovereignty, geopolitics, and the

---

[33] "Jinkō shokuryō mondai chōsakai dai ikkai sōkai ni okeru naikaku sōri daijin aisatsu," July 20, 1927, in The Cabinet of the Japanese Government, "Jinkō shokuryō mondai chōsakai dai ikkai iin sōkai giji gaiyō," *Jinkō shokuryō mondai chōsakai shorui san sōkai gijiroku*', 1927–29, National Archives of Japan Digital Archives, accessed August 20, 2019, www.digital.archives.go.jp/das/image/F0000000000000068867.

[34] Bashford, *Global Population*, 81–106; Alison Bashford, "Nation, Empire, Globe: The Spaces of Population Debate in the Interwar Years," *Comparative Studies in Society and History* 49, no. 1 (2007): 180–83.

economy that arose due to growing population density. They understood the situation as most evident in Europe, the Indian subcontinent, and East Asia – the three world regions that Ernst Georg Ravenstein had once characterized as the "global centers of population."[35] One of the American participants, Warren S. Thompson (1887–1950), later portrayed these regions, including Japan, as "danger spots in world population."[36] Thompson claimed that the population expansion in these "danger spots" would put further pressure on the land, ultimately catalyzing political and economic disasters on a global scale.[37] Put in this context, Tanaka's speech indicated the Japanese government's awareness of the global consensus forming among the population experts at the time: The earth had bounded space, and the growing population, especially in world regions with the densest populations, would imminently trigger a world crisis.

Also recognizing that the international academic discussion pointed a finger at Japan's rising population density, the Japanese government resorted to migration to relieve the pressure associated with a rising population in "Japan Proper."[38] In 1928, the government issued the Overseas Migration Cooperative Societies Law (*Kaigai Ijū Kumiai Hō*), which authorized prefectures to organize emigration projects. Responding to the law, in the 1930s, prefectures took a leading role in sending their people abroad, first to Latin America and then to Manchuria and the South Seas.[39]

The official response to the perceived population crisis was not new. In the early Meiji period, responding to the narrative of surplus people, the government mobilized the samurais to colonize Hokkaido.[40] In the mid-1880s, amid the economic depression, population expansionists argued that North America could absorb the population of impoverished Japanese farmers.[41] While this premise remained fundamentally unchanged, in the

---

[35] Bashford, Global Population, 99–100.

[36] Warren Simpson Thompson, *Danger Spots in World Population* (New York: Alfred A. Knopf, 1929).

[37] Bashford, *Global Population*, 103–6.

[38] Lu, *The Making of Japanese Settler Colonialism*, 149–233; Shiode, *Ekkyōsha no seijishi*, 226–350; Manabu Takeno, "Jinkō mondai to shokuminchi: 1920, 20 nendai no karafuto wo chūshin ni," *Keizaigaku kenkyū* 50, no. 3 (December 2000): 117–32.

[39] Chapter 4 discusses the Japanese migration to Manchuria in more detail.

[40] Lu, *The Making of Japanese Settler Colonialism*, 58–63.

[41] Sidney Xu Lu, "Eastward Ho! Japanese Settler Colonialism in Hokkaido and the Making of Japanese Migration to the American West, 1869–1888," *The Journal of Asian Studies* 78, no. 3 (2019): 521–47; Sidney Xu Lu, "Colonizing Hokkaido and the Origin of Japanese Trans-Pacific Expansion, 1869–1894," *Japanese Studies* 36, no. 2 (May 2016): 251–74, https://doi.org/10.1080/10371397.2016.1230834; Ishizaki, *Kingendai nihon no*, 69; Shiode, *Ekkyōsha no seijishi*, 112–53, 278–329.

late 1920s, the political environments specific to the period – domestic and international – exhorted government officials to consider making overseas migration an official policy.[42] Within Japan, as the politically suspected left-leaning labor and socialist movements grew in popularity, the government, especially the Home Ministry, which was also in charge of social affairs, began to consider overseas migration as an attractive option for maintaining social order. By targeting populations deemed susceptible to these movements, the government hoped overseas migration would curb the movements' further growth. Alongside this, a number of diplomatic events – notably the Allies' refusal of Japan's proposal to include a racial equality clause in the Treaty of Versailles in 1919 and the US Immigration Act in 1924 – swayed the government's opinion in favor of overseas migration. Perceiving these diplomatic incidents to symbolize the domination of the racist attitudes of white, western colonial powers within world politics, Japanese political leaders believed that Japan, as the only colored colonial power, should cultivate their own versions of colonial migration in order to take a stand in the political struggles of that time.[43]

Reflecting the government's agenda, state-endorsed emigration in the late 1920s focused on workers and farmers who were impoverished due to the economic depression, as well as on activities that could directly contribute to Japan's colonial agenda.[44] In Brazil and Manchuria, it insisted on the principle of "coexisting and coprospering" (*kyōzon kyōei*), a benevolent colonial development based on the coexistence and coprosperity of the Japanese and native populations.[45] For the Japanese state, overseas migration during this period was a device to solve the labor, social, and population problems within the metropole as well as a way to assert – what the political leaders deemed compromised – Japan's geopolitical standing in world politics via colonialism.[46] In this context, Japan's high population density conveniently justified colonial migration.

---

[42] Lu, *The Making of Japanese Settler Colonialism*, 19–20.

[43] Park, "Interrogating the 'Population Problem' of the Non-Western Empire"; Eiichiro Azuma, "Remapping A Pre-World War Two Japanese Diaspora: Transpacific Migration as an Articulation of Japan's Colonial Expansionism," *Connecting Seas and Connected Ocean Rims*, January 1, 2011, 415–39; Eiichiro Azuma, "'Pioneers of Overseas Japanese Development': Japanese American History and the Making of Expansionist Orthodoxy in Imperial Japan," *The Journal of Asian Studies* 67, no. 4 (November 2008): 1187–226.

[44] Lee, "Problematizing Population," 145. For a specific example, see Andre Kobayashi Deckrow, "São Paulo as Migrant-Colony: Pre-World War II Japanese State-Sponsored Agricultural Migration to Brazil" (PhD diss., Columbia University, 2019).

[45] Lu, *The Making of Japanese Settler Colonialism*, 206–33.

[46] Not all intellectuals endorsed the official stance. Haruna, *Jinkō, shigen, ryōdo*, 217–46.

In addition to political reasons, population density gave an economic rationale for the migration policy. The government portrayed overseas migration also as a means with which to foster what Prime Minister Tanaka called an "industrial economy." This caricature urged policymakers to see overseas migration as the twin policy of domestic migration. According to this view, overseas migration aimed to alter population density on the macro level, whereas domestic migration – encouraging people to move from more concentrated urban centers to less-dense regions within Japan Proper – was a micro-level attempt to raise the overall level of commercial and industrial productivity by "adjusting" the domestic population density to an optimal level. The logic was as follows: If the government adequately coordinated internal and overseas migration, industrialists could more efficiently deploy their business within both Japan Proper and the colonies. At the same time, migrants, as well as people who remained in Japan Proper, would have more job opportunities and, therefore, enough financial power to stimulate the Japanese economy. This additional rationale not only promoted overseas migration even further but also consolidated the official discourse in favor of internal migration.

Prime Minister Tanaka, as head of the IC-PFP, had internalized this logic when he pondered the population issues. To start with, Tanaka believed overseas migration was merely "one solution to the problem" of a growing population; therefore, he did not think it alone could dissipate the population problem in the metropole.[47] Tanaka believed overseas migration should be carried out in parallel with "policies to make Japan an industrial nation [*sangyō rikkoku*]," by which he primarily meant economic measures, including "internal migration."[48] If overseas migration could "relieve" Japan's population pressure, effective economic measures could also "absorb population."[49]

The IC-PFP Secretariat, headed by Secretary of the Cabinet Hatoyama Ichirō (1883–1959) and comprised of twenty-four bureaucrats from relevant government offices, followed Tanaka's line.[50] Topics related to

---

[47] "Jinkō shokuryō mondai chōsakai dai ikkai."

[48] Ibid.

[49] Ibid.

[50] The government offices that had representation in the IC-PFP Secretariat were: the Cabinet, Cabinet Colonial Bureau, Cabinet Resource Bureau, Ministries of Foreign Affairs, Home Ministry Bureau of Internal Affairs, Home Ministry Social Bureau, Ministry of Finance, Ministry of Agriculture and Forestry, Ministry of Commerce and Industry, Ministry of Communication, and Ministry of Railways. The Cabinet of the Japanese Government, *Jinkō shokuryō mondai chōsakai shorui kyū kanjikai gijiroku'*, 1927, National Archives of Japan Digital Archives, www.digital.archives .go.jp/das/image/F0000000000000068867, accessed August 20, 2019.

migration – domestic and overseas – and the increased economic opportunities dominated the draft of the IC-PFP Secretariat's reference plan for population policies. The draft reference plan, originally submitted by the Home Ministry's Bureau of Social Affairs to the Secretariat on August 4, included the following list of subjects that the IC-PFP Population Section should investigate to come up with "specific measures for population growth":

1. Regarding Measures to Increase Demands in Labor
2. Regarding Measures to Control the Supply and Demand in Labor
3. Regarding Measures for Domestic Migration
4. Regarding Measures for Emigrating out of the Country
5. Regarding Measures to Encourage Migration Generally
6. Regarding the Establishment of Colonial Companies
7. On the Official Administrative Organizations for Migration
8. Regarding Birth Control[51]

The list clearly reflected the government's commitment to using migration and economic development to solve the problem of population growth.

After deliberating on the bureau's draft several times, the IC-PFP Secretariat approved the modified reference plan in early October. The reference plan was passed on to both sections, and they used it to deliberate on government population policies. The discussion within the Population Section illustrated that a community of population experts, held together by policymaking, was formed alongside the government's effort to tackle "population problems." However, these policy experts, with their diverse intellectual backgrounds, presented ideas that were not always in line with the official agenda.

Population Control and Migration: The Experts' Response

For the Population Section, the IC-PFP summoned twenty-two permanent members and eleven temporary members from the pool of politicians, senior officials, bureaucrats, and population experts from fields as diverse as economics, social policy, colonial policy, statistics, medicine, and public health.[52] On October 12, 1927, the Population Section

---

[51] Jinkō Shokryō Mondai Chōsakai, "Jinkō shokuryō mondai chōsakai ni okeru naigai ijū hōsaku oyobi rōdō no jukyū chōsetsu ni kansuru hōsaku no ketsugi tōshin ni itaru keika narabini giron no yōten," February 1928, 2–5.

[52] Later, the member list became longer, including a prominent figure in colonial policy, Nitobe Inazō. The Cabinet of the Japanese Government, *Jinkō shokuryō mondai chōsakai*, 30–36.

convened its first meeting. Between then and April 1930, when the IC-PFP dissolved, the Population Section had four section meetings, twenty-four special committee meetings, six draft-making committee meetings, and three Small Committee meetings, and it officially submitted six draft reports and two resolutions in response to the government's inquiry on population matters.

Generally speaking, the Population Section followed the guidelines stipulated in the IC-PFP Secretariat reference plan,[53] and so the members at the meeting decided to work on emigration and "labor adjustment" (rōdō chōsei). This focus ultimately culminated in three draft reports, "Measures for Internal and External Emigration," "Measures Regarding the Adjustment of Supply and Demand in Labor" and "Population Measures for Various Locals Apart from Japan Proper."[54]

However, a closer look at the deliberation process presents a more nuanced picture, testifying to how the policy discussion involved complex negotiations. In many instances during the negotiations, the two groups had divided opinions over the interpretation of population – the IC-PFP Secretariat representing the voice of the government officials and the Population Section expressing the opinion of population experts. To start with, while the Population Section deliberated on emigration and labor adjustment, this did not mean they completely agreed with the Secretariat's suggestions. Some members of the Population Section's Special Committee and Draft-Making Committee were skeptical that overseas migration and domestic labor adjustment alone could "have a value, as fundamental measures, to solve the population problem."[55] Based on this view, they saw a need to come up with a more comprehensive policy that would complement these measures.

In addition, some members of the Population Section disagreed over the IC-PFP Secretariat's characterization of population measures. Specifically, they did not like that the Secretariat's draft reference plan portrayed overseas migration not only in relation to the problem of overpopulation but additionally as a geopolitical measure "to establish an international justice by realizing the principle of coexistence and coprosperity."[56] The member vociferously opposing the political overtones

---

[53] Jinkō Shokryō Mondai Chōsakai, "Jinkō shokuryō mondai chōsakai ni okeru naigai ijū hōsaku," 17.

[54] The first two draft reports were passed by the Second IC-PFP General Assembly that took place on December 15,1927, and the last one by the Third IC-PFSP General Assembly on September 27, 1928.

[55] Jinkō Shokryō Mondai Chōsakai, "Jinkō shokuryō mondai chōsakai ni okeru naigai ijū hōsaku," 47.

[56] Ibid., 47.

in this description was Nagai Tōru (1878–1973), the bureaucrat and social policy expert leading the discussions in the Population Section. At a Population Section meeting, in front of the Bureau of Social Affairs Secretary Kawanishi Jitsuzō (1889–1978), who was explaining the draft reference plan, Nagai dismissively told: "this kind of manifesto is disagreeable because it sounds like either a small nation's tale of woe or acts as the mask for an aggressor nation."[57] After discussions, the Population Section decided that its draft proposal on overseas migration would not include the aforementioned statement.[58] Some in the Population Section were clearly ambivalent about the idea that emigration might be used as a political tool to secure Japan's global status.

Finally, between the IC-PFP Secretariat and the Population Section, there was a slight discord in the understanding of which measures the government should prioritize. While the Secretariat was mainly concerned about the problems of population *quantity*, the Population Section was additionally interested in the issues of population *quality*. The deliberation process on "population control" (*jinkō yokusei*), which included the topics of birth control and eugenics, attests to this point. Whereas the IC-PFP Secretariat sidelined concerns over the subject, the Population Section spent long hours on it. The outcome of this different attitude was a stark contrast. While the IC-PFP Secretariat's reference plan put the subject of birth control at the bottom of the aforementioned list, the Population Section produced an independent policy proposal dedicated to measures for population control – one of the six draft reports officially presented by the section during its existence.

In the late 1920s, population quality problems, specifically issues touching on eugenics and birth control, were popular, yet controversial, subject matters.[59] Eugenics, a term coined by Francis Galton in 1883, was almost immediately introduced in Japan as *yūzenikkusu*, and later *yūseigaku*, "science of superior birth."[60] From the start, the intellectuals defined eugenics

---

[57] Ibid., 50.
[58] Ibid., 50.
[59] Takashi Yokoyama, *Nihon ga yūsei shakai ni naru made: Kagaku keimō, media, seishoku no seiji* (Keiso Shobo, 2015).
[60] For recent or representative works on the history of eugenic thought and movements in modern Japan, see Karen J. Schaffner, ed., *Eugenics in Japan* (Fukuoka: Kyushu Daigaku Shuppankai, 2014); Kiyoko Yamazaki, ed., *Seimei no rinri: Yūsei seisaku no keifu* (Fukuoka: Kyushu Daigaku Shuppankai, 2013); Jennifer Robertson, "Eugenics in Japan: Sanguinous Repair," in *The Oxford Handbook of the History of Eugenics*, eds. Alison Bashford and Philippa Levine (Oxford: Oxford University Press, 2010), 430–48; Yutaka Fujino, *Kōseishō no tanjō: Iryō wa fashizumu wo ikani suishin shitaka* (Kyoto: Kamogawa shuppan, 2003); Fujino, *Nihon fashizumu to yūsei shisō*; Yuehtsen Juliette Chung, *Struggle for National Survival: Eugenics in Sino-Japanese*

as a method of "racial improvement" (*jinshu kairyō*), which the Japanese could incorporate to reach a higher level of civilization.[61] This idea became more pervasive after the 1890s, when "race" (*jinshu* or *minzoku*) began to hold currency in political discussions and within the articulation of nationalistic sentiments.[62] Especially after Japan acquired Taiwan as a colony in 1895 and won the war against one of the most formidable white race countries in the Russo-Japanese War (1904–5), anthropologists, biologists, geneticists, doctors, and social scientists fervently discussed the political implications of Japan as a nonwhite empire, and actively explored the possibility of applying eugenics to raise Japan's international profile.[63] Beginning in the 1910s, events in international politics, including the aforementioned scandal at the Versailles Peace Conference in 1919, further fueled the sentiment that eugenics – now used synonymously with another neologism, "racial hygiene" (*minzoku eisei*) – was an appropriate measure for Japan to adopt to supersede the white imperial powers. Under these circumstances, publications such as the journal *Der Mensch: Jinsei* (*Humankind*), launched in 1905 by the renowned physician and medical historian Fujikawa Yū (1865–1940), offered experts in medicine, biology, psychology, anthropology, and other fields dealing with racial and population sciences an outlet for their fervor for eugenics.

With regard to birth control, an early reference in Japan's modern history appeared in the 1880s in relation to the criminalization of abortion in 1880 (see Chapter 2).[64] In the 1900s, the neo-Malthusian argument, greatly inspired by the British activist Annie Besant, began to generate public discourse on birth control.[65] However, during this period, the concept of birth control remained largely theoretical.[66] This situation began to

---

Contexts, 1896–1945 (New York: Routledge, 2002); Sumiko Otsubo and James R. Bartholomew, "Eugenics in Japan: Some Ironies of Modernity, 1883–1945," *Science in Context* 11, no. 3–4 (1998): 545–65; Yoko Matsubara, "Nihon – Sengo no yūsei hogohō toiu nano danshuhō," in *Yūseigaku to ningen shakai: Seimei kagaku no seiki wa doko e mukaunoka*, eds. Shohei Yonemoto, Yoko Matsubara, Jiro Nudeshima, and Yasutaka Ichinokawa (Kodansha, 2000), 169–236; Yoko Matsubara, "The Enactment of Japan's Sterilization Laws in the 1940s: A Prelude to Postwar Eugenic Policy," *Historia Scientiarum* 8, no. 2 (1998): 187–201; Zenji Suzuki, *Nihon no yūseigaku* (Sankyo Shuppan Kabushikigaisha, 1983).
61 Robertson, "Eugenics in Japan."
62 Kevin M. Doak, *A History of Nationalism in Modern Japan: Placing the People* (Leiden and Boston: Brill, 2007); Morris-Suzuki, *Re-Inventing Japan*; Oguma, *Tan'itsu minzoku shinwa no kigen*.
63 Sabine Frühstück, *Colonizing Sex: Sexology and Social Control in Modern Japan* (Berkeley: University of California Press, 2003); Fujino, *Nihon fashizumu to yūsei shisō*, 80–120; Oguma, *Tan'itsu minzoku shinwa no kigen*, 73–171.
64 Tenrei Ota, *Nihon sanji chōsetsu hyakunenshi* (Ningen no Kagakusha, 1976).
65 Ogino, *"Kazoku keikaku" eno michi*, 17.
66 Ibid., 12–14.

change in the late 1910s, when news about the American birth control movement arrived in Japan.[67] In particular, when Margaret Sanger, the doyenne of the American birth control movement, visited Japan in 1922, she greatly influenced the public's perception of birth control and the popular birth control movement.[68] After Sanger's Japan tour, Ishimoto Shizue – Sanger's close friend who became a leader in the Japanese birth control movement – founded the Japan Birth Control Study Group (*Nihon Sanji Chōsetsu Kenkyūkai*) in May 1922, whose participants included well-known figures such as Abe Isoo, Kaji Tokijirō, Yamakawa Kikue, and Suzuki Bunji.[69] In Kyoto, Sanger's visit urged the noted biologist and sexologist Yamamoto Senji (1889–1929) to join the birth control activism.[70] Shortly after Sanger's visit, Yamamoto published *The Critique of Mrs. Sanger's Methods of Family Limitation*, which, despite being noncommercial and academic, was phenomenally popular.[71] In January 1923, with labor activists based in Osaka, Yamamoto established the Osaka Birth Control Study Group (*Osaka Sanji Seigen Kenkyūkai*) as an organization specifically dedicated to teaching birth control methods to individuals with more than five children.[72] In February 1925, Yamamoto launched the coterie magazine *Birth Control*

---

[67] Ibid., 40–69; Karen Lee Callahan, "Dangerous Devices, Mysterious Times: Men, Women, and Birth Control in Early Twentieth-Century Japan" (PhD diss., University of California, Berkeley, 2004); Fujime, *Sei no rekishigaku*, 245–60.

[68] Sanger, the pioneer of the twentieth-century birth control movement in the United States, was spurred on by the mission to spread the "gospel of birth control" to the world. She arrived in Japan in March 1922 and stayed for a month. During this time, she delivered public lectures and participated in private meetings, where she befriended Japanese birth-control advocates. Takeuchi-Demirci, *Contraceptive Diplomacy*, 19–82; Aiko Takeuchi-Demirci, "Birth Control and Socialism: The Frustration of Margaret Sanger and Ishimoto Shizue's Mission," *Journal of American-East Asian Relations* 17, no. 3 (2010): 257–80; Ogino, *"Kazoku keikaku" eno michi*, 39; Helen M. Hopper, *A New Woman of Japan: A Political Biography of Katō Shidzue* (Boulder: Westview Press, 1996), 22–27. For the impact of Sanger's visit on the public debate, see Sujin Lee, "Differing Conceptions of 'Voluntary Motherhood': Yamakawa Kikue's Birth Strike and Ishimoto Shizue's Eugenic Feminism," *U.S.-Japan Women's Journal* 52 (2017): 3–22; Elise K. Tipton, "Birth Control and the Population Problem," in *Society and the State in Interwar Japan*, ed. Elise K. Tipton (London and New York: Routledge, 1997), 42–62; Elise K. Tipton, "Birth Control and the Population Problem in Prewar and Wartime Japan," *Japanese Studies* 14, no. 1 (1994): 54–64.

[69] The group was, however, short-lived, in part due to the Great Kanto Earthquake. Ogino, *"Kazoku keikaku" eno michi*, 45.

[70] For Yamamoto Senji's birth control activism, see Lee, "Problematizing Population," 54–80; Michiko Obayashi, *Yamamoto Senji to haha Tane: Minshū to kazoku wo aishita hankotsu no seijika* (Domesu Shuppan, 2012), 95–135.

[71] Ogino, *"Kazoku keikaku" eno michi*, 45.

[72] Ota, *Nihon sanji chōsetsu hyakunenshi*, 147.

Review (Sanji Chōsetsu Hyōron).[73] The journal grew and offered a forum where writers and readers could exchange cutting-edge ideas about birth control.

As suggested above, the socialist network – which saw birth control as a tool for liberating the working class from the chains of capitalist exploitation and poverty – provided a strong ideological backbone to the popular birth control campaign during this period.[74] However, precisely because of its link with socialism, the birth control movement had a precarious relationship with the government. Officials viewed socialism as a foreign, dissident ideology undermining state authority and thus suppressed birth control activism because it was associated with socialism. After the government issued the Peace and Preservation Law on May 12, 1925, it actively cracked down on birth control activism, claiming it was inculcating indecent ideas in the public. The police were often present at public gatherings and at times ordered participants to stop meetings on the spot. The government also censored activists' writings, using indecency as a reason.[75] As far as state authorities were concerned, the socialist birth control movement needed to be nipped in the bud; otherwise, it would agitate the public with subversive foreign ideas and disrupt political order.

Some government officials were concerned about birth control for another reason: It could pose a eugenic risk.[76] As Takata Yasuma suggested, in the 1920s, birth control was becoming increasingly popular among those perceived as biologically "superior" – the able-bodied, educated, affluent class in the cities – and their birth control practices seemed to be lowering the group's fertility rates. To the dismay of government officials, studies in the mid-1920s confirmed this point, showing fertility decline was more conspicuous among the "intellectual class" (*chishiki kaikyū*) using birth control than among people in other

---

[73] In October 1925, the journal changed its title to *Sex and Society (Sei to Shakai)*.

[74] Takeuchi-Demirci, "Birth Control and Socialism"; Lee, "Problematizing Population," 54–80, 98–109. In addition to socialism, eugenics was another important strand of thought buttressing the birth control movement during the period. See Lee, "Problematizing Population," 54–89; Lee, "Differing Conceptions."

[75] Furthermore, Kyoto University ordered Yamamoto to voluntarily resign after he spoke at the public seminar on birth control organized by Suimyakusha, which was based in the provincial Tottori City. Obayashi, *Yamamoto Senji to haha Tane*, 113–15. Toshiji Sasaki and Akinori Otagiri, eds., *Yamamoto Senji zenshū* (Chobunsha, 1979), 3: 559.

[76] For a policy discussion on birth control and eugenics during this period, see Kiyoshi Hiroshima, "Gendai nihon jinkō seisakushi shōron: Jinkō shishitsu gainen wo megutte (1916–1930)," *Jinkō mondai kenkyū*, no. 154 (April 1980): 48–49; Sugita, "*Yūsei*," "*yūkyō*."

socioeconomic groups.[77] Seeing this trend, Takata broadcast his "give birth, multiply" argument. Similarly, his colleague Teruoka Gitō (1889–1966), known as the pioneer of labor science and a proponent of social hygiene, characterized the intellectual class's practices as the "unconscientious propagation of birth control" and warned it would "lead to racial suicide."[78] Policy intellectuals were acutely aware of this "racial suicide" argument presented by population experts.

Against this backdrop, the Bureau of Social Affairs, which was drafting the IC-PFP Secretariat's reference plan, included "Regarding Birth Control" as a subject for deliberation within the Population Section, as shown in the aforementioned list.[79] However, the Secretariat's response was tepid. In the fifth Secretariat meeting on September 21, 1927, Chairperson Hatoyama said that this item could be removed because "the issue will naturally come up in the [Population Section] committee meetings anyway."[80] After the discussion, the Secretariat decided to drop the item from the original draft of the reference plan and replaced it with "Investigations Regarding the Eugenics Movement,"[81] and it was included in the final version given to the Population Section.

Why was the Secretariat hesitant about birth control? Why did it recommend eugenics instead? A plausible reason was that the term "eugenics" embraced broader meanings, including birth control, whereas the reverse was not the case. By adopting "eugenics," the Secretariat might have left more space for the Population Section to decide which topics to discuss. Another reason, which was more political, was that birth control, and population control more broadly, was a controversial topic at the time, certainly more so than eugenics. At some point in the deliberation process, Secretariat members confessed, "once the [policy recommendation for] the problem of population control is codified as a report and published in society, ... [it] could invite quite a few

---

[77] Shigeyoshi Masuda, "Sanji seigen ni kansuru chōsa" (Jinkō Shokryō Mondai Chōsakai, February 1928), 12, PDFY10121605, the Tachi Bunko Archive, National Institute of Population and Social Security Research, Tokyo (hereafter Tachi Bunko).

[78] Gitō Teruoka, "Waga kuni shusshōritsu no shakai seibutsugakuteki kansatsu," *Rōdō kagaku kenkyū* 1, no. 2 (1924): 56; Sugita, *"Yūsei," "yūkyō,"* 31–53.

[79] "Jinkō mondai nikansuru chōsa kōmoku," in The Cabinet of the Japanese Government, *Jinkō shokuryō mondai chōsakai shorui kyū kanjikai gijiroku'*, 1927, National Archives of Japan Digital Archives, 26, accessed August 20, 2019, www.digital.archives.go.jp/das/image/F0000000000000068867.

[80] "Dai sankai kanjikai giji gaiyō," in The Cabinet of the Japanese Government, *Jinkō shokuryō mondai chōsakai shorui kyū kanjikai gijiroku'*, September 21, 1927, National Archives of Japan Digital Archives, 45–57, on 50–51, accessed August 20, 2019, www.digital.archives.go.jp/das/image/F0000000000000068867.

[81] Ibid.

misunderstandings."[82] Apprehensive of the possibility that the general public would misconstrue the government's motivation for engaging with birth control, the Secretariat dropped birth control from the reference plan. Therefore, this decision could be interpreted as an attempt to deflect any potential public controversies arising from officially recommending politically contentious measures.

In turn, the Population Section poured much energy into the subject. It even set up an independent Small Committee dedicated to population control, namely, the IC-PFP Population Section Small Committee on Population Control. It had three members: Nagai Tōru, Nagai Hisomu (1876–1957, chair of the Department of Physiology at Tokyo Imperial University), and renowned economist Fukuda Tokuzō (1874–1930). Within the Small Committee, the ideas presented by Nagai Tōru and Nagai Hisomu shaped the contours of the discussion within the committee.

Nagai Tōru, after graduating from the Faculty of Law at Tokyo Imperial University in 1903, embarked on his career as a high-rank government official at the Ministry of Agriculture and Commerce and then at the Ministry of Railways until 1920.[83] Between 1920 and 1926, he became involved in social policy through his role as a managing director at the Harmonization Society (Kyōchōkai), a half-government, half-private organization established in 1919 to coordinate the labor-management relationship.[84] After gaining a doctorate in economics in 1925, Nagai Tōru established himself as one of the most pursued experts for the government's social and population policies.

Nagai Hisomu was a key figure in eugenics in this period.[85] In the 1910s, he began to present his eugenic theories widely in professional journals, popular magazines, and newspapers.[86] In 1923, he headed the Japanese Society for Sexology (Nihon Sei Gakkai), and in 1930, he headed the new Japanese Society of Racial Hygiene (Nihon Minzoku

---

[82] Jinkō Shokryō Mondai Chōsakai, "Jinkō shokuryō mondai chōsakai jinkōbu tōshin setsumei" (Jinkō Shokryō Mondai Chōsakai, April 1930), 117.

[83] For Nagai Tōru, see, e.g., Sugita, "Yūsei," "yūkyō," 116–41; Sugita, Jinkō, kazoku, seimei, 184–233.

[84] On Kyōchōkai, see Andrew Gordon, The Evolution of Labor Relations in Japan: Heavy Industry, 1853–1955 (Cambridge, MA: Council on East Asian Studies, Harvard University Press, 1985).

[85] Mitsuko Chuman, "Nagai Hisomu saikō: Yūseigaku keimō katsudō no shinsō wo saguru," in Seimei no rinri, ed. Kiyoko Yamazaki (Fukuoka: Kyushu Daigaku Shuppankai, 2013), 228.

[86] Chuman, "Nagai Hisomu saikō," 230–35.

Eisei Gakkai).[87] Alongside this, Nagai Hisomu promoted eugenics in policymaking. In the 1910s, as an HHSG member, he lobbied for an independent committee for eugenics within the HHSG.[88] At that time, his plan did not materialize. However, in the 1930s, his campaign – this time to promote a eugenic policy like the one in Nazi Germany – became more readily accepted in policymaking, so much so that it culminated in the establishment of eugenic population policies in the early 1940s.

Within the IC-PFP, Nagai Tōru acted as a core member of the Population Section from the outset, assuming a position in all the subcommittees set up within the section. Applying his expertise, he submitted private proposals for most of the issues deliberated within the section. In turn, although Nagai Hisomu was not a founding member of the Population Section, he quickly became one of the most vocal members after being appointed to the section's special committee on December 13, 1927.[89] For the Population Section's Small Committee on Population Control, the two presented draft proposals, and Fukuda came up with a revised proposal based on the two documents.[90]

Fukuda was confronted with an enormous task, because both Nagai Hisomu and Nagai Tōru had distinctive, yet strong, opinions regarding birth control and eugenics as solutions to the population problem. Trained primarily in the social sciences, Nagai Tōru maintained that the population problem required a sociological approach. He contended that population was determined by the "social scientific law," which was "calculated based on the contrast between a population and the social force – the force of a society that by itself acts like a living body."[91] Thus, the population problem "was neither the [Malthusian] problem of food nor the [Marxian] problem of employment" but the problem of the "ratio between society's productive power and the power of a population to expand."[92] This idea of population exhorted him to focus on the details of population structure (e.g., sex and age ratio) and population dynamics (e.g., fertility and mortality rates) as critical factors determining the quality of a population. It also urged him to define the population

---

[87] In 1935, *Minzoku Eisei Gakkai* was renamed *Minzoku Eisei Kyōkai*. For the role of the Society in eugenic movements in modern Japan, see Takashi Yokoyama, "Yūseigakushi niokeru nihon minzoku eisei gakkai no ichi," *Nihon kenkō gakkaishi* 86, no. 5 (2020): 197–208.

[88] Hoken Eisei Chōsakai, "Hoken eisei chōsakai dai ikkai hōkokusho," 34. Hiroshima, "Gendai nihon jinkō seisakushi shōron," 48–49.

[89] The Cabinet of the Japanese Government, *Jinkō shokuryō mondai chōsakai shorui ichi*, 40–41.

[90] Sugita, *Jinkō, kazoku, seimei*, 198.

[91] Toru Nagai, *Nihon jinkōron* (Ganshodo, 1929), 7.

[92] Ibid., 5–6, 17–18.

problem as first and foremost a societal problem that occurred primarily when the population structure and its dynamics became distorted.[93]

Based on this logic, Nagai Tōru thought the source of the current population problem was the "phenomenon of high birth, high [infant] death," which he believed represented the underdeveloped state of the "social organization" and "people's intellect."[94] To solve the problem, he believed that the state should establish social policies that would help implement institutions fostering the practice of, what he called, "childcare preservation" (*ikuji hozen*) instead of popularizing birth control. At the same time, he also acknowledged it would take time for Japanese society to mature enough to accommodate a full range of "childcare preservation" institutions. Thus, Nagai Tōru recommended birth control as an interim measure that should only be implemented until society developed fully.[95]

In turn, Nagai Hisomu's idea about population – which he categorically viewed as synonymous with biological race – was premised largely on eugenics, which he insisted on calling "racial hygiene."[96] He claimed that recent progress in genetics had clarified how parents' traits were passed down to children equally, irrespective of social groups. Thus, to improve the quality of the Japanese race, the government should strive to "select the good-quality genetic substance and exclude the bad-quality one" and "eliminate the genealogy of families, for instance, with mental illnesses, prone to tuberculosis, and prone to disability."[97] Yet, Nagai Hisomu was also skeptical about directly applying Mendelian laws of animal and plant heredity to humans. He thought that human sexual behaviors were unpredictable. Furthermore, it would be unrealistic to arrange interbreeding among different human groups within a controlled environment, as Mendelian geneticists had done with their experimental subjects.[98] For this reason, Nagai additionally relied on the biometric approach to heredity, which had its origins in the Galtonian tradition of eugenics.[99] This eclectic understanding of eugenics underpinned Nagai Hisomu's activism. In practice, he lobbied for a state policy on eugenic birth control, including sterilization, to improve the genetic composition of the Japanese race.

---

[93] Ibid., 8.
[94] Ibid., 7–8.
[95] Ibid., 415–24.
[96] Chuman, "Nagai Hisomu saikō," 230.
[97] Nagai (1913) quoted in in Chuman, "Nagai Hisomu saikō," 267. Also see Fujino, *Nihon fashizumu to yūsei shisō*, 56–62; Suzuki, *Nihon no yūseigaku*, 93–97.
[98] For details, see Chung, *Struggle for National Survival*, 43.
[99] Ibid.

Over the 1920s, Nagai Hisomu expressed his dissatisfaction that his long-term campaign for eugenic birth control was often mixed up with what he saw as frivolous popular birth control activism. What particularly bothered him was that this activism was propagating birth control among the urban intellectual class, who he thought should procreate more because of their "superior traits" (yūshū na seishitsu).[100] Thus, like Takata, Nagai Hisomu believed the birth control movement was damaging the quality of the Japanese race. Specifically, it was stimulating differential fertility, or the widening gap in fertility among different social groups, which, in this case, was symbolized by lowering fertility rates within the urban intellectual class and sustained high fertility in the lower socioeconomic groups. Borrowing from the Darwinian concept of natural selection, Nagai Hisomu called this "reverse selection" (gyaku tōta) and characterized it as "those with inferior traits" (retsuaku na soshitsu no mono) dominating those with "superior traits."[101] He saw the question of "reverse selection" as a pressing population problem but also observed that the current debate neglected this issue as it concentrated on the Malthusian problem of overpopulation. To rectify the situation, Nagai Hisomu became convinced that the government should intervene in current birth control practices more proactively. He believed the government should establish eugenic birth control as a national policy and work toward implementing a more regulated birth control initiative targeting "those with inferior traits" to solve the problems of population quantity *and* quality.

With this in mind, Nagai Hisomu submitted a draft proposal to the Small Committee. The proposal was titled "Draft of a Report on Eugenic Problems," which clearly demonstrated his conviction that a policy on eugenic birth control was necessary to tackle population issues. The preamble defined the "population problem" as "not simply an issue of population quantity but also something that intends to improve population quality from a eugenic point of view."[102] Following this definition, three out of the nine recommendations in the proposal were about reproductive practices and were aimed to prevent the process of "reverse selection." Specifically: (1) the establishment of "appropriate institutions enabling consultations on marriage, childbirth, and birth control"; (2) the enforcement of "control over distribution, sales, and

---

[100] Sugita, *Jinkō, kazoku, seimei*, 193.
[101] Ibid., 192–93.
[102] Jinkō Shokryō Mondai Chōsakai, "Jinkō shokuryō mondai chōsakai jinkōbu," 41. Also see Sugita, *Jinkō, kazoku, seimei*, 190–99; Fujino, *Nihon fashizumu to yūsei shisō*, 121–31; Hiroshima, "Gendai nihon jinkō seisakushi shōron," 50–59.

advertisement for instruments, pharmaceuticals, and other materials assisting contraception"; and (3) the promotion of "research on various institutions from the eugenic viewpoint."[103]

In contrast, Nagai Tōru's draft proposal, titled "Measures Regarding Population Regulation," mirrored his view that the population problem was a problem of population itself, caused by the "current situation of high birth, high death, as well as high marriage and high divorce rates," and thus claimed the state should implement policies aiming to regulate population structures and dynamics.[104] Reflecting his focus on "childcare preservation," the proposal stressed the necessity for implementing "social institutions protecting women giving birth and preserving childcare" and opportunities to teach men and women about population problems.[105] While Nagai Hisomu's proposal put eugenic birth control at the fore, Nagai Tōru's draft hid the statement, placing it almost casually in the middle of a paragraph and merely suggesting the government should encourage "investigation and research" on birth control.[106]

Fukuda, in consultation with Nagai Tōru and Nagai Hisomu, made great efforts to merge the two draft proposals into the final draft report. Having looked at both documents, Fukuda came up with a neutral title: "On Various Measures for Population Control." After many discussions, the three agreed on the following broad definition of population control:

> Unlike so-called birth control, population control does not only refer to the control of the number of a population, it even includes positive meanings, such as the decrease in mortality rates and the prolongation of average life expectancy. Furthermore, it does not only aim to solve the problem of the population number but to improve the quality of the population.[107]

Based on this understanding, the draft report presented a total of nine policy recommendation items, which were, on the whole, more sympathetic to Nagai Tōru's social policy approach than Nagai Hisomu's eugenic recommendations. Echoing the sentiment of the HHSG (see Chapter 2), many of these items were about measures for the promotion of maternal and infant health, public health in rural farming villages and cities, and the prevention of tuberculosis. Only two items were

---

[103] Jinkō Shokryō Mondai Chōsakai, "Jinkō shokuryō mondai chōsakai jinkōbu," 40–41.
[104] Ibid., 39.
[105] Ibid., 38–40.
[106] Ibid., 39.
[107] Ibid., 118.

specifically on birth control. With regard to eugenics, it adopted the view presented in Nagai Tōru's proposal, only recommending "research on various institutions relevant from the viewpoint of eugenics."[108] The draft report was approved at the IC-PFP general assembly on December 19, 1929.

By passing "On Various Measures for Population Control," the IC-PFP implicitly endorsed population control by means of social policy – and to a lesser extent eugenics – as a solution to the population problem. However, the IC-PFP did not make any further efforts to act on this policy recommendation.[109] Consequently, no population control policies came about as a result of the exercise. This was a stark contrast to the draft report on migration submitted by the Population Section, which, as mentioned above, directly corresponded to the Overseas Migration Cooperative Societies Law.

Official Research on the "Population Problem"

While failing to make an actual policy, the Population Section's deliberation efforts made a significant mark in the history of population science in modern Japan: It paved the way for institutionalizing policy-oriented population research.

As mentioned in the previous chapters, the analysis of population figures comprised a significant part of the daily work of official bureaucrats by this period. However, policy-oriented population research was conducted on a project or ad hoc basis. The need to set up an official, permanent institution dedicated to coordinated, policy-oriented population studies was addressed early on by the members of the Population Section's Special Committee. Among them, committee member Nitobe Inazō – by now established as a prominent colonial-policy scholar and undersecretary-general of the League of Nations – took a first step. Nitobe, also present at the abovementioned World Population Conference, submitted a written opinion piece to the special committee in the spring of 1929 in favor of a permanent research organization for population issues.[110] Others agreed with Nitobe.[111] Based on Nitobe's written

[108] Ibid., 56–57.

[109] Fujino, *Kōseishō no tanjō*, 42; Fujino, *Nihon fashizumu to yūsei shisō*, 131.

[110] The Cabinet of the Japanese Government, *Jinkō shokuryō mondai chōsakai shorui jūroku daigokai sōkai giji sokkiroku*, March 27, 1930, National Archives of Japan Digital Archives, 38–44, accessed February 20, 2020, www.digital.archives.go.jp/das/image/F0000000000000068880.

[111] The Cabinet of the Japanese Government, *Jinkō shokuryō mondai chōsakai shorui jūroku*, 38.

piece, Nagai Tōru drafted "The Proposal Concerning the Establishment of a Population Research Institute."[112]

The draft recommended the government set up a permanent research organization named the Population Research Institute (*Jinkō Kenkyūsho*), as either a national institute or a public interest corporation, with the objectives of conducting research on population problems and making recommendations in response to government inquiries.[113] The Population Research Institute would consist of "experienced academic specialists" and "bureaucrats in the relevant fields" acting as councilors, as well as a small number of researchers. They would conduct studies on: (1) the population composition, distribution, and dynamics; (2) eugenics and other topics relevant to population control; (3) specific measures on overpopulation; and (4) other population policies and theories.[114] The members of the institute would be obliged to present and publish their research findings, organize lectures and seminars, and join the "international councils on population" and dispatch representatives to its general meetings.[115]

Based on Nagai Tōru's proposal, the Population Section made the resolution, "Matters Concerning Setting Up a Permanent Research Organization Specialized in Population Problems," which the IC-PFP passed and submitted to the government on March 29, 1930. The resolution claimed that, given the complex nature of population problems, the government would "take a wrong course" and "make errors in setting the standards for policy measures" for population problems if it did not have a permanent research institution supplying up-to-date data and analysis.[116] It also stressed the advantage of a permanent institution from the international viewpoint. An official institution, according to the resolution, would act as a collaborative partner internationally, liaising between the Japanese government and international population organizations, such as the International Union for the Scientific Investigation of Population Problems (est. 1928) that was a result of the World Population Conference.

The government initially took up the resolution and secured a budget to set up a permanent population research institute for the 1931 financial

[112] "Jinkō kenkyūsho secchi nikansuru kengian (Nagai iin shian)," April 10, 1929, in The Cabinet of the Japanese Government, "Dai nikai jinkōbu tokubetsu iinkai shōiinkai giji gaiyō," *Jinkō shokuryō mondai chōsakai shorui roku jinkōbu tokubetsu iinkai gijiroku'*, 1927–30, National Archives of Japan Digital Archives, 295–322, on 301–2, accessed August 22, 2019, www.digital.archives.go.jp/das/image/F0000000000000068870.
[113] Ibid., 301.
[114] Ibid., 301–2.
[115] Ibid., 301–2.
[116] Sugita, *Jinkō, kazoku, seimei*, 188.

year. However, the plan fell through due to the cabinet's resignation on April 13, 1931 after Prime Minister Hamaguchi Osachi was shot.[117] Nevertheless, the IC-PFP's call was not entirely futile. It eventually led to the establishment of the Foundation-Institute for Research of Population Problems (*Zaidan Jinkō Mondai Kenkyūkai*, hereafter IRPP) on November 22, 1932. Though not directly a government body, in many ways the IRPP was a successor organization to the IC-PFP Population Section, de facto acting as a policy research institute and a government inquiry body on population matters.[118] The Population Section's campaign in the late 1920s laid a foundation for establishing an institution specialized in policy-relevant population discussion and research.

In addition to lobbying for a permanent research institute, the Population Section itself contributed to the policy-oriented research on population problems. Specifically, the Home Ministry Bureau of Social Affairs conducted birth control research during the policy deliberation process. The study not only supplied materials for discussion to the Population Section but significantly also buttressed the official policy on birth control in later years.

The Home Ministry Bureau of Social Affairs began investigations on birth control in 1922 – upon Sanger's visit to the country.[119] However, the work conducted at the time concentrated on the collection and analysis of publications on birth control.[120] In the late 1920s, for a policy deliberation within the IC-PFP, the Bureau of Social Affairs conducted a more thorough investigation. First, it prepared a reference list on birth control research that had been conducted internally. The list, dated January 1928, was used by Bureau of Social Affairs Secretary Kawanishi when he presented at the Population Section Special Committee meeting on July 13, 1928, which brought up birth control and eugenics for discussion for the first time within the Population Section.[121]

Along with this, Bureau of Social Affairs bureaucrat Masuda Shigeki conducted research on the birth control movement that justified why the Bureau of Social Affairs had recommended birth control research in its draft of the IC-PFP Secretariat's reference plan. Masuda, who had long engaged with population issues for his work on social work and labor policies, adopted eclectic approaches to the topic. For the

---

[117] Ibid.
[118] Sugita, "*Yūsei*," "*yūkyō*," 188–89.
[119] Jinkō Shokryō Mondai Chōsakai, "Jinkō mondai nikansuru yoron," January 1928.
[120] Ibid.
[121] Jinkō Shokryō Mondai Chōsakai, "Jinkō shokuryō mondai chōsakai jinkōbu," 33–34.

internal classified report, "An Investigation on Birth Control," dated February 1928, Masuda first conducted a review of the world history of "modern thoughts regarding birth control," starting from the publication of Thomas Malthus's *An Essay on the Principle of Population* in 1798, and studied the evolution of birth control movements in Britain, the United States, Holland, and Norway, as well as international movements since the mid-nineteenth century.[122] He then gathered information on the popular birth control movement in Japan since Sanger's visit in 1922. He collected names, addresses, and the details of the services provided by individuals running birth control clinics and selling birth control products, including abortifacients, in the cities of Tokyo, Osaka, Chiba, Nagoya, and Kobe, as well as in Chiba and Shizuoka Prefectures. Additionally, Masuda examined various birth control methods, including surgical methods such as hysterectomy. Based on the report, Masuda explained the situation surrounding birth control at the meeting of the Population Section Special Committee on July 13, 1928, standing next to his boss, Kawanishi.[123]

Masuda's report generally maintained a neutral take on birth control, concentrating mostly on giving the factual data he had gathered through his investigation. Yet, from time to time, certain views shaped the report, especially when he gave analysis and policy recommendations. Of these, one echoed the concern over "reverse selection" addressed by Nagai Hisomu. Elaborating on the argument in favor of birth control from the viewpoint of population problems, Masuda warned that officials in Japan would have to consider the following two points if they were planning to endorse birth control policies: (1) "the fact that the birth rate is declining among the upper class," and (2) "the [fact that] population growth … refers to the drastic growth of the people in low classes."[124] Following this comment, Masuda showed statistical data on fertility among different social groups in Japan, such as the table showing the number of births per every thousand in the "rich" and "poor" areas within Tokyo's Yotsuya-Ward, and suggested that a process of "reverse selection" had a jumpstart in Japan, at least in cities.[125]

In conjunction with this, Masuda introduced the birth control initiative in Holland as an example of a successful state-led birth control campaign. He praised the initiative for not only "decreas[ing] the mortality and infant mortality rates significantly" but for also making

---

[122] Masuda, "Sanji seigen ni kansuru chōsa," 1.
[123] Jinkō Shokryō Mondai Chōsakai, "Jinkō shokuryō mondai chōsakai jinkōbu," 34.
[124] Masuda, "Sanji seigen ni kansuru chōsa," 12.
[125] Ibid., 12–19.

the rate of natural population growth in Holland "one of the best in the world," thus projecting the message that a birth control initiative with a strong government presence could tactfully adjust the population size and quality to an optimal state.[126] Based on this information, he made five recommendations toward the end of the report, which included government leadership in "the establishment of a consultation clinic catering to the women of the lower class and poverty class with many children who, because of their situations, wish to practice birth control."[127] The report clearly projected the view that a government-led birth control program was integral to social work and that specifically targeting the lower classes would adequately circumvent the process of "reverse selection."

Though produced by a mid-rank bureaucrat, Masuda's report had a lasting impact on the government's attitude toward birth control.[128] It not only underpinned the official discourse on the subject but also provided a blueprint for the government's birth control policy.[129] The aforementioned Population Section Small Committee dedicated to population control decided to incorporate two of the policy recommendations presented in Masuda's report which recommended, respectively, that the government should "establish appropriate institutions enabling consultations on marriage, childbirth, and birth control" and "enforce control over distribution, sales, and advertisement for instruments, pharmaceuticals, and other materials assisting contraception."[130] Of these, the latter was taken up by the Special Committee on Racial Hygiene (*Minzoku Eisei nikansuru Tokubetsu Iinkai*), established on June 24, 1930 within the HHSG following the dissolution of the IC-PFP in April 1930. This special committee – which Nagai Hisomu was instrumental in founding – submitted a draft proposal calling for the control of "harmful" (*yūgai*) contraceptives.[131] The Home Ministry adopted the proposal and issued the Ordinance for the Control of Harmful Contraceptive Devices on December 27, 1930 as Ministerial Ordinance No. 40 (enacted on January 6, 1931).[132] Following Masuda's original report, the ministerial

---

[126] Ibid., 3, 28–30.
[127] Ibid., 72.
[128] Hiroshima, "Gendai nihon jinkō seisakushi shōron," 53–54.
[129] Ibid., 53.
[130] Jinkō Shokryō Mondai Chōsakai, "Jinkō shokuryō mondai chōsakai jinkōbu." Also see Sugita, *Jinkō, kazoku, seimei*, 185–99. Yokoyama, *Nihon ga yūsei shakai ni naru made*, 198–218.
[131] Hoken Eisei Chōsakai, "Hoken eisei chōsakai dai jūyonkai hōkokusho" (1930), 11.
[132] Hidebumi Kubo, *Nihon no kazoku keikaku shi: Meiji, Taisho, Showa* (Shadan Hōjin Nihon Kazoku Keikaku Kyōkai, 1997), 44.

ordinance defined harmful contraceptive devices to include "contraceptive pins," and "other contraceptives designated by the home minister to cause harm from the viewpoint of hygiene." In the 1930s, Masuda's report set the tone for the official attitude toward birth control as a population measure at a time when population became redefined as a valuable resource to be mobilized for Japan's engagement in a total war.

Conclusion

In the 1920s, with the increased availability of demographic data, "population problems" became a topic of public discussion. The heightened public interest in "population problems" exhibited an emerging consensus that a distorted population trend could be a source of economic, political, and social problems. High-rank bureaucrats and politicians followed public debate's logic. They portrayed "population problems" as dovetailing with a wide range of interlinked issues, such as food, industry, employment, poverty, space, and race, and thought they should be tackled with government policies.[133] Through public and policy discussions, "population problems" became shorthand for a myriad of issues associated with the population trend that were subjected to government intervention.

The establishment of the IC-PFP as the first-ever official research committee specializing in population matters in this context symbolized the burgeoning of a new mode of interplay between science and the governing of Japan's population. It was set in motion by the belief that rational population management required government policies as well as an independent institution dedicated to policy discussion and research. Reflecting the all-encompassing concept of "population problems," the IC-PFP singlehandedly took charge of coordinating policy-oriented population research under one roof. Yet, recognizing that the population problems required intervention from many disciplinary angles, the IC-PFP also summoned population experts from diverse medical and scientific fields. As a consequence, the IC-PFP – reifying the government's commitment to solving "population problems" through policies – helped foster population research as an institutionally-based, multidisciplinary endeavor. It also produced a new kind of population specialists, an amorphous community of bureaucrats, scientists, and medical researchers acting as policy experts, who were united by the effort to conduct policy-oriented population research and to advise the government on population issues based on the research.

---

[133] Sugita, *Jinkō, kazoku, seimei*, 189–90.

Overall, the new mode of science-governing interplay built around the IC-PFP was productive. The population research conducted under the aegis of the IC-PFP directly fed policy discussion, which then yielded policy recommendations on the topics initially raised by the government. Yet, the actual deliberation process also highlighted elements of discord in this relationship, as exhibited in the discussions on birth control, eugenics, and population control within the IC-PFP Secretariat and the Population Section. But, the dissonance did not automatically mean this interplay was unproductive. In the 1930s, as Japan's entry into total war heightened official interests in eugenic population management, "On Various Measures for Population Control" made a significant mark in history.[134] It paved the way for institutionalizing eugenic and social policy measures with a specific aim to "improve" the quality of the Japanese race and population. From the government's point of view, even the elements of discord yielded a productive outcome.

The 1930s saw the rise of another conception of population – as a "resource" – which in turn shaped the official narrative. Amid total war, the policy agenda informed by this understanding addressed issues of population quantity, quality, and movement, but specifically focused on the relationship between people and the intangible yet emotionally charged concept of "national land" (*kokudo*).

---

[134] Sugita, *Jinkō, kazoku, seimei*, 199; Hiroshima, "Gendai nihon jinkō seisakushi shōron," 56.

4   National Land Planning
Distributing Populations for the Wartime Nation-State-Empire

For the bureaucrats and experts working on population problems in the metropole, total war (1937–45), triggered by the outbreak of war with China in 1937, marked a watershed moment. The Konoe Fumimaro cabinet's call for general mobilization to construct a "new order" in East Asia changed the official treatment of population issues in a number of ways. First, officials now redefined a large population as an asset and thus adopted a population growth policy.[1] Second, the government's demand for a "high-quality" population during the war emphasized the significance of applying eugenics and racial hygiene to the official mobilization scheme.[2] Third, the government established the Ministry of Health and Welfare (MHW) in 1938 and made it the administrative office in charge of "regulating and utilizing" the population as a valuable "resource" for the nation at war.[3] Finally, in 1939, the government founded the Institute of Population Problems (IPP) within the MHW as an official institute dedicated to population studies.[4] The official effort to tackle the population issues in connection with the war culminated in the cabinet approval of the General Plan to Establish the National Population Policy *(Jinkō Seisaku Kakuritsu Yōkō)* on January 22, 1941. Toward the end of the war, the Japanese government had established what Ogino

[1] Hiroyuki Takaoka, "Senji no jinkō seisaku," in *Kazoku kenkyū no saizensen jinkō seisaku no hikakushi: Semegiau kazoku to gyōsei,* ed. Hiroshi Kojima and Kiyoshi Hiroshima (Nihon Keizai Hyoronsha, 2019), 101.

[2] Matsubara, "The Enactment of Japan's Sterilization Laws"; Christiana A. E. Norgren, *Abortion before Birth Control: The Politics of Reproduction in Postwar Japan* (Princeton: Princeton University Press, 2001), 52–59; Yutaka Fujino, *Nihon fashizumu to yūsei shisō;* Fujime, *Sei no rekishigaku;* Oguma, "Tsumazuita junketsu shugi"; Kiyoshi Hiroshima, "Gendai nihon jinkō seisaku shi shōron: 2- kokumin yūseihō ni okeru jinkō no shitsu seisaku to ryō seisaku," *Jinkō mondai kenkyū,* no. 160 (October 1981): 61–77.

[3] Ogino, "'Kazoku keikaku' eno michi," 113; Fujino, "Kōseishō no tanjō".

[4] Kōseishō Jinkō Mondai Kenkyūsho, ed., *Jinkō mondai kenkyūsho no ayumi: 40-shūnen wo kinen shite* (Kōseishō Jinkō Mondai Kenkyūsho, 1979).

Miho once called the "system of managing the population under the war," in which population bureaucrats and experts played a pivotal role.[5]

With the emphasis on population increase and the improvement in population quality, the General Plan to Establish the National Population Policy was instituted in tandem with the eugenic and social policies established during the war, which aimed to primarily promote the health and welfare of women and children. Partly corroborating the Foucauldian theory of biopolitics, and partly the portrayal of wartime Japan as a "fascist welfare state," the wartime population policy significantly reified the Japanese effort to enhance the reproductive capacity of the "population of Japan Proper" (*naichi jinkō*) and the colonial subjects, now called *gaichi jinkō* ("population of the outer territories"), for the eternal existence of Japan as a nation-state-empire.[6] Yet, the "system of managing the population under the war" was far more pervasive, generating various ways in which the population was articulated and controlled in relation to mass mobilization. One of these ways, which has hitherto enjoyed less attention in historical inquiries, was to regard the population as a *composition* and to manage the population's quantity and quality by pursuing a balance in its composition through the population movement.[7] In the late 1930s and early 1940s, this way of discerning the population and population management manifested in the debate over a "comprehensive population distribution planning" policy, which surfaced as a mass mobilization measure accountable for one of the most important national policies in the total war: "national land planning" (*kokudo keikaku*).[8]

---

[5] Ogino, *"Kazoku keikaku" eno michi*, 112.

[6] For a study that draws from Foucauldian theory, see the introduction. For a representative study on fascist welfare, see Takaoka, *Sōryokusen taisei to "fukushi kokka"*; Gregory J. Kasza, "War and Welfare Policy in Japan," *The Journal of Asian Studies* 61, no. 2 (May 2002): 417–35; Fujino, *Nihon fashizumu to yūsei shisō*. For selected case studies, see Sunho Ko, "Managing Colonial Diets: Wartime Nutritional Science on the Korean Population, 1937–1945," *Social History of Medicine* vol. 34, no. 2 (2021): 592–610; Yoneyuki Sugita, "Toward a National Mobilization: The Establishment of National Health Insurance," in *Japan's Shifting Status in the World and the Development of Japan's Medical Insurance Systems*, ed. Yoneyuki Sugita (Singapore: Springer, 2019), 93–125.

[7] The exception was the works of Hiroyuki Takaoka and Kiyoshi Hiroshima, which this chapter is highly indebted to. Takaoka, "Senji no jinkō seisaku"; Takaoka, *Sōryokusen taisei to "fukushi kokka"*; Hiroshima, "Gendai nihon jinkō seisaku shi shōron."

[8] Today, the term *kokudo keikaku* is primarily associated with the postwar "national comprehensive development planning" project (*zenkoku sōgō kaihatsu keikaku*). However, as this chapter will show, it had roots in wartime state planning. For works that depict continuities between the wartime *kokudo keikaku* and postwar comprehensive development, see Eric G. Dinmore, "'Mountain Dream' or the 'Submergence of Fine Scenery'? Japanese Contestations over the Kurobe Number Four Dam, 1920–1970,"

This chapter is about the policy discussions and research on "population distribution" that emerged in the process of establishing a "national land plan" as a state policy. The population work for "national land planning" effectively illustrates the mode of interaction between science and state-led population management during the war in the context of fascist imperialism.[9] First, it shows the ways population experts imagined the population as a distributable "ethnic group/ race" (*minzoku/jinshu*) and "resource" (*shigen*) and how they directly interacted with the Japanese state's attempts to manage its population for the sake of fascist imperialism.[10] Second, it illustrates that population science under a dictatorship was, to borrow the words of Sang-hyun Kim, "actively mobilized by the state ... to materialize the vision of a self-reliant political economy."[11] Under a fascist regime, Japan invested in population research because demographic calculation was perceived to be fundamental for the construction of an economically and politically contained imperium, the Greater East Asia Co-Prosperity Sphere.[12]

One outcome of the state mobilization of population research was the formalization of population studies as an officially endorsed field of inquiry. Another was the expanding role of technical and research bureaucrats in policy-oriented population research. For the most part, these bureaucrats, employed to undertake the state-sanctioned population research, diligently completed the tasks assigned to them. However, a closer look at their research practices also suggests that the knowledge

Water History 6, no, 4 (December 2014): 315–40; Eric G. Dinmore, "Concrete Results?: The TVA and the Appeal of Large Dams in Occupation-Era Japan," *The Journal of Japanese Studies* 39, no. 1 (January 2013): 10–12; Takashi Mikuriya, *Seisaku no sōgō to kenryoku: Nihon seiji no senzen to sengo* (Tokyo Daigaku Shuppankai, 1996).

[9] Fascist imperialism, according to historian Louise Young, describes the "synergy and interdependence between imperial expansion and the development of the fascist programmes throughout the nation-state-empire," "When Fascism Met Empire in Japanese-Occupied Manchuria," *Journal of Global History* 12, no. 2 (July 2017): 280.

[10] Shigeo Kato, "Senjiki nihon no kagaku to shokuminchi, teikoku," *Rekishi hyōron* 832 (August 2019): 36–46. For this argument, see Jean-Guy Prévost, *Total Science: Statistics in Liberal and Fascist Italy* (Montreal: McGill-Queen's University Press, 2009), 11–14.

[11] Sang-Hyun Kim, "Science and Technology: National Identity, Self-Reliance, Technocracy and Biopolitics," *The Palgrave Handbook of Mass Dictatorship*, eds. P. Corner and J. H. Lim (London: Palgrave Macmillan, 2016), 82.

[12] Therefore, state mobilization of population research took place in the same discursive space that gave rise to the "New Order for Science and Technology" (*kagaku kijutsu shintaisei*), formulated by the Konoe cabinet in 1941 to establish state coordination of scientific and technological activities for rational resource management in Japan Proper and its colonies. Moore, *Constructing East Asia*; Mizuno, *Science for the Empire*; Oyodo, *Gijutu kanryō no seiji sankaku*, 142–86.

created by population studies was founded upon unstable epistemological grounds, despite the assertion of certainties demanded by the political regime.

From Burden to Valuable Resource: The Population Phenomena in the 1930s

When government officials and population experts raised the issue of "population problems" during policy discussion in the early 1930s, the message they projected had changed little from that of the late 1920s. Their perspective was firmly locked onto the problem of "overpopulation." The only difference: Policymakers were now wearily tracing the rising discourse of unemployment triggered by the Wall Street Crash of 1929.[13] Attributing the unstable economic situation that had been in place since the Great Kanto Earthquake of 1923 to the intensification of the leftist labor movements, these policymakers were worried that mass unemployment, in conjunction with the ever-growing population, might ultimately result in a political crisis. Population experts described this official concern using a blanket term: "unemployment problem" (*shitsugyō mondai*).[14]

Though not so obvious at first, the issue of overpopulation implicit within the "unemployment problem" was very much a rural economy issue – one specifically linked to the rural community's inability to absorb a surplus population.[15] Well before the early 1930s, the population growth rate in "the countryside" (*gunbu*) was already high, far exceeding that of "the urban area" (*shibu*).[16] This trend continued throughout the 1920s, with the rates gradually increasing: from 17.33 per 1,000 population in 1925 to 18.09 in 1930 in the countryside and from 5.60 to 6.92

---

[13] Michiya Kato, "Hidden from View?: The Measurement of Japanese Interwar Unemployment," *Annual Research Bulletin of Osaka Sangyo University*, no. 1 (December 2008): 77–103.

[14] Tōru Nagai, *Nihon Jinkōron*, 58–59.

[15] Penelope Francks, *Rural Economic Development in Japan: From the Nineteenth Century to the Pacific War* (London and New York: Routledge, 2006); Ann Waswo and Yoshiaki Nishida, *Farmers and Village Life in Twentieth-Century Japan* (London: RoutledgeCurzon, 2003); Ann Waswo, "Japan's Rural Economy in Crisis," in *The Economies of Africa and Asia in the Inter-War Depression*, ed. Ian Brown (London: Routledge, 1989), 115–36.

[16] In population studies at the time, the "urban area" conventionally included the major cities of Tokyo, Osaka, Nagoya, Kyoto, Kobe, and Yokohama, as well as cities with populations of 100,000 and over. See, e.g., Minoru Tachi and Masao Ueda, "Taisho 9-nen, taisho 14-nen, showa 5-nen, showan 10-nen dōfuken betsu oyobi shibunbetsu hyōjunka shusshōritsu, shibōritsu oyobi shizen zōkaritsu," *Jinkō mondai kenkyū* 1, no. 1 (April 1940): 21–28.

in cities.[17] In turn, the rural economy was enduring enormous hardships due to the post–World War I (WWI) depression and the destabilization of economy after the Great Kanto Earthquake. It was also crushed by the volatile prices of rice and silk cocoons, the two major profit-making agricultural products for farmers.[18] Throughout the 1920s, the countryside was increasingly feeling the pressure of a growing population.

This was the backdrop when the worldwide depression of the early 1930s struck the rural economy in Japan.[19] From the Malthusian point of view, the economic depression brought a tangible population crisis to the countryside, obliterating the already precarious balance between population growth and subsistence growth. The effect of the collapse of the population-subsistence ratio was felt the harshest in the northern region of Aomori, Iwate, Miyagi, Akita, Yamagata, and Fukushima. The region additionally suffered from severe crop failures caused by cold summers in 1931 and 1934. Yet, in the face of this, the population kept expanding.[20] A scholar studying the diet in a village in Aomori Prefecture in 1934 lamented that villagers were so desperate that they were subsisting on rotten potatoes and sake lees.[21] In part to solve this dire situation, throughout the first half of the 1930s, families in the region sent more sons out to work in diversified industries and more daughters off to places, both in and outside of the region, in search of jobs as factory workers, entertainers, waitresses, and prostitutes than ever before.[22]

In this context, experts raised concerns that the "unemployment problem" might further deepen the population crisis in the already debilitated rural community. They feared that the countryside, thus far a major supplier of labor force in cities, might lose an outlet for its "surplus population" due to the economic depression and that the surplus population would disrupt the political order. Official concerns

---

[17] Ibid., 21–22; Minoru Tachi, "Wagakuni chihōbetsu jinkō zōshokuryoku ni kansuru jinkō tōkeigakuteki ichikōsatsu' (ge)," *Jinkō mondai* 2, no. 1 (June 1937): 217–38. Also see Minoru Tachi, "Showa 12 nen zenkoku, toshibu, gunbu oyobi rokudai toshi jinkō dōtai hikaku," 1937, PDFY090803054, Tachi Bunko.

[18] Yoshio Ando, ed., *Showa keizaishi* (Nihon Keizai Shinbunsha, 1994).

[19] Dietmar Rothermund, *The Global Impact of the Great Depression 1929–1939* (London: Taylor and Francis, 1996), 115–19.

[20] Tachi and Ueda, "Taisho 9-nen, Taisho 14-nen," 24–25.

[21] Shiro Aoshika, "Tohoku chihō no kyōsaku nituite," *Tokyo-shi nōkaihō*, no. 21 (December 10, 1934): 7–8.

[22] "Miyagi-ken dekasegi ni kansuru chōsa," n.d., PDFY09110678, Tachi Bunko. The document was a carbon copy of the meticulously handwritten chart showing the figures of migrant workers from Miyagi Prefecture between 1933 and 1935. The author is unknown, but it was written on official manuscript paper produced by Miyagi Prefecture.

peaked after the attempted military *coup d'état* by junior army officers on February 26, 1936. High-ranking government officials considered overpopulation, particularly in the farming villages, to be behind the incident, acting as a factor in the political radicalization that caused the attempted coup.[23] From the perspective of government officials, it seemed obvious that the post–Depression countryside had turned into a problem region due to the growing pressure caused by the expanding "surplus population."

The government responded to the crisis by further promoting overseas migration.[24] However, in the early 1930s, Latin America, which by then had become a major destination for the officially endorsed migration project, was becoming less attractive to Japanese emigrants.[25] During this period, Brazil, the major recipient country in the region, became less welcoming to the Japanese because Getúlio Vargas's nationalist government, which came into power in 1930, imposed assimilation and exclusion policies on the quickly expanding Japanese migrant communities.[26] Under these circumstances, Manchuria loomed on the horizon as a new promised land.[27] The formation of Manchukuo as Japan's puppet state in 1932 additionally gave the government hope that it could mobilize rural populations to turn Manchuria into an important site of colonial development.[28] After the Hirota cabinet approved a program

[23] Takaoka, "Senji no jinkō seisaku," 102.
[24] Another official measure linked with overseas migration was regional development, and Tohoku was the target region. The Tohoku Development and Promotion program, launched in response to the famine caused by 1934 crop failure, was approved by the Imperial Diet in 1936. Atsushi Kawauchi, "Jinkō to Tohoku: Senjiki kara sengo ni okeru Tohoku 'kaihatsu' tono kanren de," in *Tōhoku chihō "kaihatsu" no keifu: Kindai no sangyō seisaku kara higashi nihon daishinsai made*, ed. Takenori Yamamoto (Akashi Shoten, 2015), 1–17; Makoto Okumura, "Tōhoku chihō kaihatsu no rekishi," *Toshi keikaku* 61, no. 2 (April 2012): 5–10.
[25] Lu, *The Making of Japanese Settler Colonialism*, 222–29.
[26] Shiode, *Ekkyōsha no seijishi*, 336–38; Toake Endoh, *Exporting Japan: Politics of Emigration toward Latin America* (Urbana: University of Illinois Press, 2009), 32–34.
[27] Shinichi Yamamuro, *Manchuria under Japanese Dominion* (Philadelphia: University of Pennsylvania Press, 2006); Prasenjit Duara, *Sovereignty and Authenticity: Manchukuo and the East Asian Modern* (New York: Rowman and Littlefield Publishers, 2004), 41–86; Louise Young, *Japan's Total Empire: Manchuria and the Culture of Wartime Imperialism* (Berkeley and London: University of California Press, 1998), 3–52; Sandra Wilson, "The 'New Paradise': Japanese Emigration to Manchuria in the 1930s and 1940s," *International History Review* 17, no. 2 (1995): 121–40. In tandem with this, Mongolia – which often appeared in association with Manchuria, as in the expression *manmō* – was imagined as terra incognita.
[28] Azuma, "'Pioneers of Overseas Japanese Development'"; Yoshihisa Tak Matsusaka, *The Making of Japanese Manchuria, 1904–1932* (Cambridge, MA: Harvard University Asia Center, 2003); Young, *Japan's Total Empire*, 53.

for the mass colonization of Manchuria in 1937, local and prefectural organizations arranged a systematic emigration of farmers to Manchuria to establish "branch villages" (*bunson*).[29] Posters, travel journals, and historical writings stressed the image of Manchuria as a vast and empty frontier, urging many Japanese to dream of migration as an opportunity to materialize a vision of the future that they thought would be impossible to achieve at home.[30] For Japanese officials, the emigration of Japanese farmers to Manchuria would buttress what historian Louise Young once called "social imperialism."[31] They were convinced that migration was an effective social policy that would relocate a myriad of domestic problems associated with "overpopulation" to Manchuria while also fostering Japan's colonial development.[32]

Population experts, especially social scientists working on the rural community, were behind the official migration program to Manchuria.[33] One of the most prominent was the agrarian economist Nasu Shiroshi (1888–1984). Nasu was affiliated with the agrarian movement led by Katō Kanji, a right-wing activist who, in the 1920s, ran schools for rural youth in Ibaraki and Yamagata to realize a farm colonization in Korea and Manchuria.[34] In the early 1930s, Nasu argued in front of government officials that agricultural migration to Manchuria was an effective way to relieve the population pressure of the resource-poor metropole and simultaneously give hope to the farming villages hardest hit by the depression.[35] Using his status as a well-reputed academic, in February 1932, Nasu and his colleague Hashimoto Denzaemon consulted with the Guangdong Army (or Kwantung Army, in Japanese *Kantōgun*), additionally justifying migration on the grounds of security.[36] Going along with

[29] Young, *Japan's Total Empire*, 336.
[30] Duara, *Sovereignty and Authenticity*, 62.
[31] Young, *Japan's Total Empire*, 12–13.
[32] Matsusaka, *The Making of Japanese Manchuria*.
[33] Lu, *The Making of Japanese Settler Colonialism*, 187–90; Young, *Japan's Total Empire*, 352–98.
[34] For a more recent work referring to Kanji Katō and Japanese migration to Manchuria, see Yasumasa Ishibashi, "Mobilizing Structures in Manchuria Agricultural Emigration in Imperial Era: Idea and Practice of Kanji Kato as a 'Mediator'" [in Japanese]. *Korokiumu*, no. 6 (June 2011): 111–34.
[35] Lee, "Problematizing Population," 148–58, 171–78; Takaoka, *Sōryokusen taisei to "fukushi kokka,"* 108–10.
[36] Shinnosuke Tama, *Sōsenryoku taiseika no manshū nōgyō imin* (Yoshikawa Kobunkan, 2016); Sandra Wilson, "Securing Prosperity and Serving the Nation: Japanese Farmers and Manchuria, 1931–33," in *Farmers and Village Life in Twentieth-Century Japan*, eds. Ann Waswo and Yoshiaki Nishida (New York: Routledge, 2003), 156–74; Kyōji Asada, "Manshū nōgyō imin seisaku no ritsuan katei," in *Nihon teikokushugika no manshū imin*, ed. Manshū Iminshi Kenkyūkai (Ryukei Shosha, 1976), 7–8.

government officials, agrarian population experts stressed that migration was simultaneously a panacea for the domestic problem of "overpopulation" and a tool for the imperial project to turn Manchukuo into Japan's "life line."[37]

While the migration program continued into the late 1930s, the official discourse on population problems changed dramatically after the outbreak of war with China in 1937. With the rising demand for labor in the war industry, the narrative of an "unemployment problem" dissipated and was replaced by an argument that stressed the problem of labor shortage. Linked to this, the problem of declining fertility surged as a policy agenda, as mass conscription had a tangible effect on birth rates starting in the latter half of 1938.[38] The changing political situation in 1938 further exhorted government officials to reconsider population in a different light. In that year, the first Konoe Fumimaro cabinet (est. June 1937) redefined the war with China as a prolonged conflict, and on April 1, 1938, issued the National General Mobilization Law to mobilize the population for the construction of a "national defense state" (kokubō kokka). The demand for total mobilization intensified even more when Konoe issued a communiqué about the Chinese government in November 1938, which stated that the new goal of the current conflict was world peace realized through the "construction of a new order" for "eternal stability in East Asia." This was immediately followed by another, issued a month later, which stated that the friendship, military collaboration, and economic cooperation of Japan, Manchuria, and China would be ideal for the construction of a "new order" in East Asia. In this political context, a large population size supported by high birth rates, which hitherto had been seen as a socioeconomic menace, was quickly redefined to represent "national power" (kokuryoku).[39]

Also behind the change in the official discourse was an additional understanding of population that had gradually become dominant in policymaking since the interwar period: Population was a valuable national resource. This idea emerged shortly after WWI, when the term "resource" (shigen) entered the official lexicon.[40] WWI exposed Japan's shortcomings as a small island nation that was poor in resources. This fostered the consensus, especially within the Army, that the government

[37] Matsusaka, The Making of Japanese Manchuria, 214–23.
[38] Takaoka, "Senji no jinkō seisaku," 103.
[39] Jinkō Mondai Kenkyūsho, "Kokudo keikaku toshiteno jinkō haichi (yohō) shōwa 15 nen 8 gatsu," August 1940, PDFY09111754, Tachi Bunko.
[40] Jin Sato, "Motazaru kuni" no shigenron: Jizoku kanō na kokudo wo meguru mouhitotsu no chi (Tokyo Daigaku Shuppankai, 2011), 66–68; Dinmore, "A Small Island Nation Poor in Resources."

should invest in resource management to prepare the country for future conflicts.[41] This view led to the launch of the Cabinet Bureau of Resources in 1927 as the official organization charged with the investigation, management, and mobilization of resources.[42]

The bureau defined resources broadly. According to Bureau Chief Matsui Haruo, "every kind of source contributing to the existence and prosperity of an organization" fell into the category of "resource."[43] However, partly because of the army's involvement in the bureau, Matsui's seemingly neutral take on the term was full of political overtones.[44] Indeed, "resource," as defined by Matsui, referred to materials that could be utilized for the expansion of military power and war industries, and the "organization" was not just any institution, it was the Japanese state preparing for a future war. This interpretation of resource simultaneously gave rise to the idea that the population, too, could be a type of resource. This articulated what Matsui called *jinteki shigen* ("human resource"), and he insisted that a population, like any other type of resource, could be mobilized for national prosperity. The notion of population as "human resource" was an aggregate of people whose capability was defined not only by size but also by its qualitative values, such as spirit, morality, and physical strength.[45] In effect, "human resource" for Matsui was human power that directly enhanced the nation's economic and military capabilities.

In the late 1920s and early 1930s, this formulation of population failed to become a mainstream narrative within the policy-oriented population debate, which was overly focused on the problem of "overpopulation." However, the situation changed in the wake of total war. The concept of population as "human resource" gained currency in a political environment in favor of mass mobilization.[46] The tendency became prominent, especially after October 1937, when the government merged the existing Cabinet Bureau of Resources and the Planning Agency to create the

---

[41] Yasuo Mori, *"Kokka sōdōin" no jidai: hikaku no shiza kara* (Nagoya: Nagoya Daigaku Shuppankai, 2020); Michael A. Barnhart, *Japan Prepares for Total War: The Search for Economic Security, 1919–1941* (Ithaca: Cornell University Press, 2013), 64–76.

[42] For the establishment of the Cabinet Bureau of Resources, see Sato, *"Motazaru kuni" no shigenron*, 68–77; Mikuriya, *Seisaku no sōgō to kenryoku*; Toshiaki Yamaguchi, "Kokka sōdōin kenkyū josetsu: Dai ichiji sekai taisen kara shigenkyoku no seiritsu made," *Kokka gakkai zasshi* 92, no. 3–4 (1979): 266–85.

[43] Cited in Sato, *"Motazaru kuni" no shigenron*, 73.

[44] Yamaguchi, "Kokka sōdōin kenkyū josetsu."

[45] Sato, *"Motazaru kuni" no shigenron*, 73.

[46] Tsukada Ippo, *Kokka sōdōinhō no kaisetsu* (Shūhōen shuppanbu, 1938), 22–23. Also see Aiko Kurasawa, *Shigen no sensō: "Daitōa kyōeiken" no jinryū, butsuryū* (Iwanami Shoten, 2012).

Cabinet Planning Board (CPB) as the government office charged with resource management and total mobilization. The CPB explicitly understood "human resource" as manpower, a determining factor for the military and labor force being able to sustain the wartime economy and national defense.[47] This idea of population directly shaped the National General Mobilization Law drafted by the CPB. The law stipulated that "human resource," juxtaposed with "material resource" (*butteki shigen*), would be subject to "controlled management" (*tōsei unyō*) in times of emergency so that the state could fully take advantage of its capabilities for national defense. For the rest of the war, this conceptualization buttressed the central government's mobilization schemes, such as conscription, the migration of workers, and mass evacuation.[48]

Alongside the rise of "human resource" in the official discourse, the meanings assigned to the growing rural population changed. The surplus population in the countryside, hitherto perceived suspiciously as a seed of political unrest, was now seen positively, as an asset directly assisting the Japanese state's struggle to win the battle. In parallel, the farming community became described as the primary supplier of a healthy and morally sound "human resource" appropriate for serving the Japanese nation-state-empire. Needless to say, this view did not simply emerge out of a vacuum but was strongly informed by the antimodern, antiwestern agrarianism endorsed by activists such as Katō that came to hold currency under the wartime fascist regime.[49] The ideology denounced cities for fostering western values of decadence, individualism, and liberalism, while romanticizing the farming community as a source of Japan's national identity and power. When applied to the wartime population debate, the ideology manifested itself in criticism that blamed cities for causing the decline in people's physical and mental constitutions and blamed the urban lifestyle for the fertility decline. At the same time, the ideology lent itself to the argument in favor of protecting the farming population by means of social policy.[50] The wartime demand for "human resource," compounded with agrarianism, invited policymakers to revise their views on farmers. At the same time, the positive view reinforced the existing tendency to regard farmers as a primary target for policy interventions.

[47] Aiko Kurasawa, *Shigen no sensō*; Paul H. Kratoska, "Labor Mobilization in Japan and the Japanese Empire," in *Asian Labor in the Wartime Japanese Empire: Unknown Histories*, ed. Paul H. Kratoska (London: Routledge, 2005), 3–21.
[48] Takaoka, "Senji no jinkō seisaku," 117–18.
[49] Thomas R. H. Havens, "Kato Kanji (1884–1965) and the Spirit of Agriculture in Modern Japan," *Monumenta Nipponica* 25, no. 3/4 (1970): 295–322.
[50] Takaoka, *Sōryokusen taisei to "fukushi kokka,"* 205.

Wartime Population Policies: Creating a Large and Robust Population for the Nation at War

The concept of "human resource" amid war highlighted the significance of certain demographic phenomena in the policy discussion.[51] As already suggested above, fertility decline was attracting the most attention. In fact, even prior to the war, population experts were warning that the birth rate – falling after it peaked in 1920 at 36.19 per 1,000 population – heralded a contracting and aging population in the future.[52] But, a significant dip in the rates in 1938, from 30.61 the previous year to 26.70 per 1,000 population, was a significant blow to government officials.[53] They were now worried that Japan would have a less mobilizable "human resource" in the near future. As a MHW document succinctly summarized, the "lack of a population is a lack of military force and workforce," and the country at war would certainly suffer from the consequences.[54] Due to this logic, government officials singled out fertility decline as a policy agenda.

In addition to declining fertility, the "lowering" of the Japanese people's physical strength caused concern among government officials. The problem of compromised physical strength had been addressed by military health officers since the 1910s (see Chapter 2). In the 1930s, Army Ministry Medical Affairs Director Koizumi Chikahiko (1884–1945) brought the argument to the frontlines of policy discussions.[55] Pointing to the rising number of men failing the physical examination for conscription due to tuberculosis and substandard muscle and bone strength, Koizumi warned that "physical aptitude" (*tai'i*) in Japan was in crisis.[56] He then pointed out that some countries in western Europe, confronted by a similar challenge after WWI, tried to rectify the situation by setting up a government office specialized in nurturing "people's power" (*minryoku*) by means of public health and suggested Japan should follow a similar path.[57] Based on this logic, he urged his seniors at the Ministry of Arms to lobby the government to found what he called the "Hygiene Ministry" (*Eiseishō*).[58]

---

[51] Yuriko Sakurada, "Senji ni itaru 'jinteki shigen' wo meguru mondai jōtai: Kenpei kenmin seisaku tōjō no haikei," *Nagano daigaku kiyō*, no. 9 (March 1979): 41–55.

[52] Teruoka, "Waga kuni shusshōritsu."

[53] Kōseishō Jinkōkyoku, "Shusshōritsu yori mitaru genka no jinkō mondai," 5–6, March 1942, PDFY090212123, Tachi Bunko.

[54] Ibid., 8.

[55] Takaoka, *Sōryokusen taisei to "fukushi kokka,"* 26–56.

[56] Hiroyuki Takaoka examined the data related to the military physical examination and concluded that the "fact" about the lowering level of national physical strength Koizumi presented was "clearly a fiction" that he "intentionally" came up with by manipulating the data; Takaoka, *Sōryokusen taisei to "fukushi kokka,"* 43.

[57] Ibid., 26.

[58] Ibid., 26–27.

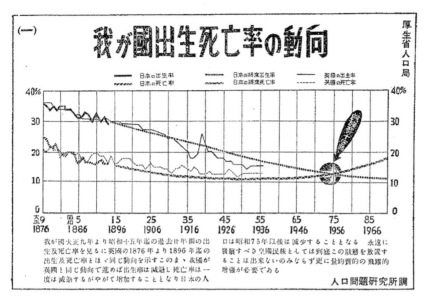

Figure 4.1 The trend of birth and death rates in our country. A poster published by the IPP in 1942. The caption states how Japan was following England's path, and if the trend continued, the Japanese population would start shrinking in 1956. Toward the end, the text below the graph states: "Not only can our imperial race not ignore this situation for our eternal development, but also it needs drastic and further strengthening of population quality and quantity."
Source: *Jinkō mondai kenkyū*, 3, no. 6 (June 1942): 31.

What fueled official anxiety even more was the dire state of maternal and infant health in the countryside. This became apparent in the investigation into the demographics and health in approximately 134 villages that the Home Ministry Sanitary Bureau had conducted since 1918 as a follow-up to the HHSG (see Chapter 2). The study, published in 1929, clearly pointed out high stillbirth and child mortality rates in the countryside.[59] It pointed out that the ten-year average rate of stillbirth in the 7 and 77 villages studied by the bureau and local authorities, respectively, were 2.35 and 2.66 per 1,000 population, which exceeded the national average (2.18) and the average in cities (1.85).[60] The ten-year average rate of child mortality in all villages in the study was 16.2 per 100

[59] Naimushō Eiseikyoku, *Nōson hoken eisei jicchi chōsa seiseki* (1929).
[60] Ibid., 33.

live births, which was more than the national average of 13.7.[61] This trend alarmed population bureaucrats and experts because these figures revealed that the countryside, supposedly the source of strong, youthful, and high-quality "human resource," was actually inundated with issues, which could easily lead to fertility decline *and* falling physical strength, the two biggest demographic problems of the day. They were concerned that high child mortality in rural areas symbolized the imminent future loss of Japan's "national power."[62]

The wartime government came up with specific measures in response to these concerns. To accommodate the request from the Army Ministry, the Konoe cabinet authorized the establishment of the MHW, which materialized on January 11, 1938.[63] The MHW stated that its missions included the improvement of physical strength and maternal and child health to address issues related to fertility decline. In 1939, the MHW assigned the newly established Bureau of Society's Life Section (*Shakai-kyoku Seikatsuka*) to look into matters concerning population problems, and on August 1, 1941, it launched the Population Bureau.[64] In 1939, the government also set up the Institute of Population Problems (IPP) within the MHW as a permanently based official institution dedicated to population studies and policymaking.[65] As historian Fujino Yutaka once suggested, the "policy of cultivating and mobilizing 'human resource'" under the "fascist regime" urged the wartime government to institutionalize the health and welfare administration and research dedicated to population matters.[66]

Between 1940 and 1941, the IPP was involved in drafting a proposal for population policies, which culminated in the cabinet's approval of the key wartime population policy, the General Plan to Establish the National Population Policy (*Jinkō Seisaku Kakuritsu Yōkō*, hereafter General Plan for Population, GPP) on January 22, 1941. The GPP was a direct response to the Outline of a Basic National Policy issued on July 26, 1940 by the second Konoe cabinet (est. July 22, 1940). The outline confirmed the Konoe cabinet's commitment to the total mobilization for

---

[61] Ibid., 40–42.

[62] See, e.g., the poster "Shusshōritsu no teika suru kuni wa horobiru," n.d., in PDFY09121667, Tachi Bunko. In response to the report, the HHSG set up the Special Committee Regarding the State of Hygiene in Farming Villages in 1930 and examined policy measures intended to promote rural health. Hoken Eisei Chōsakai, "Hoken eisei chōsakai dai 15 kai hōkokusho" (April 1931).

[63] Takaoka, *Sōryokusen taisei to "fukushi kokka,"* 63–70; Fujino, *Kōseishō no tanjō*, 55; Fujino, *Nihon fashizumu to yūsei shisō*, 266–67.

[64] Takaoka, "Senji no jinkō seisaku," 104.

[65] Kōseishō Jinkō Mondai Kenkyūsho, ed., *Jinkō mondai kenkyūsho no ayumi*, 1–2.

[66] Fujino, *Kōseishō no tanjō*, 9.

establishing a "new order in Greater East Asia." It also exhorted Minister of Foreign Affairs Matsuoka Yūsuke to pronounce that Japan's political goal was to establish a "Greater East Asia Co-Prosperity Sphere" – the term he coined to refer to an economically and politically integrated area in Asia under Japanese leadership – to fend off western imperial intervention and materialize world peace.[67] On the topic of population, the outline characterized a large and high-quality population as "a driving force for the execution of the national policy" and stated that the government should strive to "establish a permanent policy for population increase, for the improvement in the quality, and for the physical strength of the nation's people." Following the outline, on August 1, the cabinet decided that the MHW, CPB, Ministry of Agriculture and Forestry, and Ministry of Colonial Affairs would draw up a proposal to establish the population policy, while the Home Ministry, Army Ministry, Navy Ministry, and Ministry of Commerce and Industry would act as the main ministries involved in deliberations on the policy.[68]

The GPP, which was made as a result of the interministerial collaboration, stated that it should act as a guide to establish a "fundamental and perpetual population policy" for the "construction and eternal and healthy development of a Greater East Asia Co-Prosperity Sphere."[69] It further explained that the population policy should achieve the following four objectives: (1) "to ensure our population's eternal development," (2) "to surpass other countries in terms of the population's growth power and population quality," (3) "to acquire the required military and labor force for a high national defense state," and (4) "to appropriately deploy populations to secure Japanese leadership vis-à-vis other East Asian races." The GPP further presented the following four categories for policy measures: (a) "measures for population growth," (b) "measures for strengthening population quality," (c) "the preparation of relevant materials," and (d) "the establishment of organizations."[70]

Responding to the outline, the GPP endorsed pronatalism.[71] It stated that a tangible goal of the current policy should be to increase the

---

[67] Jeremy A. Yellen, *The Greater East Asia Co-Prosperity Sphere: When Total Empire Met Total War* (Ithaca: Cornell University Press, 2019); Kousuke Kawanishi, *Daitōa kyōwaken: Teikoku nihon no nanpō taiken* (Kodansha, 2016).

[68] Japan Center for Asian Historical Records (JACAR) Ref. B02030544800, Shina-jihen kankei ikkei dai 15-kan (A-1-1-0-30_015) (Gaimushō Gaikō Shiryōkan).

[69] Takaoka, "Senji jinkō seisaku no saikentou," 160–73.

[70] Jinkō Mondai Kenkyūsho, "Jinkō seisaku kakuritsu yōkō," March 1941, PDFY091105017, Tachi Bunko.

[71] For how pronatalism dominated fascist rhetoric in another national context, see Carl Ipsen, *Dictating Demography the Problem of Population in Fascist Italy* (Cambridge: Cambridge University Press, 1996), 173–84.

Figure 4.2 Birth rates within the Greater East Asia Co-Prosperity Sphere. A propaganda poster published by the IPP. The caption states: "The birth rate in Japan Proper is the lowest among the fellows in the Greater East Asia Co-Prosperity Sphere. We need to supersede other countries in terms of population growth power and quality."
Source: *Jinkō mondai kenkyū* 3, no. 6 (June 1942): 35.

"population of Japan Proper" (*naichijin jinkō*) to 100 million by 1960.[72] To realize this objective, the GPP proposed the government strive to lower the average age of marriage down by approximately three years and to raise the average number of children per married couple up to five. It further stipulated that these pronatalist measures should be accompanied by others that aimed to lower general mortality by approximately 35 percent over the next two decades. Together, these measures would ensure the growth and perpetual development of the "population of Japan Proper," thus enabling the population to perform at its full capacity as a "driving force" behind the national mission – so the outline stated.

In addition to pronatalism, eugenics also acted as a backbone for the GPP.[73] The GPP recommended that the government should "strengthen the physical and mental traits required for national defense and labor" and recommended the "diffusion of eugenic thought" and a thorough implementation of the National Eugenic Law (*Kokumin Yūsei Hō*), which was issued in May 1940. This eugenic clause in the GPP came in tandem with the MHW's efforts to popularize eugenics.[74] From its inception, the MHW had an independent Section of Eugenics within the Division of Prevention. After the government issued the National Eugenic Law and the National Physical Strength Law (*Kokumin Tairyoku Hō*) in 1940, the MHW instigated the "healthy soldiers, healthy citizens" (*kenpei kenmin*) movement. This campaign, organized under Koizumi, the new minister of health and welfare, promoted eugenic health and educational initiatives as well as medical research in the metropole and the colonies on topics such as the eradication of tuberculosis, venereal disease control, sterilization, and psychosomatic disorders in order to produce a "physical robust, intellectually sharp, and determined … imperial Japanese population (*kōkoku jinkō*)."[75] The GPP placed these measures at the center of wartime population policy.

Though initially only a guideline, the GPP's status was elevated when Japan's attack on Pearl Harbor in December 1941 turned into a full-blown war involving the Allied Forces. Discourses on race and racism dominated the war, and the government leaders portrayed the

---

[72] Takaoka, "Senji jinkō seisaku no saikentou," 161–65.
[73] Rihito Yasuda, "Kindai nihon ni okeru jinkō seisaku kōsō no ichi danmen (II)," *Kokusai bunkagaku*, no. 32 (March 2019): 155–79; Yoko Matsubara, "Nihon ni okeru yūsei seisaku no keisei" (PhD diss., Ochanomizu University, 1998).
[74] Yokoyama, *Nihon ga yūsei shakai ni naru made*, 198–201, 253–71; Mitsuko Chuman, "Nagai Hisomu saikō."
[75] Takaoka, *Sōryokusen taisei to "fukushi kokka,"* 228; Fujino, *Nihon fashizumu to yūsei shisō*, 343–69.

population problem even more explicitly as a matter of Japanese leadership in the colored people's racial struggle against white, western imperialism, which, in the specific context of Japan's effort to construct a "new order" in East Asia, entailed a struggle that could be overcome through cooperation among the five races in the region (Koreans, Manchurians, Mongolians, Han Chinese, and Japanese).[76] Under these circumstances, in February 1942, the Konoe cabinet requested the newly founded Advisory Council for the Construction of Greater East Asia (*Dai Tōa Kensetsu Shingikai*) come up with "population and race policies to accompany the construction of Greater East Asia." The advisory council's response, a policy proposal titled "The Population and Race Policy Accompanying the Construction of the Greater East Asia," stated the main goals of the population policy were to "expand and strengthen the Yamato race" and recommended the government implement the measures introduced in the GPP.[77] Following the proposal, in November 1942, the government founded the Ministry of Health and Welfare Research Institute (MHW-RI) Department of Population and Race and ordered the new institute to examine the GPP in light of the new policy.[78] After this, official activities for population and race policies converged more than ever before.

This characterization of population policy – as synonymous with the policy aiming to strengthen the physical and mental capabilities of the Japanese race – was widely shared among high-rank government officials during the war.[79] It focused on the corporeal aspect of a population and therefore endorsed eugenic, health, and reproductive measures as solutions to the problems of both population quantity and quality. Yet, this was not the only rationale that buttressed wartime population policy.[80] Another important rationale was summed up in the expression "population distribution," which allowed contemporaries to expand the scope of their definition of the "population of Japan Proper": in the context of Japan's struggles to develop a multiethnic empire with a highly controlled economic system. It also exhorted wartime policy intellectuals and policymakers to ask how the "population of Japan Proper," as "human resource," could be best deployed across the Greater East Asia Co-Prosperity Sphere to

---

[76] John W. Dower, *War without Mercy: Race and Power in the Pacific* (London: Faber, 1986), 262–90. For a more recent work on the racial assimilation in Japanese empire, see Hanscom and Washburn eds., *The Affect of Difference*.

[77] Takaoka, "Senji no jinkō seisaku," 116–17.

[78] With the establishment of the MHW-RI, the IPP ceased to exist. It was revived after the war.

[79] "Jinkō seisaku kakuritsu yōkō no kettei," *Jinkō mondai kenkyū* 2, no. 2 (1941): 56–57.

[80] Takaoka, *Sōryokusen taisei to "fukushi kokka,"* 178–80.

maximize Japan's imperial power. Questions surrounding "population distribution" surged when the government pondered over the population problem in relation to its grand wartime state planning scheme: "national land planning."

Distributing Populations for the Greater East Asia Co-Prosperity Sphere

As MHW IPP staff were writing drafts of the GPP, their colleagues in the Cabinet Planning Board (CPB) – another office assigned to draw up population policy – were also engaged in population issues. Reflecting the CPB's role as the cabinet's war planning and mobilization body, the CPB staff contextualized population problems in terms of the wartime state's ultimate planning scheme: "national land planning."

"National land planning" (*kokudo keikaku*) was conceived sometime in the fall of 1939 as a comprehensive state planning scheme designed for Konoe's "new order" movement.[81] It was first discussed in the National Land Planning Study Group, which Konoe's close advisor Gotō Ryūnosuke created within the Showa Research Association (*Showa Kenkyūkai*).[82] Representing the voice of pro-fascist, anti-capitalist "new order" supporters, Gotō claimed the top-down comprehensive state planning ensured by technocratic management was an ideal foundation for the self-sufficiency of the Japan-Manchuria-China Bloc. Responding to Gotō's call, in January 1940, the association submitted the "Memorandum on National Land Planning," which triggered policy deliberations within the CPB. The appointment of Hoshino Naoki as the head of the CPB at the inauguration of the second Konoe cabinet in July 1940 gave the policy initiative a boost, since Hoshino had already headed a similar project in Manchuria. The government proclaimed that the "establishment of a national land development plan aimed to expand a comprehensive national power throughout Japan, Manchuria, and China" would be a core policy item in the aforementioned Outline of

---

[81] Recent works on *kokudo keikaku* in the total war include Janis Mimura, *Planning for Empire: Reform Bureaucrats and the Japanese Wartime State* (Ithaca: Cornell University Press, 2011); Janis Mimura, "Japan's New Order and Greater East Asia Co-Prosperity Sphere: Planning for Empire," *The Asia-Pacific Journal: Japan Focus* 9, no. 3 (2011): 1–12; Shinichi Yamamuro, "Kokumin teikoku, nihon no keisei to kūkanchi," in *Kūkan keisei to sekai ninshiki*, ed. Shinichi Yamamuro (Iwanami Shoten, 2006), 19–76; Mikuriya, *Seisaku no sōgō to kenryoku*; Takashi Mikuriya, "The National Land Planning and the Politics of Development," *The Annals of Japanese Political Science Association* 46 (1995): 57–76.

[82] Yamamuro, "Kokumin teikoku," 65; Saburo Sakai, *Showa kenkyukai: Aru chishikijin shūdan no kiseki* (Chuokoron-sha, 1992).

a Basic National Policy. On September 24, 1940, the cabinet approved the General Plan to Establish National Land Planning (*Kokudo Keikaku Settei Yōkō*, hereafter the General Plan for Land, GPL), which was drafted by the CPB and based on the Outline of a Basic National Policy.

In a narrow sense, the national land planning delineated by the GPL was an economic policy endorsing self-sufficiency, a means to enhance national productive power via a careful planning of what Ramon H. Myers once called the "enclave economy" of the Japan-Manchuria-China Bloc.[83] Yet, it was not just a narrowly conceived and managed economic scheme.[84] "National land planning" was as much a policy of resource economics and national defense as a political technology for constructing a "new order" in East Asia. It involved state bureaucrats' active participation in the comprehensive development and management of resources in relation to "national land" (*kokudo*), an ideologically laden, emotive concept denoting the topographical landmass, the geopolitical concept of space, and the source of Japan's spiritual identity.

A key mandate of national land planning was to adopt a rational approach for seeking an optimal geographical relationship between the "national land" and resources to reach a higher level of efficiency.[85] To attain this goal, the GPL stressed that the resources acquired within the Greater East Asia Co-Prosperity Sphere should be distributed "in a controlled manner" and "in relation to the national land," and assigned the CPB to administer the controlled coordination of resources.[86]

To fulfill the mandate, the GPL adopted the expansive definition of resources expressed by the CPB since the 1920s. They included natural resources (e.g., ore, trees, and water), energy, humanmade institutions, systems such as the industrial system, transportation, cultural and welfare facilities, and, finally, the population. Among these different kinds

---

[83] Ramon H. Myers, "Creating a Modern Enclave Economy: The Economic Integration of Japan, Manchuria, and North China, 1932–1945," in *The Japanese Wartime Empire, 1931–1945*, eds. Peter Duus, Ramon H. Myers, and Mark R. Peattie (Princeton: Princeton University Press, 1996), 136–70. Also see Janis Mimura, "Economic Control and Consent in Wartime Japan," in *The Palgrave Handbook of Mass Dictatorship*, eds. Paul Corner and Jie-Hyun Lim (London: Palgrave Macmillan, 2016), 157–69; Yasuyuki Hikita, "Daitoakyōeiken ni okeru tōsei keizai," in *"Teikoku" nihon no gakuchi dai 2 kan "teikoku" no keizaigaku*, ed. Shin'ya Sugiyama (Iwanami Shoten, 2006), 2: 257–302.

[84] For different interpretations of "national land planning," see Mikuriya, "The National Land Planning and the Politics of Development," 58; Dinmore, "A Small Island Nation Poor in Resources," 59; Mimura, *Planning for Empire*, 11–12.

[85] Janis Mimura, "Technocratic Visions of Empire: Technology Bureaucrats and the 'New Order' for Science-Technology," in *The Japanese Empire in East Asia and Its Postwar Legacy*, ed. Harald Fuess (Munich: Indicium Verlag GmbH, 1998), 97–118.

[86] Yamamuro, "Kokumin teikoku," 65.

of resources, the GPL regarded population as particularly critical, thus spending a substantial amount of space elaborating on what it called "population planning," or the designs for population policies.[87]

As part of national land planning, the primary objective of "population planning" was to raise efficiency through a rational coordination of population and the "national land." The GPL based on this principle stressed "population distribution" as the chief means for attaining the goal. It stipulated "population planning" should aim for "an appropriate distribution of the population according to regions and professional abilities" and included "comprehensive population distribution planning" (sōgōteki jinkō haibun keikaku) in the list of the nine most important policy items for national land planning. "Comprehensive population distribution planning," according to the GPL, consisted of the following four interlinked measures: (1) coordination of urban populations, (2) distribution of populations divided by occupational categories, (3) distribution of populations divided by regions, and finally, (4) "comprehensive migration." In practical term, this entailed the movement of primarily Japanese people within the Japan-Manchuria-China Bloc, or more broadly, the amorphous sphere of imperial Japan's reach. However, it was not simply an extension of the existing state-endorsed migration scheme. The aim of the existing migration program was to solve the domestic problem of overpopulation by "relieving" the population pressure in the metropole. The "comprehensive population distribution planning" in the GPL was a population *growth* policy realized through a careful coordination of populations vis-à-vis Japan's military strategy and the industrial adjustment within the Bloc.[88] These two migration schemes had different fundamental premises for the "population problem" that necessitated migration.

Having said this, the argument for the "comprehensive population distribution planning" had roots in a number of overlapping discursive sites that thrived in the 1920s as Japan was struggling to build its international reputation as the only nonwestern, industrial colonial power. Among these, two stood out. One was the field of social sciences and social policy that engaged with the population problem as an economic – specifically labor – issue, and the other was geopolitics. As for the first,

---

[87] Kyoko Kondo, "Kokudo keikaku to jinkō no shiten no hensen," *Tōkei* 62, no. 12 (December 2011): 17–26.
[88] Minoru Tachi, "Jinkō seisaku no tachiba yori mitaru kokudo keikaku ni kansuru jakkan no kihonteki mondai shiken," *Shōkō keizai* 11, no. 1 (January 1941): 81–114. While the GPL's primary focus was on the redistribution of the Japanese population, the idea of "population distribution" at times was expansive, including the labor migration of other ethnic subjects within the Japan-Manchuria-China Bloc.

social scientists serving the IC-PFP Population Section recommended migration in the name of "labor adjustment" (see Chapter 3). Nearly a decade later, social scientists discussed migration again, but this time to tackle the problem of labor shortage and the declining quality of the workforce arising from the rapid expansion of the munitions industry.[89] The renowned economist Ōkouchi Kazuo argued that these labor problems were inhibiting the expansion of industrial productivity and suggested the government establish social policies aimed at controlling the supply of the workforce as "human resource."[90] Partly in response to this kind of argument, between 1938 and 1939, the government issued a number of legislations to manage the labor market, including the amended Work Placement Law (*Shokugyō Shōkai Hō*) in April 1938 that nationalized the work placement scheme. In this context, government officials redefined work placement as a government initiative to "deploy labor appropriately."[91]

Corresponding to this trend, social policy specialists examined population distribution as a wartime labor policy, calling it a "deployment of the workforce." The Labor Problem Study Group, established in February 1939 and consisting mainly of CPB bureaucrats, put forward the "quantitative deployment of labor force" as a specific measure for the wartime economy.[92] Taking up Ōkouchi's idea that the wartime labor policy should be a "production policy that seizes workers as its object," the study group argued that the policy should address the question of "how to distribute the labor force effectively ... in relation to the maintenance and expansion of productivity as well as national defense."[93] The GPL took up this idea. It explained that one of the policy's objectives should be "an appropriate distribution of the population according to ... professional abilities." The "population distribution" in the GPL resonated with the narrative of the "deployment of the workforce" that prevailed in the policy discussions on wartime economy.[94]

Another field endorsing "population distribution" for the GPL was geopolitics.[95] Geopolitics, originally formulated by Friedrich Ratzel,

---

[89] Takaoka, *Sōryokusen taisei to "fukushi kokka,"* 133–34.
[90] Ibid., 133–37.
[91] Ibid., 132.
[92] Ibid., 137.
[93] Ibid., 137.
[94] The scope for the "deployment of the workforce" in the GPL was directly tied to the labor issues in Manchuria, including recruitment, skills, and high turnover. Paul H. Kratoska ed., *Asian Labor in the Wartime Japanese Empire: Unknown Histories* (London: Routledge, 2005).
[95] For geopolitical thinkers' engagement in national land planning, see Yamamuro, "Kokumin teikoku," 60–69.

Rudolf Kjellén, and Karl Haushofer, was popularized in Japan in the latter half of the 1920s by figures such as the geographer Iimoto Nobuyuki.[96] Envisioned at a time when Japan's international standing was becoming increasingly precarious, geopolitics was depicted in Japan as a theory that justified Japanese imperialism as a colored race's struggle against western domination in global politics.[97] After Japan withdrew from the League of Nations in 1933 and began to explore an alternative way to ensure world peace through Pan-Asianism, the academic field called Greater East Asian Geopolitics (*Daitōa Chisekigaku*) gained political power.[98] Scholars in the field claimed that Japan, as a country endowed with a special relationship between land and people due to its unique geographical location, was in a fortunate position from which to overhaul the world order currently predicated upon the white-centric Westphalian system. The proponents of Greater East Asian Geopolitics also argued for a construction of a borderless and inclusive *Lebensraum* in East Asia, united by moral values arguably specific to Eastern philosophies, including altruism and filial piety.[99] Beginning around 1940, geographers striving to establish the field of Japanese Geopolitics (*Nihon Chiseigaku*) also promoted the view.[100] Under the Konoe government, their arguments legitimated the Greater East Asia Co-Prosperity Sphere, as well as the national land planning that aimed to materialize it.

In national land planning, the geopolitical idea of "race/ethnicity/people," encapsulated in the term *minzoku*, buttressed the population distribution policy.[101] Applying the metaphor of "blood and soil" that had been originally presented by Haushofer, Japanese geopolitical thinkers claimed a race (= "blood") to be a crucial geopolitical actor that maintained a mutually exclusive relationship with the land (= "soil"). Geopolitical thinker Ezawa Jōji equated the "land" with *kokudo*.[102] According to Ezawa, people would become *minzoku* by living in the *kokudo*. However, *kokudo* for *minzoku* did not represent a mere physical space but

[96] Atsuko Watanabe, *Japanese Geopolitics and the Western Imagination* (Cham: Palgrave Macmillan, 2019), 154.

[97] Haruna, *Jinkō, shigen, ryōdo*, 177–94, 215–61.

[98] Yamamuro, "Kokumin teikoku." For the works explaining how Pan-Asianism legitimated the imperial order in East Asia, see Cemil Aydin, "Japan's Pan-Asianism and the Legitimacy of Imperial World Order, 1931–1945," *The Asia-Pacific Journal: Japan Focus* 6, no. 3 (March 2008): 1–33; Eri Hotta, *Pan-Asianism and Japan's War 1931–1945* (New York: Palgrave Macmillan, 2007).

[99] Watanabe, *Japanese Geopolitics*, 187–218.

[100] Yoichi Shibata, *Teikoku nihon to chiseigaku: Ajia taiheiyō sensōki ni okeru chiri gakusha no shisō to jissen* (Osaka: Seibunsha, 2016).

[101] Morris-Suzuki, *Re-Inventing Japan*, 32.

[102] Watanabe, *Japanese Geopolitics*, 199–214.

the "basis of communal affects," the "externalization of the *minzoku*'s worldview and … collective experiences."[103] Ezawa claimed the relationship between *kokudo* and *minzoku* therefore was intimate and powerful precisely because the power to expand the *Lebensraum*'s boundary resided in the mutually affective interactions occurring within the relationship.[104] It was this geopolitical formulation of *minzoku* and *kokudo* that made "population distribution" an urgent matter for national land planning. Geographer Iwata Kōzō emphasized national land planning should be a plan to attain an "appropriate" (*tekisetsuna*) relationship between the people and *kokudo*.[105] The GPL incorporated this argument when it depicted population distribution. It suggested, with the "population distribution … according to regions," that the state would guarantee an "appropriate" relationship between the population and *kokudo* and fuel the limitless expansion of the self-sufficient Greater East Asia Co-Prosperity Sphere as *Lebensraum*.[106]

The geopolitical concerns over race addressed by the GPL made it apparent that the GPL's policy was indeed part and parcel of the general wartime population policy embodied in the GPP.[107] Both were premised on the idea that the population policy should facilitate the expansion and perpetuation of the Japanese population as the leading race in Asia. Both incorporated the logic ingrained in the Outline of a Basic National Policy, in particular, that the farming population as the source of Japan's "national/ethnic/racial power" (*minzokuryoku*) should be protected through governmental policies.[108] The GPL's and GPP's population policies were synonymous in so far as they both aimed to enhance Japan's "racial power."

At the same time, the GPL approached the subject matter differently from the GPP. In contrast to the GPP, whose population measures were informed primarily by the biological model of population-as-race, an economic and geopolitical rationale buttressed the GPL. Furthermore, in part because the GPL concentrated on resource distribution, the GPL population measures endorsed a much more structural understanding of population. Population seen in this light was built on the axis of quality

---

[103] Cited in Ibid., 201.

[104] Ibid., 200–201.

[105] Kōzō Iwata, "Kokka sōryokusen to sōgō kokudo keikaku," *Chiri kyōiku* 33, no. 5 (February 1941): 1–13.

[106] Yamamuro, "Kokumin teikoku," 60–69.

[107] Strictly speaking, the population policy delineated in the GPL was a constitutive element of the GPP, as the former was supposed to contribute to the general wartime population policy summarized in the GPP.

[108] Masayasu Kusunoki, "Jinkō mondai to kokudo keikaku," *Ikai jihō*, no. 2365 (January 1940): 12.

and quantity and made up of individuals with multiple social attributes. This notion of population further consolidated perspective on population quality and quantity that was different from the one that prevailed in the GPP. In contrast to the majority of the population quality measures in the GPP that focused on people's genetic, physical, and mental constitutions, the population quality described by the GPL was shaped by a balance in the composition of social segments that defined the population. Similarly, while population quantity applied to the eugenic and health measures in the GPP exhorted pronatalism as a strategy for population expansion, population quantity in the GPL, focusing on the ratio of the population numbers in relation to the places of domicile and work, implied that a rational coordination of people's location according to "regions" and "occupations" was the most effective means to increase a population's size.

This distinctive approach was most visible in the specific measures the MHW and CPB came up with to tackle the issue of declining fertility among the farming population. While both recognized that the fertility decline among the farmers could weaken the "racial power" of the Japanese, the countermeasures they came up with were different. The MHW, which was involved in drafting the GPP, recommended health and welfare measures (e.g., the prevention of infant diarrhea and the expansion of maternity facilities). In turn, the CPB policymakers, when drafting the GPL, endorsed a controlled migration of farmers between the countryside and cities as well as between the metropole and colonies. To support this measure, the CPB applied the theory established in the early 1930s by the renowned social scientist Ueda Teijirō, who showed a correlation between fertility decline and the movement of people from the countryside to the cities.[109] In concrete terms, this meant the CPB policymakers deliberated on the migration and work deployment measures to "secure a certain percentage of the population of [Japanese] farmers in farming," which should be based on the sum of the farmers in "Japan Proper" and those in the Japan-Manchuria-China Bloc.[110] After much discussion, they settled on 40 percent as the necessary figure.[111] The different solutions presented by the CPB and MHW in part mirrored the different perspectives of the GPP and GPL, and the different

---

[109] Takaoka, Sōryokusen taisei to "fukushi kokka," 108–9; Kingo Tamai and Naho Sugita, "Nihon ni okeru jinkō no 'ryō' 'shitsu' gainen to shakai seisaku no shiteki tenkai: Ueda Teijirō kara Minoguchi Tokijirō e," Keizaigaku zasshi 3, no. 1 (September 2015): 25–40.

[110] Dai Yonkai Jinkō Mondai Zenkoku Kyōgikai, "Kigen nisen roppyakunen kinen dai yonkai jinkō mondai zenkoku kyōgikai ni kafu seraretaru seifu shimon ni taisuru tōshin," November 15, 1940, PDFY09111747, Tachi Bunko.

[111] For the details about how the number was ascertained through the policy discussion, see Takaoka, Sōryokusen taisei to "fukushi kokka," 210–15.

perspectives within the emerging fields of policy-oriented health and the social sciences specializing in population matters.

As a policy initiative, wartime national land planning was a failure.[112] While the plan was initially moving ahead quickly, in the end, the CPB went only as far as to produce the Proposal for the Outline of the Yellow Sea and Bo Hai National Land Planning in March 1943 and to distribute the Rough Draft of the Proposal and the Proposal for the Outline of Central Planning to various government offices in October 1943 as a policy guideline. The CPB ceased to exist in November 1943, when it was absorbed by the new Ministry of Munitions. Reasons for the policy failure were multifaceted, but internal politics was among the most conspicuous. The struggle for leadership over the wartime economy led to the accusation that communism had infiltrated the CPB, which ultimately led to the arrest of three CPB research bureaucrats for violating the Peace Preservation Law.[113] National land planning was directly influenced by the dissolution of the CPB after this incident.

In contrast, the "population distribution" policy, originally presented in the GPL, survived in the GPP. The GPP depicted "population distribution" measures as an effective means of achieving one of its stated objectives, namely, "to appropriately deploy populations to secure Japanese leadership vis-à-vis other East Asian races." It then presented the following two as part of "measures for strengthening population quality":

(1) [The government should] plan for the rationalization of the population composition and distribution based on national land planning, in particular, [it should] plan for the dispersal of urban populations by means of evacuation. To achieve this, [the government should] do its utmost to disperse factories, schools, and other institutions in provinces.

(2) In view of the fact that the farming village is the most superior provider of military and work force, [the government should] do its utmost to maintain the population of farming communities of Japan Proper at a certain level and to keep 40% of the population of Japan Proper across Japan, Manchuria, and China for farming.

Later, population distribution comprised a core principle in the Population and Race Policy Accompanying the Construction of the Greater

---

[112] Bureaucrats were not very effective in coordinating economic activities despite the overbearing presence of bureaucratic rationality in rhetoric. Mimura, *Planning for Empire*, 138–69.

[113] Ibid.

East Asia. The proposal contained two clauses, which reiterated the items in the GPP.[114] Following the proposal, the agricultural policy established on June 24, 1942 by the Advisory Council for the Construction of Greater East Asia stated explicitly that at least 40% of the population of Japan Proper should be comprised of a farming population at all times. In the same month, the government issued a general plan for war mobilization, which in effect banned the building of new factories in the four major industrial areas within Japan Proper. As the war intensified, the population policy initially designed for national land planning became integrated into general war mobilization policies.[115]

The process of making a population policy for national land planning not only highlighted the centrality of the notion of population distribution in the wartime national policy, but it also underlined the increasingly important role scientific investigations played in policymaking: They were conducted by technical and research bureaucrats who specialized in population issues.

Research Bureaucrats for National Land Planning

Albeit a failure as a policy, national land planning highlighted a critical aspect of wartime statecraft: It relied on the brainpower and footwork of bureaucrats. At the top were elite bureaucrats such as Kishi Nobusuke (1896–1987), who drew up national land planning as the ultimate wartime mobilization scheme. They were "reform bureaucrats," a new generation of state administrators who were defined by their proactive and managerial function and engaged in coordinating work within production and strategic planning.[116] They belonged to a line of what historian Laura Hein called "reasonable men" with "powerful words," many of whom spent their formative years at the University of Tokyo where they were exposed to the Marxist social sciences and social movements of the 1910s and 1920s.[117] These reform bureaucrats thrived in the post-WWI industrial capitalist society, in which the technological advances engendering complex and expensive systems and the perceived decline in liberal capitalism led to an increased demand for a controlled economy and a strong state. During the war, they tried to materialize their technocratic vision of state organization through national land planning. To implement national land planning, they applied the political power derived from the close network

---

[114] Takaoka, "Senji no jinkō seisaku," 116–17.
[115] Ibid., 117–18.
[116] Mimura, *Planning for Empire*, 12.
[117] Hein, *Reasonable Men, Powerful Words*.

of politicians, officers in the Army and Navy, and industrialists and used the state's power to manage and coordinate industries both in the metropole and for the Japan–Manchuria–China Bloc.

Working side by side with these elite managerial technocrats were the technical bureaucrats called *gijutsu kanryō*, also known as *gikan*,[118] the title given to career bureaucrats who served the government through their medical, scientific or technical expertise. The category was established in the Meiji period, when the new government's commitment to building a modern nation-state with a technologically enhanced industry and military instigated the training of technically competent bureaucrats. However, technical bureaucrats remained a minority within the state bureaucracy. The demand for them increased in the 1930s, after a report by the Cabinet Bureau of Resources on the poor state of scientific research triggered the move to establish governmental and semigovernmental institutions dedicated to the promotion of science.[119] After the National General Mobilization Law, technical bureaucrats were sought after even more. Specifically, the Konoe cabinet mobilized them for its "New Order for Science and Technology" (*kagaku gijutsu shintaisei*), formulated in 1941 to establish the state coordination of scientific and technological activities for rational resource management in both the metropole and its colonies.[120] In the first half of the 1940s, technical bureaucrats strove to consolidate their status in state bureaucracy by stressing their role as the vanguards of cutting-edge techno-science and by promising Japanese Empire's self-sufficiency through their involvement in the scientific distribution of natural resources, labor, and capital.

Overlapping with technical bureaucrats was the category of bureaucrats specializing in fundamental research. Known by various titles, such as "research staff" (*kenkyūin*), "fieldworker" (*chōsain*), or "research bureaucrat" (*kenkyūkan* or *chōsakan*), these research bureaucrats, like technical bureaucrats, were civil servants with scientific expertise and often with a technocratic worldview. However, in contrast to technical bureaucrats, whose expertise was concentrated in highly technical and applied fields such as engineering and medicine, many research bureaucrats had backgrounds in social science.[121] Moreover, while technical bureaucrats were expected to stay in the same ministry

---

[118] Kashihara, *Meiji no gijutu kanryō*; Moore, *Constructing East Asia*; Mizuno, *Science for the Empire*, 19–68; Oyodo, *Gijutu kanryō no seiji sankaku*.
[119] Oyodo, *Gijutu kanryō no seiji sankaku*, 142–44.
[120] Ibid., 142–86.
[121] For economists mobilized for the war effort as research bureaucrats, see Makino, *Senjika no keizagakusha*; Hein, *Reasonable Men, Powerful Words*, 77–82.

for their entire career, research bureaucrats tended to be hired on a fixed-term basis for a specific project. Thus, many research bureaucrats moved between projects within the same ministry or worked on secondment for a fixed-term technical project organized by another ministry. Depending on the project, public intellectuals and scholars would also be recruited as temporary researchers serving for specific government ministries or other government organizations. In turn, some research bureaucrats, who were affiliated with external organizations accountable for official inquiries, were involved in drafting policy recommendations. In a nutshell, research bureaucrats contributed to state affairs by investigating issues specific to their areas of expertise, mainly for policymaking.[122]

In the national land planning population policy, Tachi Minoru (1906–72) took central stage as a research bureaucrat.[123] Tachi was a product of the University of Tokyo's social sciences that generated the "powerful men" mentioned above. He studied economics at the university between 1926 and 1929 and learned about population problems there.[124] Upon graduation, for over a year he continued his studies with Hijikata Seibi (1890–1975), the soon to be chair of the Department of Economics at the university. After serving as a commissioned editor for Nihon Hyōronsha publishing house for a little over three years, in 1933 Tachi was appointed by the recently founded Foundation-Institute for Research of Population Problems (IRPP) to serve as visiting staff. He then became a permanently based "research bureaucrat" (kenkyūkan) at the IPP when it was established in 1939. In 1942, he became the director of the Division of Population Policy Research of the MHW-RI Department of Population and Race. Shortly after the war, in May 1946, he became the director of the Department of General Affairs at the revived IPP, while still serving as a statistician for the MHW from 1947 on. From the time he assumed the directorship at the IPP in 1959 until his death in 1972, Tachi led population studies in Japan.

Prior to full-scale war with China, Tachi undertook research that became relevant to national land planning in later years. In the mid-1930s, he studied the Tohoku population as a member of the IRPP

---

[122] What has been described was a general tendency. Many research technocrats, in fact, shared the qualities ascribed to the gikan.

[123] Another important figure was Tokijirō Minoguchi (1905–83). For Minoguchi, see Tamai and Sugita, "Nihon niokeru jinkō no 'ryō' 'shitsu' gainen," 25–40.

[124] "Ko Tachi Minoru shochō no ryakureki to gyōseki," Jinkō mondai kenkyū 123 (July 1972): 44–62.

research staff, engaging with the question of population distribution.[125] During the war, Tachi collated and compared vital statistics in cities and rural areas, drawing on Mizushima Haruo's demographic work on six major cities (Tokyo, Osaka, Nagoya, Kyoto, Kobe, and Yokohama).[126] At the IPP, along with his colleague Ueda Masao, Tachi compiled standardized birth, death, and population growth rates in every prefecture.[127] These studies became vital for engaging with the most pressing question for wartime population distribution policy: What percentage of the "population of Japan Proper," especially the farmers, should be relocated without eroding the population's ability to expand?

Tachi began to express his opinions on population problems and policies publicly from the mid-1930s onward. He argued that the "population problem" had changed significantly in recent years. Amid the rise of racial struggles, it changed from being an "economic problem" (*keizai mondai*) to a "racial problem" (*minzoku mondai*).[128] Tachi then defined population as something that "organically composes a race or a nation, just like cells compose a biological body."[129] In the early 1940s, he suggested the Japanese "race population problem" (*minzoku jinkō mondai*), related to the construction of "new East Asia," was a problem of population quantity *and* quality, and policymakers should take into account the following elements of population: (1) as "military power," (2) as "members required for the industry," and (3) as "required for racial [growth]."[130] Tachi's understanding of population problems was eclectic, predicated on the idea of population as an organic body and a sociological entity. This multifarious formulation of population informed his engagement with population studies and policies in the late 1930s and early 1940s.

Through national land planning, research bureaucrats such as Tachi became a critical cog in the machine driving the Japanese state's effort to expand the boundary of its nation-state-empire. At the same time,

[125] Jinkō Mondai Kenkyūkai, "Tōhoku chihō jinkō nikansuru chōsa kōmoku," n.d., PDFY09110655, Tachi Bunko; Toshimichi Odauchi, Shigeki Masuda, and Minoru Tachi, "Tōhoku chihō jinkō nikansuru chōsa taiyō," March 7, 1935, PDFY09110671, Tachi Bunko; "Tōhoku chihō no jinkō nikansuru chōsa," March 1935, PDFY09110675, Tachi Bunko.

[126] Tachi, "Shōwa 12 nen zenkoku." For Mizushima Haruo's statistical activities during this period, see Kenichi Ohmi, "Mizushima Haruo ra no shokuminchi seimeihyō kenkyū ni miru dainiji sekai taisen zen, senchū no igaku kenkyū saikō," Nihon kenkō gakkai zasshi 86, no. 5 (September 2020): 209–223.

[127] Tachi and Ueda, "Taisho 9-nen, taisho 14-nen," 21–28.

[128] Tachi, "Wagakuni chihōbetsu jinkō zōshokuryoku."

[129] Ibid., 3.

[130] Ibid., 6.

their research helped to establish population studies as a policy science, despite the policy itself failing to materialize.[131]

Population Studies for National Land Planning

Since the 1910s, official investigation into demographic trends and problems was gradually becoming, more important in policymaking. After the war with China broke out, the government invested in population research more directly and created the IPP in 1939. In parallel, the Japan Society for the Promotion of Scientific Research (*Nihon Gakujutsu Shinkōkai*) launched the Eleventh Special Committee in October 1939, which additionally promoted population research as a branch of the committee's specialization, "racial science" (*minzoku kagaku*).[132] On June 19, 1941, experts in racial and population sciences founded the Japan National Racial Policy Study Group (*Nihon Minzoku Kokusaku Kenkyūkai*) as officially a nonofficial organization studying racial and population policies. The group acted as a policy think tank working alongside the MHW Population Bureau.[133] By the time the Konoe cabinet approved the GPL, population organizations both in and outside the government had long been fostering policy-oriented population research, creating foundations for population studies to thrive as a policy science.

Under these circumstances, population studies accountable for national land planning took place in three overlapping sites. The first was the CPB, charged with national land planning. Within the CPB, high-rank officials widely shared the idea that fundamental research, including demographic research, was a prerequisite for the government to actualize the vision of total state planning predicated upon a rational management of resources.[134] However, they also judged the existing research was organized haphazardly by different ministries, and this was preventing efficient planning.[135] Thus, in the wake of total war, the CPB

---

[131] For the idea of demography as a policy science, see, e.g., Dennis Hodgson, "Demography as Social Science and Policy Science," *Population and Development Review* 9, no. 1 (1983): 1–34.

[132] Takaoka, *Sōryokusen taisei to "fukushi kokka,"* 184–87. The English translation of Nihon Gakujutsu Shinkōkai today is "Japan Society for the Promotion of Science," but I use the translation adopted at that time.

[133] Takaoka, *Sōryokusen taisei to "fukushi kokka,"* 224.

[134] "Chōsa kenkyū renmei setsuritsu yōkō (Showa 17-nen 8-gatsu 28-nichi kakugi kettei)" in Kokudo Keikaku Kenkyūsho, *Kokudo keikaku kenkyūsho tsūshin*, no. 2, December 20, 1942, PDFY09111729, Tachi Bunko.

[135] Kikakuin Dai Ichibu, "Kokudo keikaku honkakuteki settei no hōhō ni tsuite (dai ichibunsatsu) hi (fu dai go jun gunjiteki kenchi ni motozuku kenkyū mondai," n.d., PDFY09111768, Tachi Bunko.

decided to coordinate the research by making it in-house. It requested the government approve the employment of additional research staff, which was realized in 1937 with the CPB hiring fourteen new personnel members.[136] Along with this, specifically for population research, the CPB created an independent Population Group within the First Department Third Section and recruited six research bureaucrats.[137] After the approval of the GPL, the CPB officially made scientific research a part of its administrative work for national land planning.[138]

The MHW IPP was the second site where national land planning population research was conducted. The research began in 1940, after the government published the Outline of a Basic National Policy. While drafting the GPP, IPP research staff collected data and examined subjects they saw as relevant to national land planning.[139] In October 1940, the IPP made a confidential report, the "General Plan for the Population Deployment as National Land Planning."[140] The content of the report fed into the policymaking process and was reflected in the GPP and Rough Draft of the Proposal Outlining Central Planning of 1943.

Finally, the abovementioned research organizations were where population studies related to national land planning thrived during this period. Among them, the IRPP occupied central stage. It hosted the Fourth National Conference on Population Problems between November 14 and 15, 1940 in response to the official inquiry made by Minister of Health and Welfare Kanemitsu Tsuneo.[141] Following the conference, on December 18, 1940, the IRPP set up the National Land

---

[136] JACAR, Ref.A14100539800 Kōbunruishū, dai 61-pen, Showa 12-nen, dai 4-kan, shokkan-2, kansei-2 (naikaku 2), Cabinet Privy Council Bureau of Law-making, Hiranuma Kiichiro, 1937, "Dajōruiten dai 2-hen, Meiji 4-nen k Meiji 10-nen, dai 85-kan, 1937," "Kikakuin chōsakan no tokubetsu nin'yō ni kansuru ken wo sadame," accessed July 29, 2019, www.digital.archives.go.jp/das/image/M0000000000001764902.

[137] Naikaku Kikakuin, "Kokudo keikaku jimu buntan ni kansuru ken," March 27, 1942, PDFY090226027, Tachi Bunko.

[138] Kikakuin Dai Ichibu, "Kokudo keikaku honkakuteki settei no hōhō ni tsuite."

[139] See Jinkō Mondai Kenkyūsho, "Kokudo keikaku ni kanshi jinkō seisakujō kōryo subeki shutaru jikō sankō shiryō (gokuhi 6-bu nouchi dai 2-gō)," June 24, 1942, PDFY090212096, Tachi Bunko; "Daitōa kensetsu shingikai daisanbukai tōshin'an setsumei shiryō no uchi sangyōbetsu oyobi chiikibetsu haichi ni okeru jinkō baransu (shi'an) (gokuhi 100-bu no uchi dai 12-gō)," April 13, 1942, PDFY090212097, Tachi Bunko; "Daitōa kensetsu shingikai daisanbukai tōshin'an setsumei shiryō no uchi wagakuni jinkō no toshi shūchū to tohi zōshokuryoku (hi)," April 11, 1942, PDFY090212098, Tachi Bunko.

[140] Jinkō Mondai Kenkyūsho, "Kokudo keikaku toshiteno jinkō haichi keikaku yōkō'an Showa 15-nen 10-gatsu hi," October 1940, PDFY09111757, Tachi Bunko.

[141] Tsuneo Kanemitsu, "(Shimon) Kigen nisen roppyakunen kinen daiyonkai jinkō mondai zenkoku kyōgikai. Kokudo keikakujō jinkō seisaku no kenchi yori kōryo

Planning Section Group within its National Population Policy Committee to make the national land planning population research more permanently based in the organization.[142] Comprised of members from the military, academia, and government offices, and headed by Director of the IRPP Sasaki Yukitada, the section group was a high priority within the IRPP.[143]

Population studies conducted in these sites was integral to policymaking.[144] The research design drawn up by the CPB for its Population Group, for instance, confirmed population studies' utility for national land planning. The topics the CPB assigned to the group included "relationship between supply and demand in populations," "regional distribution of physical strengths according to racial groups," and "the effect of population concentration and movement (organized by the place of origin and the destination)," which clearly resonated with the demographic goals of national land planning. In turn, these goals directly shaped the objectives of the population research conducted under the aegis of the CPB.[145] For instance, to correspond with national land planning's goal for "the optimal location of the Japanese race vis-à-vis other races across the Greater East Asia Co-Prosperity Sphere," the CPB stated that its population research, "from the perspective of population expansion and welfare," aimed to "adequately deploy the population of Japan Proper according to occupations and from the viewpoint of national missions, such as guiding various races in East Asia, promoting industries, development of resources, and the defense of the Greater East Asia Co-Prosperity Sphere."[146] Population research clearly interacted with the CPB's administrative and policymaking activities in the state planning scheme.

Behind this research arrangement was a trust in population studies within the government administration. The CPB valued the demographic

---

subeki ten ni tsuki sono kai no iken wo tou" November 14, 1940, PDFY09111748, Tachi Bunko; Tachi Minoru, "Dai yonkai jinkō mondai zenkoku kyōgikai ni kafu seraretaru seifu shimon ni taisuru tōshin'an yōkō (an) oyobi koreni taisuru kisō iin ikensho," October 10, 1940, PDFY09111739, Tachi Bunko; Dai Yonkai Jinkō Mondai Zenkoku Kyōgikai, "Kigen nisen roppyakunen kinen dai yonkai jinkō mondai zenkoku kyōgikai ni."

[142] Jinkō Mondai Kenkyūkai Jinkō Kokusaku Iinkai Kokudo Keikaku Bunkakai, "Jinkō kokusaku iinkai kokudo keikaku bunkakai (dai ikkai kaigō)."

[143] Ibid.

[144] "Chōsa kenkyū renmei setsuritsu yōkō." n.d. c.1940, PDFY090226030, 1.

[145] Naikaku Kikakuin, "Kokudo keikaku jimu buntan ni kansuru ken," March 27, 1942, PDFY090226027, Tachi Bunko.

[146] Kikakuin Daiichibu, "Kokudo keikaku honkakuteki settei no hōhō ni tsuite," n.d. c. 1940, PDFY09111769.

work because it firmly believed that current population research was fully equipped to provide what Sheila Jasanoff once called the "serviceable truth," the knowledge that "satisfies tests of scientific acceptability and supports reasoned decision making."[147] Concretely speaking, the CPB officers were convinced that the demographic knowledge about the population composition produced by population research would effectively assist the government's decisions regarding a coordinated distribution of the population of Japan Proper, because the idea that mathematical calculation and analysis would reveal the objective truth about the nation had by then reached a firm consensus in the scientific and policy fields. They further believed that population distribution based on this demographic knowledge would help the Japanese to assume the leading role in the construction of a "new order" in East Asia, first by fostering a rational arrangement of economic activities in the Japan–Manchuria– China Bloc and second by ensuring the construction of a hierarchical power structure between the Japanese and other races within the Greater East Asia Co-Prosperity Sphere.[148] For the CPB officers, demographic knowledge was key to the political maneuverings of the Japanese state and empire at war.

It was in this environment that Tachi thrived as a research bureaucrat engaged in national land planning population studies. Quickly building his reputation within the government and among his colleagues in the 1930s, Tachi was involved in national land planning population work in the CPB, IPP, and IRPP. At the CPB, Tachi was employed on a temporary basis to work in the Population Group and to assume a supervisory role for research on "the form of the decentralization of manufacturing industries and the limits of the urban population."[149] At the IPP, he was involved in drafting the "General Plan for the Population Deployment as National Land Planning." While there, he was also a member of the IRPP National Land Planning Section Group and drafted a policy recommendation document for the Fourth National Conference on Population Problems.[150] In the early 1940s, Tachi established his name as an

[147] Sheila Jasanoff, *The Fifth Branch: Science Advisors as Policymakers* (London and Cambridge, MA: Harvard University Press, 1990), 250.
[148] Jinkō Mondai Kenkyūsho, "Kokudo keikaku toshite no jinkō haichi (yohō) showa 15-nen 8-gatsu"; Kōseishō, "Rōmu dōtai chōsa teiyō," December 1939, PDFY090226050, Tachi Bunko.
[149] Naikaku Kikakuin, "Kokudo keikaku jimu buntan."
[150] See Tachi, "Dai yonkai jinkō mondai zenkoku kyōgikai ni"; Mondai Kenkyūkai Jinkō Kokusaku Iinkai Kokudo Keikaku Bunkakai, "Jinkō kokusaku iinkai kokudo keikaku bunkakai (dai ikkai kaigō)," December 18, 1940, PDFY09111751, Tachi Bunko.

expert in national land planning population research by moving agilely between the three institutions and between his roles as bureaucrat, population expert, and policy advisor.

Tachi's population research for national land planning was motivated by his desire to come up with new planning schemes in alignment with the demands of the "new order" movement and therefore entirely different from the planning work of earlier eras. First, he claimed "comprehensive migration" in the GPL was not the same as the existing migration scheme, arguing that the latter, aiming to relieve population pressure, was based on a Malthusian, "liberalist concept."[151] In contrast, "comprehensive migration" was combined with a controlled economy and a migration program that engaged with geopolitical concerns. Second, the government should consider forming "blocs" in the process of implementing population deployment "by regions." However, unlike an earlier idea, the "blocs" in national land planning should not "foster a mechanical formation of a population group." Instead, they should form *"Lebensraum."*[152] Third, the "distribution of populations divided by occupational categories" should not be equated with a preexisting work placement scheme.[153] It should raise "industrial productivity," but it should not be done at the cost of "consuming the human resource." For this reason, it should be complemented with welfare measures.[154] Fourth, the "dispersal of industries" – "dispersing" factories around the nation to "adjust" the population ratio between cities and the countryside – should be conducted with caution.[155] Fifth and finally, policymakers should factor in the "human aspect," which had been neglected in the existing planning schemes from which national land planning evolved. For this reason, they should consider building cultural and welfare institutions as a population measure for national land planning. This was important from the viewpoint of racial prosperity.[156] As Tachi saw it, population measures for national land planning were a "new order" planning policy because they addressed geopolitical and economic concerns as combined factors and maximized the population's potential in the three domains he elaborated on above – military power, labor force, and the source of "racial power." This was the reason why they were in no way the same as prewar liberalist population measures.

---

[151] Tachi, "Jinkō seisaku no tachiba yori mitaru kokudo keikaku", 94.
[152] Ibid., 101.
[153] Ibid., 102–7.
[154] Ibid., 102.
[155] Ibid., 107–8.
[156] Ibid., 108–12.

Tachi's population work based on this philosophy was wide ranging. He compiled vital statistics and analyzed the patterns of child mortality, age, and gender composition.[157] He also compiled materials indicating the numbers for the "working populations of Japan Proper" in commerce, heavy industries, ore industries, fisheries, transportation, civil service, freelance work, and housemaid and butler work.[158] Furthermore, reflecting the centrality of the metropole's farming population for wartime population policy, he also engaged with the question of how the countryside could act as, what he called, the "imaginary hinterland," a land supplying populations to cities without destroying its own population's capacity to grow.[159] At the same time, Tachi tried to collect demographic materials concerning other strategically important population groups (e.g., Manchurians, Koreans, and the Taiwanese).[160]

Tachi's work directly contributed to national land planning. The results from the research on the distribution of people in the countryside versus cities were directly useful for the government when trying to decide to what extent it should work toward "dispersing populations of overextended cities ... in relation to the dispersion of industries into regions" and "develop the city in a way that it can retain reproductive and growth power" and "prevent industrialization from lowering the population's power in regional small- to mid-size cities."[161] He also used vital statistics to calculate the "excess labor" among the women of Japan Proper and the maximum number of the women mobilizable for war industries.[162]

[157] Jinkō Mondai Kenyūsho, "Showa 15-nen kokusei chōsa kekka ni motozuku danjo-kakusaibetsu jinkō no suikei," n.d. c.1940, PDFB5041EST40A, Tachi Bunko.

[158] Jinkō Mondai Kenkyūsho, "Kokudo keikaku toshite no jinkō haichi (yohō) Shōwa 15-nen 8-gatsu." Also see Kōseishō, "Rōmu dōtai chōsa teiyō"; "Tōhoku rokken sangyō daibunrui betsu yūgyō jinkō senbunhi saikō chiiki," n.d. PDFY09110679, Tachi Bunko.

[159] Minoru Tachi and Masao Ueda, "Toshi jinkō hokyūgen toshite no 'kasōteki haichi' no kettei ni kansuru ichi kōsatsu," *Jinkō mondai kenkyū* 2, no. 2 (February 1941): 33–43; Minoru Tachi, "Jinkō saibunpai keikaku no kiso toshite mitaru jinkō zōshokuryoku no chiikiteki tokusei," *Jinkō mondai kenkyū* 3, no. 2 (February 1942): 1–40.

[160] "1. Manshū teikoku kokusekibetsu jinkō shirabe [hoka]," 1937, PDFY090803051, Tachi Bunko; "[1] 'Dainiji taisen shuyō kōsenkoku jinkō kōseizu' setsumei," n.d. c.1942, PDFY090803053, Tachi Bunko; Minoru Tachi, "Sorenpō jinkō ni kansuru shuyō tōkei tekiyō [gokuhi]," May 1, 1945 PDFY09110621, Tachi Bunko; "1. Manshū teikoku kokusekibetsu jinkō shirabe (Shōwa 11-nen matsu)," 1937, PDFY09111703, Tachi Bunko; Jinkō Mondai Kenkyūkai, "Waga kuni jinkō ni kansuru shuyō tōkei bassui," August 1938, PDFY090803056, Tachi Bunko.

[161] Tachi, "Shōwa 12-nen zenkoku"; "Dai yonkai jinkō mondai zenkoku kyōgikai ni."

[162] Minoru Tachi, "Joshi dōin ni kansuru shiryō," April 27, 1944, PDFY09110638, Tachi Bunko.

Vital statistics was also used to ascertain how many people within "Japan Proper" should be relocated between 1943 and 1960 (in the two periods divided by the year 1950).[163] The data calculated from these works was used to produce the "General Plan for the Population Deployment as National Land Planning." The document estimated that 85,579,000 should be the minimum population required for Japan Proper in 1950 "for the development of the Japanese race." Of these, 49,074,000 should be of a "productive age" and at least 35,269,000 workers should be strategically deployed to various industries within Japan Proper. In addition, a minimum of 19,686,000 Japanese people should be based in the Greater East Asia Co-Prosperity Sphere, of which 8,111,000 should be in "China Proper" (*shina hondo*), 6,885,000 in Manchuria, 2,200,000 in Korea, and 2,390,000 in the area covering French Indochina, Thailand, Dutch East India, and the Philippines.[164] Later, in 1942, for the work the IPP conducted in response to the inquiry made by the Third Section of the Advisory Council for the Construction of Greater East Asia, Tachi recalculated figures for a strategic distribution of the population of Japan Proper. The document concluded that a minimum of 9,410,000 additional people in "Japan Proper" would need to be relocated between 1940 and 1950 to the area covering Korea, Taiwan, Manchuria, "China Proper," French Indochina, Thailand, Burma, the Philippines, the Dutch East Indies, Australia, and New Zealand, and of those, 6,330,000 should be dedicated to agriculture.[165] The documents became the basis for the recommendations made in the aforementioned Rough Draft of the Proposal Outlining Central Planning of 1943.[166]

---

[163] Tachi, "'Toshi haichi nikansuru jinkō shisakuteki mokuhyō' hōkoku shiryō," June 19, 1943, PDFY090226036, Tachi Bunko.

[164] Jinkō Mondai Kenkyūsho, "Kokudo keikaku toshite no jinkō haichi keikaku yōkō'an Showa 15-nen 10-gatsu hi," October 1940, PDFY09111757, Tachi Bunko. For context, see Satoshi Nakano, *Japan's Colonial Moment in Southeast Asia 1942–1945: The Occupiers' Experience* (Abingdon: Routledge, 2019).

[165] Jinkō Mondai Kenkyūsho, "Daitōa kensetsu shingikai daisanbukai tōshin'an setsumei shiryō no uchi sangyōbetsu oyobi chiikibetsu haichi ni okeru jinkō baransu (shi'an) (gokuhi 100-bu no uchi dai 12-gō) April 13, 1942, PDFY090212097, Tachi Bunko." As Mariko Tamanoi has pointed out, this type of source should be read bearing in mind the politics and human agency engrained in the practice of classifying the ethnonational categories. Mariko Tamanoi, "Knowledge, Power, and Racial Classifications: The 'Japanese' in 'Manchuria,'" *The Journal of Asian Studies* 59 (May 2000): 248–76.

[166] Jinkō Mondai Kenkyūsho, "Kokudo keikaku toshite no jinkō haichi keikaku yōkō'an Showa 15-nen 10-gatsu hi."

As such, population studies conducted by research bureaucrats quickly became institutionalized as the war progressed. Reflecting the government's trust in population research, the government employed population experts for national land planning and assigned them to provide data on population distribution, which was strategically important for the execution of the wartime national policy. Tachi, as one of the most prominent research bureaucrats in this context, duly responded to the role ascribed to him and produced demographic knowledge that policymakers could utilize readily. The total war fostered a specific form of population studies conducted by research bureaucrats.

However, the political influence on population studies did not end there. National policy also shaped the studies profoundly by exhorting researchers to focus on certain demographic subjects that were particularly pertinent to Japan's political struggles. In turn, by orienting itself to the policy debate, demographic studies crystallized the racial and gender stereotyping within the characterization of the target population groups in the debate. Consequently, the demographic subjects appearing in the population research were depicted in gendered and racialized terms.

Gendered and Racialized Demographic Subjects

The population research Tachi was involved in was significant, not only because it provided applicable demographic data for policymaking, but also because it elaborated on the demographic subjects who were perceived as threats to the Konoe cabinet's "sacred mission" to construct a "new order" in East Asia. In the context of national land planning, in which the "sacred mission" was depicted in terms of ethnonational struggles, the identified demographic subjects were also depicted as racialized national groups.[167]

Among them were the populations of western countries vying for power in Asia, in particular the Soviet Union (USSR). Caricaturing the population as "a basis of national power," population experts showed interest in the Soviet population, especially after the Nomonhan Incident of 1939, in which the devastating defeat in the military confrontation with the USSR dealt the Japanese Army a serious blow. They were particularly concerned that the Soviets would prevent Japan from

---

[167] Dower, *War without Mercy*, 263–65. A more recent work sheds light on the experience of mixed-race people, which has hitherto been hidden due to the focus on the narrative of "race war." W. Puck Brecher, "Euraasians and Racial Capital in a 'Race War,'" in *Defamiliarizing Japan's Asia-Pacific War*, eds. W. Puck Brecher and Michael W. Myers (Honolulu: University of Hawai'i Press, 2019), 207–26.

completing the "sacred mission" with USSR's expansive landmass and population. Koya Yoshio, Tachi's colleague at the MHW-RI and one of the most influential technical bureaucrats specializing in racial science (see Chapter 6), claimed the USSR was formidable not only because of its vast landmass but also because of its demographic composition, which was biased toward children and youth thanks to high fertility. In contrast, the Japanese population was meager in size and getting old due to the fertility decline. Comparing the demographic trend of the two countries, Koya warned that the "racially young" Soviets would soon take over Japan's position as the ruler of Asia.[168] As Koya saw it, fecundity represented racial vitality and political force, thus the "racially younger" and larger populations of the neighboring countries in Asia, enabled by fecundity, necessarily jeopardized the Japanese influence in Asia.

Population research internalized this logic as it collected the Soviet demographic data in the early 1940s. The MHW-RI Department of Population and Race compiled data about Soviet statistics on birth, death, and natural population growth rates and on the population composition by class, age, and occupation, along with those of other western countries participating in the current war (the United States, England, Germany, and Italy).[169] In 1943, Tachi, as a member of the department's research staff, prepared confidential notes showing estimates of the recent population trends in the USSR. For the work, he used the census data from 1897 – since the time of Imperial Russia – population estimates calculated by the USSR and the South Manchurian Railway, and vital statistics produced by the scholar Tachi called "Kuczynski."[170]

---

[168] Yoshio Koya, *Kokudo, jinkō, ketsueki* (Asahi Shinbunsha, 1941), 218; "Kokudo keikaku to jinteki shigen," *Ishi kōron bessatsu*, no. 1475 (November 2, 1940), PDFY09111808, Tachi Bunko.

[169] Kōseishō Kenkyūsho Jinkō Minzokubu, "Shuyō kōsenkoku jinkō tōkei tekiyō," Jinkō mondai kenkyū shiryō (Kōseishō Kenkyūsho Jinkō Minzokubu, May 10, 1943), PDFY090212071, Tachi Bunko.

[170] Minoru Tachi, "Sovietto [*sic*] renpō saikin no jinkō nikansuru suikei shiryō (miteikō) (ichi) (hi)," November 16, 1943, PDFY09110603, Tachi Bunko; Minoru Tachi, "Sovietto [*sic*] renpō saikin no jinkō nikansuru suikei shiryō (miteikō) (ni no tsuika) danjo nenreibetsu jinkō kōsei oyobi nenreibetsu zettai shōmō heiryoku no suikei (ni) (hi)," November 25, 1943, PDFY09110604, Tachi Bunko; Minoru Tachi, "Sovietto [*sic*] renpō saikin no jinkō nikansuru suikei shiryō (miteikō) (ni) danjo nenreibetsu jinkō kōsei oyobi nenreibetsu zettai shōmō heiryoku no suikei (hi)," November 20, 1943, PDFY09110605, Tachi Bunko; Minoru Tachi, "Sovietto [*sic*] renpō saikin no jinkō nikansuru suikei shiryō (miteikō) (san) 1939-nen hatsu niokeru danjo shakai kaikyūbetsu sangyōbetsu jinkō no suikei (hi)," November 26, 1943, PDFY09110606, Tachi Bunko; Minoru Tachi, "Sovietto [*sic*] renpō saikin no jinkō nikansuru suikei shiryō (miteikō) (san no kaitei) 1939-nen hatsu niokeru danjo shakai kaikyūbetsu

Tachi's findings about the current state of the Soviet demography were more modest than the alarmist view presented by Koya and other colleagues earlier in the decade. He estimated that the Soviet population had actually decreased from 173,549,000 in 1940 to 171,812,000 in 1943, and would even further decrease to 162,898,000 if soldiers' deaths from the current war were counted.[171] He attributed the population contraction to the drastic fertility decline in the 1930s, which occurred despite pronatalist policies.[172] Tachi also carried out a covert study on the population capable of engaging in (re)productive activities and concluded that "the capacity of the USSR to mobilize human resources has reached a limitation. [Yet] it would not be impossible to expand military mobilization [therefore] we should not see this as a considerable obstacle for [the Soviets] securing a production force."[173] Compared to the rhetoric of racial scientists that magnified the racial power of the Soviets, Tachi's evaluation of the Soviet demography was soberer. However, Tachi's study also implied that the Soviets were still capable of undermining Japan's "sacred mission." In this way, Tachi's population research consolidated the image of the Soviets as a potential threat to Japan's political project in Asia.

If the Soviets were perceived as an external threat, Koreans were depicted as a demographic subject destabilizing the Japanese endeavor from within. From the onset of the Japanese annexation of Korea, Japanese-language literary and medical writings pathologized Koreans as prone to crime and depicted this "proclivity" as a factor that would undermine Japanese colonial rule in Korea.[174] This view, informed by racism, continued into the 1920s within the discussion of "overpopulation." Confronted with an independent movement, Japanese colonial

sangyōbetsu jinkō no suikei (hi)," December 18, 1943, PDFY09110607, Tachi Bunko; Minoru Tachi, "Sovietto [sic] renpō saikin no jinkō nikansuru suikei shiryō (miteikō) (san no kaitei) no tsuiho (hi)," December 19, 1943, PDFY09110608, Tachi Bunko; Minoru Tachi, "Sorenpō genzai niokeru jinteki dōin jōkyō no hantei nikansuru shiryō (miteikō) (Tachi kenkyūkan shaken) (gokuhi)," December 10, 1943, PDFY09110609, Tachi Bunko. Due to the lack of materials, I was unable to confirm who exactly Kuczynski was, but it was most likely Robert René Kuczynski (1876–1947), who was a renowned demographer at the time.

[171] See appendix in Tachi, "Sovietto renpō saikin no jinkō nikansuru suikei shiryō (miteikō) (ichi) (hi)."

[172] Ibid., 10–13. David L. Hoffmann, "Mothers in the Motherland: Stalinist Pronatalism in Its Pan-European Context," *Journal of Social History* 34, no. 1 (September 2000): 44.

[173] Tachi, "Sorenpō genzai niokeru jinteki dōin jōkyō no hantei nikansuru shiryō (miteikō) (Tachi kenkyūkan shaken) (gokuhi)," 8.

[174] See Jin-kyung Park, "Husband Murder as the 'Sickness' of Korea: Carceral Gynecology, Race, and Tradition in Colonial Korea, 1926–1932," *Journal of Women's History* 25, no. 3 (2013): 116–40.

officers viewed "overpopulation" as a potential catalyst for further political tension that could jeopardize Japan's colonial rule in the peninsula. At the same time, in the context of the 1920s, in which Japan itself had a growing population and was relying more and more on rice imported from Korea, the population growth in Korea heralded a future crisis in the relationship between the Government-General of Korea and the metropolitan government.[175] Japanese colonial officials thought the expanding Korean population would erode their effort to build a sustainable relationship between colonial Korea and the metropole.[176]

However, in wartime, the official attitude toward Koreans changed slightly. The demand for "human resource" and the rhetoric of racial harmony among the five races that was propagated by the wartime Japanese government in support of Konoe's "new order" movement served to shift Japanese views on Koreans from exclusionary racism to what historian Takashi Fujitani once called "polite racism," a subtle form of discrimination that is tactfully cloaked in a narrative of equality and inclusion.[177] For instance, Korean males were now allowed to vote and conscripted to serve the Japanese state – as Japanese subjects – as soldiers in the name of *naisen ittai* (harmony between the Japanese and Koreans).[178] At the same time, in the *koseki*, they remained *gaichijin*, "people of outer Japan."[179] As historian Oguma Eiji once argued, the kind of racism fostered by the imperative of the war turned Koreans into "a national resource as a 'Japanese,' but at the same time, not 'Japanese.'"[180]

Though part of the war mobilization effort, the policy debate and research on population that was accountable for national land planning was surprisingly mute when it came to polite racism's inclusion or equality logic. First, reflecting the legal definition of Koreans as belonging to "outer Japan," the population debate and research regarded the Korean population as a separate entity from the "population of Japan Proper."[181] Furthermore, the idea of a hierarchical difference between the Japanese

---

[175] Park, "Interrogating the 'Population Problem' of the Non-Western Empire."

[176] On the Japanese colonial rulers based in Korea, see Jun Uchida, *Brokers of Empire: Japanese Settler Colonialism in Korea, 1876–1945* (Cambridge, MA: Harvard University Press, 2011).

[177] Takashi Fujitani, *Race for Empire Koreans as Japanese and Japanese as Americans during World War II* (Berkeley: University of California Press, 2011), 40.

[178] Ibid., 40–75; Oguma, *"Nihonjin" no kyōkai*, 417–57; Makiko Okamoto, "Ajia taiheiyō sensō makki niokeru chōsenjin, taiwanjin sanseiken mondai," *Nihonshi kenkyū*, no. 401 (January 1996): 53–67.

[179] Endo, *Koseki to kokuseki*, 188–215.

[180] Oguma, *"Nihonjin" no kyōkai*, 457.

[181] Jinkō Mondai Kenkyūsho, "Daitōa kensetsu shingikai daisanbukai tōshin'an setsumei shiryō no uchi sangyōbetsu oyobi chiikibetsu haichi ni okeru jinkō baransu (shi'an) (gokuhi 100-bu no uchi dai 12-gō)."

and Koreans shaped the research agenda more strongly than the rhetoric of equality. The secretary of the CPB's First Division, Konuki Hiroshi, bluntly stated at the first meeting of the National Population Policy Committee on December 18, 1940 that the population problem under national land planning was a "problem of the Korean people" – specifically the question of how the Japanese could manage the uncontrollably fecund and "inferior" Koreans.[182] Konuki argued that population research based on this point should assist the policy work that "reappraises *hakkō ichiu*," the political slogan propagated under the Konoe cabinet that endorsed Japanese rule rather than egalitarian brotherhood in Asia.[183] Following Konuki's comments, the CPB demanded that, from the viewpoint of military affairs, its population research should respond to concrete questions about how to allocate population groups for Japanese imperial rule and for the "new order" in East Asia, and Koreans featured prominently in these questions.[184] Taken together, population research was expected to recognize the line between the Koreans and the "population of Japan Proper" when estimating population distribution figures. This expectation was clearly premised on the idea of racial differences between the Japanese colonial ruler and its colonial subjects.

Significantly, the Koreans who appeared in these questions referred specifically to unassimilated Koreans.[185] In the official discussion, unassimilated Koreans were described in condescending ways, as uncivilized, antisocial, criminal, promiscuous, and, last but not least, fecund. So, when the CPB prepared a research agenda for its population studies, it also requested research staff to address these questions: "How should the government respond to the growing population of the unassimilated Koreans ... in the event that the population of 'Japan Proper' declined due to the effects of war?" "How much should the government allow the migration of Korean laborers in Japan, given that Koreans are known for their 'custom of antisociality and miscegenation,' 'criminality,' and 'the danger of their lowering the living standard'?" "How could the policy avoid racial frictions between Koreans and the local populations in the event of Korean migration to Manchuria and China?"[186] For the

[182] Jinkō Mondai Kenkyūkai Jinkō Kokusaku Iinkai Kokudo Keikaku Bunkakai, "Jinkō kokusaku iinkai kokudo keikaku bunkakai (dai ikkai kaigō)."

[183] Ibid.

[184] Kikakuin Dai Ichibu, "Kokudo keikaku honkakuteki settei no hōhō ni tsuite," 23–25.

[185] Michael A. Weiner, *Race and Migration in Imperial Japan*, Sheffield Centre for Japanese Studies/Routledge Series (London: Routledge, 1994).

[186] Kikakuin Dai Ichibu, "Kokudo keikaku honkakuteki settei no hōhō ni tsuite," 23–25.

population research accountable for national land planning, the rhetoric of vulgar racism that prevailed in the earlier decades shaped the questions.

The CPB formed these questions in the specific context of the 1930s and early 1940s in which the intensified mobilization of Koreans for Japan–Manchuria–China Bloc fueled official anxiety over the allegedly indolent yet recalcitrant unassimilated Koreans. In the metropole, the civic effort to assimilate Koreans surged in the mid-1920s, when thousands of Koreans were massacred by the police following the Great Kanto Earthquake of 1923.[187] Japanese officials provided support for the effort and during the mid-1930s nationalized it. In December 1940, as the Korean workers were recruited by coercion, the official assimilation effort was further systematized with the launch of the Central Harmonization Association (*Chūō Kyōwakai*).[188] However, the Korean community resisted this by protesting.[189] In turn, implicating the protests with labor and communist activism, Japanese officials understood that it could disrupt the controlled economy.[190] At the same time, for the officials who were cognizant of the declining fertility among the "population of Japan Proper," the image of unassimilated yet fecund Koreans signified Japan's weakened political leverage it could exploit to rule Asia. Under the circumstance, the "Korean problem" was translated in population research as a question of how to accurately calculate the ratio of the expanding Korean population to the Japanese in order to help diffuse political tensions.

Though himself not so central to the population research specifically tackling the "Korean problem," Tachi was well positioned for such research. Before joining the IPP, between December 1938 and April 1939, Tachi had a four-month stint as a temporary editor at the Central Harmonization Association. Perhaps because of this work experience, he had access to confidential data about the distribution of Korean populations and crimes committed by Koreans from the metropole.[191]

---

[187] Weiner, *Race and Migration*, 156.
[188] *Chūō Kyōwakai* promoted the cultural and racial assimilation of Koreans and policed dangerous ideologies among Koreans in the metropole. Weiner, *Race and Migration*, 154–65. For the response to the assimilation policy in Korea, see, e.g., Uchida, *Brokers of Empire*, 353–93.
[189] Weiner, *Race and Migration*, 165–86.
[190] Brandon Palmer, *Fighting for the Enemy: Koreans in Japan's War, 1937–1945* (Washington: University of Washington Press, 2013), 4.
[191] E.g., "Naichi zaijū chōsenjin mondai gaikyō hi," April 28, 1939, PDFY100916017, Tachi Bunko; "Chōsenjin bunpu zu (Showa 13-nen matsu genzai 799.865 nin) hi," n.d. c.1938–39, PDFY100916015, Tachi Bunko.

The documents Tachi gathered from the association consolidated the racist attitude toward Koreans that was dominate in the policy discussion on population. The demographic research on the "problem of Koreans living in Japan Proper" – which was submitted in April 1939 as a confidential document to a meeting called the Round-Table Meeting of the Board of Trustees for Population Problems – pointed out the surge in the number of Koreans in the metropole in the early Showa period (1926–89), in particular after 1932: from 143,000 in 1926 to 799,878 in 1938.[192] It also stated that there was a "significantly higher crime rate" of Koreans compared to Japanese by showing the crime rate of 4.8% among the Koreans and 2.2% among the people of Japan Proper and by listing that 10,699, 6,290, 3,003, and 1,037 Koreans were arrested in 1938 for gambling, theft, assault, and fraud, respectively.[193] The document mentioned that the majority of Koreans were originally "illiterate" but "many have become educated and cultivated" in recent years. But, instead of interpreting this positively, the document warned that this trend, coupled with the decreasing number of Japanese workers in the metropole due to military conscription, might lead to a "serious antagonism between the Japanese and Koreans at work."[194] Following the narrative in the policy discussion, the document Tachi collected also portrayed Koreans as fecund, criminal, and politically suspect.

However, for the IPP, with which Tachi was primarily affiliated, the data he collected from the Central Harmonization Association showing the criminality of the Koreans was less valuable than the data on the Japanese in Korea. In the study the IPP carried out in 1940, which became the basis for the aforementioned "General Plan for the Population Deployment as National Land Planning," the IPP studied the deployment, composition, and physical quality of the population of Japan Proper in Korea.[195] Based on this study, it concluded that at least 10% of the total population on the Korean Peninsula should be colonist from the population of Japan Proper.[196] Yet, the fact that the IPP did not use the data from the Central Harmonization Association did not necessarily mean the IPP study was devoid of racism. In fact, to the contrary, the same condescending view on the Korean

---

[192] "Naichi zaijū chōsenjin mondai gaikyō hi," 1.
[193] Ibid., 5–6.
[194] Ibid., 3–4.
[195] Jinkō Mondai Kenkyūsho, "Kokudo keikaku toshite no jinkō haichi keikaku yōkō'an Showa 15-nen 10-gatsu hi."
[196] Ibid., 13.

people dominated the IPP policy document. To justify the 10% mark, the IPP contended that Koreans "grow expansively even though their quality is inferior," thus a certain ratio of the Japanese was required to control the Korean population.[197] Though not as explicit as in the policy debate, the population research conducted for national land planning also incorporated the racist characterization of Koreans as a dangerous demographic subject in relation to the Japanese political endeavor.

While population research adopted a racial category to examine the level of threat certain demographic subjects posed to Japanese imperial rule, it used a gender classification to examine how the government could further strengthen national power. Gender classification – the analysis of demographic trends by categorizing a population by sex – was by then an established practice in population studies. Similar to age, population scholars thought the ratio of men to women would reveal fundamental qualities of a given society and simultaneously influence the population composition profoundly. With this premise in mind, the IPP readily sorted the population data of Japan Proper by sex to ascertain the most rational way to distribute the population.[198]

Still, in the context of total war, in which mobilization was categorically a gendered affair, studying the population through the category of sex was more than simply routine work. It also embodied a gender ideology that shaped Japan's war effort: the ideology that magnified men's contribution to the warring state through their productive and military prowess and women's through their reproductive and assistive functions.[199] Thus, to analyze the male population, the IPP considered the men's roles primarily as soldiers and workers (including farmers) and used the framework of the "population of productive age" (*seisan nenrei jinkō*) to calculate the balance between the men deployed in the military and others mobilized as workforce.[200] Based on this perspective, the IPP claimed that 23,104,000 out of the aforementioned 35,269,000 workers to be distributed to industries in Japan Proper by 1950 should be men. It further concluded that 92% of the male workers – precisely 21,683,000 – should be "people of a productive age."[201] For the number of Japanese expats in the Greater East Asia Co-Prosperity Sphere,

---

[197] Ibid., 16.
[198] Ibid.
[199] Andrea Germer, Vera Mackie, and Ulrike Wöhr, *Gender, Nation and State in Modern Japan* (London: Taylor and Francis Group, 2014).
[200] Jinkō Mondai Kenkyūsho, "Kokudo keikaku toshite no jinkō haichi keikaku yōkō'an Showa 15-nen 10-gatsu hi," 19–22, 29–30.
[201] Ibid., 21–22.

one-third of the total population should be men of a "productive age, at least at the beginning."[202]

In turn, the IPP was less concerned with applying the concept of "population of productive age" to analyze the female population. Instead, it was more preoccupied with women's capacity to enhance the "reproductive power" (*seisanryoku*) of the population at large.[203] However, it also recognized the importance of women in the metropole as workers filling the void created by conscripted men.[204] In the end, the IPP took a compromised stance: It defined women's participation in work as primarily "harmful for the population growth" but also argued it could be encouraged insofar as it did not damage their reproductive capacity.[205] Based on this position, the IPP calculated the maximum percentage of women permitted to work without "harming" their reproductive capacity. After the investigation, the IPP concluded that the ratio of female workers to the total workforce in the metropole should not exceed more than 17% in the manufacturing industry and 10% in mining.[206]

Parallel to this, the IPP also recommended that 12,165,000 Japanese women should migrate overseas by 1950 to support the Japanese rule of the Greater East Asia Co-Prosperity Sphere. However, in contrast to their male counterparts, it did not recommend this by mobilizing the category of "population of productive age." While we cannot entirely cross out the possibility that this was an error or oversight, at the very least, it resonated with the gendered image of female expats, who, as respectable daughters and wives, helped the men dispatched to colonies to engage in productive activities as farmers, workers, merchants, colonial officers, etc. Like "continental brides" who were systematically sent to Manchuria in the late 1930s, their primary function was defined less by productive work than by their "reproductive power." Using this power, they were expected to maintain a balanced growth of the population of Japanese empire builders, as well as to maintain the expat community's racial purity by giving birth to the next generation of pure-bred

---

[202] Kōseishō Jinkō Mondai Kenkyūsho, "Kokudo keikaku toshite no jinkō haichi keikaku yōkō," 1940, PDFY09111756, 22.

[203] Jinkō Mondai Kenkyūsho, "Kokudo keikaku toshite no jinkō haichi keikaku yōkō'an Showa 15-nen 10-gatsu hi."

[204] Yuri Horikawa, "Senji dōin seisaku to kikon josei rōdōsha: Senjiki ni okeru josei rōdōsha no kaisōsei wo meguru ichi kōsatsu," *Shakai seisaku* 9, no. 3 (2018): 128–40; Regine Mathias, "Women and the War Economy in Japan," in *Japan's War Economy*, ed. Erich Pauer (London and New York: Routledge, 1999), 65–84.

[205] Jinkō Mondai Kenkyūsho, "Kokudo keikaku toshite no jinkō haichi keikaku yōkō'an Showa 15-nen 10-gatsu hi," 20.

[206] Ibid.

Japanese.[207] This assumption was inscribed in the way the IPP research depicted the category of sex.

The research based on this way of gendering the population was directly in lines with a number of social policy measures established in the 1930s, which aimed to promote health and welfare for women and children.[208] These measures, integrating the logic of the gendered division of labor, portrayed wartime social reforms categorically as gendered work. According to these measures, men were leading the fight for the prosperity of the nation-state-empire at the *front* – at the war front and at the colonial frontier – as productive workers and robust soldiers. In contrast, women were supporting the men at *jūgo* – "the back of the gun" – by keeping themselves as healthy as possible and by serving Japanese imperialism through their domestic and reproductive capabilities.[209] Women's contributions should be done primarily through their role as wives and mothers, and secondarily as workers.[210] Population research mirrored this logic found in social policy. By perpetuating the logic, research solidified the gender norms that assigned leadership roles to men while confining women to the reproductive domain and an assistive position in productive labor. And this logic was behind the IPP research's caricature of the "population of Japan Proper" as a gendered demographic subject.

[207] Sidney Xu Lu, "Japanese American Migration and the Making of Model Women for Japanese Expansion in Brazil and Manchuria, 1871–1945," *Journal of World History* 28, no. 3 (2017): 439–40; Sidney Xu Lu, "Good Women for Empire: Educating Overseas Female Emigrants in Imperial Japan, 1900–45," *Journal of Global History* 8, no. 3 (November 2013): 436–60.

[208] Sugita, "*Yūsei*," "*yūkyō*," 54–84; Sugita, *Jinkō, kazoku, seimei to shakai seisaku*, 108–60. Furthermore, outside the government, the Imperial Gift Foundation Aiiku-kai, established in 1934 through an imperial gift to commemorate the birth of the crown prince Akihito in 1933, set up "model villages" and explored ways to improve conditions for mothers and children in rural areas. Osamu Saito, "Bosei eisei seisaku ni okeru chūkan soshiki no yakuwari: Aiikukai no jigyō wo chūshin ni," in *Senkanki nihon no shakai shūdan to nettowāku: Demokurashī to chūkan dantai*, ed. Takenori Inoki (NTT Shuppan, 2008), 359–79; Saito, "Senzen nihon ni okeru nyūyōji shibo mondai"; Naoko Yoshinaga, "The Modernization of Childbirth and the Indoctrination of Motherhood in Prewar Japan: The 'Aiiku-Son' Project of Imperial Gift Foundation 'Aiiku-Kai'" [in Japanese], *Tokyo daigaku daigakuin kyoikugaku kenkyuka kiyo* 37 (1997): 21–29.

[209] Mikiyo Kano, *Onna tachi no jūgo* (Chikuma Shobō, 1995). For a creative interpretation of women's contributions at *jūgo*, see Annika A. Culver, "Battlefield Comforts of Home," in *Defamiliarizing Japan's Asia-Pacific War*, eds. W. Puck Brecher and Michael W. Myers (Honolulu: University of Hawai'i Press, 2019), 85–103.

[210] At the flipside of the coin was what Janis Matsumura calls the "moral panic" among the officials toward overly sexualized female workers and soldiers' wives who were suspected of disrupting social order due to their promiscuity and criminality. Janice Matsumura, "Unfaithful Wives and Dissolute Labourers: Moral Panic and the Mobilisation of Women into the Japanese Workforce, 1931–45," *Gender & History* 19, no. 1 (2007): 78–100.

However, for research bureaucrats, population research was not simply an intellectual exercise. In the case of Tachi, it was also grounded in his day-to-day administrative activities and conditioned by the epistemological challenges posed by the research. What kinds of activities supported his research? What did the process that was shaping knowledge about the demographic subjects under study involve?

Precarious Research Practices

To start with, the population research conducted by Tachi involved much paperwork, in part due to the CPB's administrative demands. For instance, the CPB First Department ordered its research groups to compile lists of the relevant academic publications for each of the research subjects they were in charge of, mark the materials with a level of confidentiality (secret, top secret, confidential, military resource secret, military resource partially secret), and submit a report regularly so it could compile a monthly reference catalog.[211] This meant Tachi, as a member of the staff at the CPB First Department's Population Group, must have been consumed with this laborious documentation work. The work could be particularly cumbersome for a subject such as population, which dovetailed with wide-ranging fields – from genetics, racial hygiene, and obstetrics-gynecology to statistics and macroeconomics.[212]

However, for Tachi, as a population expert, compiling demographic data was a more central focus than the above activity. As previously mentioned, Tachi was engaged in work that transformed demographic data into knowledge directly useful for national land planning. However, the process to generate "useful" knowledge was not always smooth. To the contrary, Tachi stumbled over challenges along the way. In terms of deskwork, there were two issues. The first pertained to methodology. For instance, when Tachi collaborated with his colleagues, Ueda Masao and Kubota Yoshiaki, at IPP on the study of populations in the Greater East Asia Co-Prosperity Sphere, they ran into problems calculating population dynamics because the method of census taking was not standardized. Some countries within the area simply had no system of collecting population data, while others that did adopted vastly different methodologies. As Tachi saw it, there were roughly six different data collection methods: (1) "modern" census; (2) "unmodern" census; (3) the method combining 1 and 2 but taking corrective actions; (4) calculation of a total

---

[211] See Kikakuin Daiichibu, *Kokudo keikaku shiryō mokuroku geppō* 2, no. 10, October 1942, PDFY090226029.
[212] Ibid., no. 4, April 1942, PDFY090226028.

population based on a partial census taking; (5) estimation; and (6) "a so-called simple guess."[213] Tachi and his colleagues had to grapple with the essentially incomparable data collection methods before even beginning to attempt to tabulate the "population of the Greater East Asia Co-Prosperity Sphere." To tackle the issue, in the end, they decided to do their best while largely relying on the *Annual Report of the Statistics of the Greater Japanese Empire*, which was slow to reflect the quickly evolving political reality that determined the population boundaries in Asia.[214]

The additional issue that troubled Tachi was linked to the slippery geographical definition of the Greater East Asia Co-Prosperity Sphere. As suggested above, this sphere was indeed an ideological construct that justified the expansion of the Japanese Empire. Thus, by definition, its boundaries were kept elusive. However, population research aiming to determine the optimal ratio between the population and landmass required knowledge about the sphere's clear-cut boundaries. Tachi and his colleagues were compelled to grapple with the tensions between the conditions created by Japan's political goal and scientific demands. Their solution was to take a compromising approach. In the aforementioned study with Ueda and Kubota, Tachi first stressed the sphere's amorphous, boundless, and expansive character, defining it as "the bounds toward which the power of the Japanese Empire reaches."[215] At the same time, recognizing that the study needed a clear understanding of the size of the sphere in order to calculate population density, the researchers simply decided to make up a working definition. They proposed what they called "the Greater East Asia Co-Prosperity Sphere and its adjacent area," which referred to "the area between 60 and 180 degrees east longitudinally but also included the islands of Hawaii simply because 40 percent of its population was Japanese."[216] Furthermore, they depicted it as excluding the USSR, British India, Afghanistan, Iran, Australia, New Zealand, and New Caledonia. However, they also mentioned that the areas of British India, Australia, New Zealand, and New Caledonia could be included depending on the context.[217] In other words, the researchers drew a flexible boundary, responding to the shifting understanding of what constituted the Japanese Empire and its population. As this case indicates, the process of making numerical facts for the Japanese empire-nation-state involved much tinkering along the way.

---

[213] Minoru Tachi, Masao Ueda, and Yoshiaki Kubota, "Tōakyōeiken jinkō ryakusetsu (zanteikō) (1)," *Jinkō mondai kenkyū* 3, no. 10 (October 1942): 3.
[214] Ibid., 4–9.
[215] Ibid., 20.
[216] Ibid., 4.
[217] Ibid., 20–21.

To make it more complex, population studies during the period had to reconcile the multiple understandings of population/race presented in its adjacent field of racial science.[218] Among these was, for instance, the idea that Japanese and Koreans were *konwa minzoku* ("a mixture of races," literally translated), which served to blur the boundaries between the Japanese, colonial subjects, and other races in Asia.[219] Komai Taku (1886–1972), professor of genetics at Kyoto Imperial University, claimed in 1942 that the Japanese were a *konwa minzoku*, comprised of the Ainu, the Chinese, and Koreans, while Koreans were made up of two or three culturally similar but biologically different racial groups.[220] By providing a creative interpretation of the link between race, culture, and history, which stressed the racial affinity between the Japanese and Koreans yet simultaneously insisted on the former's cultural superiority, racial science justified Japanese leadership in the geopolitical project of, and for, the Greater East Asian race's liberation from white dominance.[221] However, the emphasis on racial affinity led to tensions in Tachi's demographic study, though it primary relied on the legal definition of race and population that showed a clearer boundary between different racial groups.

Tachi believed a challenge of this kind could be overcome by technical means, by improving the methods for collecting population data. Thus, in the early 1940s, he participated in the movement within the government to reform the administrative infrastructure for the collection of vital statistics.[222] In the private draft proposal he authored on June 23, 1942, Tachi made a wish list for the "Greater Imperial Japanese Government" to act upon, which included compiling vital statistics for: "(1) the residents in Japan Proper grouped according to the categories in the civil registrations (*minseki*) and additionally nationwide, by prefectures or by cities with more than a population of 100,000," and "(2) the population of Japan Proper in Japan's colonies and in foreign countries classified by regions."[223] Additionally, the IPP proposal for national land planning, the drafting of which Tachi was involved in, stressed the

[218] Kingsberg Kadia, *Into the Field*; Hyun, "Racializing Chōsenjin"; Hoshino, "Racial Contacts across the Pacific"; Sakano, *Teikoku nihon to jinrui gakusha*; Morris-Suzuki, "Debating Racial Science."

[219] Soyoung Suh, *Naming the Local: Medicine, Language, and Identity in Korea since the 15th Century* (Cambridge, MA: Harvard University Asia Center Publications Program, 2017), 71–104.

[220] Hyun, "Racializing Chōsenjin," 501.

[221] Hoshino, "Racial Contacts across the Pacific."

[222] See Shakaiyoku Hokenbu, "Jinkō dōtai tōkei ni taisuru kibō jikō," n.d. c.1942, PDFY090803214, Tachi Bunko.

[223] Minoru Tachi, "Jinkō dōtai tōkei kaizen seibi ni kansuru kibō jikō (shi'an)," June 23, 1942, PDFY090803217, Tachi Bunko.

need to set up a system to comprehensively register every person of the metropole, which would facilitate the process of "deploying populations according to regions and of sending the population of Japan Proper out of the country." Based on this claim, the proposal recommended that the government should set up a National Registration Bureau (*Kokumin Tōroku Kyoku*) within the central government and National Registration Offices across the country, which would be in charge of registering every individual's "social status, skills, whereabouts, and other personal details" and, in case of immigration, would "train the migrants so that their activities could bring the best effect in their respective destinations."[224] Tachi believed the reform, promoting more methodical ways of collecting data about the "population of Japan Proper," was at least a first step toward solving the challenges he was confronted with in his demographic studies.

The demographic work Tachi and other research bureaucrats undertook for national land planning was premised on the assumption that the geographic and racial boundaries of the research subjects were evidently clear. However, at times, they struggled in their research activities, precisely because they were confronted with uncertainties surrounding this very assumption. As a way to overcome these challenges, they made concessions. At the same time, they resorted to a technical fix. The everyday research practices of these population bureaucrats exhibited how precarity was woven into the ways in which demographic knowledge was created and stabilized. At the same time, they showed how population experts grappled with the problem of uncertain knowledge in the context of the wartime state's policymaking, which persistently demanded clear-cut answers.

Conclusion

There was little doubt that research bureaucrats such as Tachi recognized that their work during the war made population studies into a policy science. They had many reasons to think this way. In the late 1930s, the status of population studies was raised within the government, as the political exigency of the war demanded the mobilization of people as "human resource." The government founded the IPP as the official research institution dedicated to policy-oriented population studies. On a smaller scale, the CPB established the Population Group for a similar purpose. The IRPP evolved into a professional organization that

[224] Jinkō Mondai Kenkyūsho, "Kokudo keikaku toshite no jinkō haichi keikaku yōkō'an Showa 15-nen 10-gatsu hi," 34.

the government turned to for expert advice. The conditions under total war – the fascist welfare rationale, the drive to control the economy, and the aspiration for imperial expansion, as well as the geopolitical concerns that surfaced as a result of this – accelerated the development of population studies. They also promoted a specific form of policy-oriented population studies that was conducted by bureaucrat-experts.

In turn, population bureaucrats like Tachi responded to their ascribed roles by presenting demographic knowledge that directly aided the state goal. The demographic knowledge, based on a specific formulation of population – a deployable resource, synonymous with race, and the subject of biopolitical, economic, and geopolitical strategies – supported Japan's engagement with imperial fascism from within. However, everyday scientific work also indicated how the process to produce this demographic knowledge required layers of negotiations. Consequently, compromise was part of the knowledge production process.

The state mobilization of population studies in the late 1930s and early 1940s had implications far exceeding the context of total war. Despite the political change after Japan's surrender in 1945, the practice of policy-oriented population research by technical or research bureaucrats survived, strongly influencing the trajectory of the field of population studies in years to come.

5   Birth Control Survey
Visualizing a Productive Japanese Population for Postwar Reconstruction

Japan's surrender on August 15, 1945, marking the end of World War II (WWII), immediately brought about changes in the political frameworks that upheld the population discourse and studies until then – a notable one was the Japanese territory's significantly reduced landmass. Japan had lost Taiwan, Korea, and other recently acquired colonial possessions. Manchukuo, once a promised land for Japan's "surplus people," was dissolved. Karafuto and the islands north of Hokkaido turned into a site of contention in the post-WWII diplomatic relations with the Soviet Union.

The idea of overpopulation as a source of socioeconomic problems returned with these changes. Commentators stressed that Japan, a nation already poor in resources and now with a much smaller territory, was simply unable to sustain its growing population. Even worse for Japanese policymakers, the old option of sending people overseas to solve the problem of the expanding population was now less viable because overseas migration was tainted by its association with fascism and military aggression.[1] This situation created favorable ground for birth control promoters, and the spread of birth control brought about a birth control survey boom.

The survey fervor emerged in Japan in the late 1940s and lasted for about a decade. This boom was in part supported by media organizations such as *Mainichi Newspaper*, which set up an in-house research group specializing in population issues and organized its own birth control survey. The government also played a pivotal role in this boom. In response to policy debates about the possible implementation of birth

---

[1] Edward A. Ackerman, *Japan's Natural Resources and Their Relation to Japan's Economic Future* (Chicago: University of Chicago Press, 1953), 161. Despite the argument, overseas migration still took place, albeit in a limited way compared to the previous eras. Lu, *The Making of Japanese Settler Colonialism*; Pedro Iacobelli, *Postwar Emigration to South America from Japan and the Ryukyu Islands* (London: Bloomsbury Academic, 2017).

control as population management, the Ministry of Health and Welfare (MHW) ordered the Institute of Population Problems (IPP) to conduct birth control surveys. In the early 1950s, these surveys provided material for policy debate, leading to a cabinet decision to make birth control a national policy in 1951 and the official announcement in 1954 that made birth control a population control measure. From the mid-1950s onward, the surveys further supported state-endorsed family planning initiatives. The official birth control survey continued until the late 1950s, when birth rates plummeted and overpopulation was no longer a policy item.

This chapter considers birth control surveys as a form of social survey and contextualizes the official involvement in the survey boom in terms of the interplay between science and the governing of Japan's population that occurred in the post-WWII political environment. The social survey, according to historian of modern China Tong Lam, was a "mode of knowledge production" directly linked to "China's transformation from a dynastic empire to a modern nation-state."[2] In the case of Japan, the official birth control survey did not thrive when Japan was transformed into a modern state but after WWII, when policymakers discussed population control via birth control as a critical condition for the "reconstruction" (*fukkō*) of the war-torn nation, a national slogan that emerged amid the occupation and constantly shifting regional geopolitics.[3] In this context, a birth control survey clarifying the demographic influence of people's sexual and reproductive behaviors was inherently a "political practice."[4] By providing knowledge that helped the government to better discern and manage the newly repackaged post-WWII "Japanese population," the survey helped facilitated Japan's political reconstruction more efficiently.

Partly due to the important role assigned to birth control, the knowledge produced by the official survey and research resonated with the political undertones of the reconstruction effort. First, going along with the official reconstruction efforts that involved securing the boundaries of the Japanese population based on the notion of racial homogeneity, the official birth control surveys did not look at race, despite the pervasiveness of race in Japanese population discourse of the time.[5] Second,

[2] Lam, *A Passion for Facts*, 2–3.
[3] Barak Kushner and Sherzod Muminov, eds., *The Dismantling of Japan's Empire in East Asia: Deimperialization, Postwar Legitimation and Imperial Afterlife* (London: Routledge, 2017); Toyomi Asano, ed., *Sengo nihon no baishō mondai to higashi ajia chiiki saihen* (Jigakusha Shuppan, 2013).
[4] Lam, *A Passion for Facts*, 2.
[5] Kristin A. Roebuck, "Orphans by Design: 'Mixed-Blood' Children, Child Welfare, and Racial Nationalism in Postwar Japan," *Japanese Studies* 36, no. 2 (2016): 191–212; Roebuck, "Japan Reborn," 103–84.

in harmony with the domestic- and economic-centric developmentalist framework that shaped the government's engagement with reconstruction, the survey research classified demographic data by region and socioeconomic status, two categories experts adopted when assessing a country's modernization and development achievements, or, in the case of post-WWII Japan, in postwar reconstruction.[6] In other words, the central premises buttressing post-WWII Japan's reconstruction exercise – the notion of Japan's ethnically homogenous population and the developmentalist logic upholding an introspective perspective on the socioeconomic and regional hierarchy within Japan – provided the basic framework for the birth control survey research. In turn, the knowledge about the population produced by the research informed the government's attempt to regulate fertility for the sake of reconstruction.

The coproduction of birth control survey research and the government's engagement with reconstruction did not occur only on the epistemological level; it also tangibly shaped science surrounding the research and governing of Japan's population associated with the reconstruction. To illustrate this, I look at Shinozaki Nobuo (1914–98), a colleague of Tachi who spearheaded the IPP's birth control survey research. Shinozaki was not just a rank-and-file technical bureaucrat. He would have been a younger member of the elite scientific circle of what Miriam Kingsberg Kadia termed "men of one age," the generation of human scientists who were at the prime of their careers during Japan's transwar period (1930s–60s).[7] In the 1950s, when he headed the birth control research, Shinozaki headed the establishment of a professional organization for population science, while also actively participating in the family planning initiative that unfolded in the half-government, half-private New Life Movement. Because of his multiple and often blurred identities as technical bureaucrat, population expert, and birth control campaigner, Shinozaki was able to effectively thread together the sites of science making and population management for post-WWII reconstruction. Specifically, Shinozaki used his birth control campaign to collect data for his policy-relevant survey research, and the research he conducted justified birth control policy as a population control measure that was tied to the reconstruction. Parallel to this, Shinozaki's popular campaign and research acted as basis for his involvement in the creation of a specific community of population

---

[6] For comparison, see Malcom Thompson, "Foucault, Fields of Governability, and the Population–Family–Economy Nexus in China." *History and Theory* 51, no. 1 (2012): 42–62.
[7] Kingsberg Kadia, *Into the Field*, 1.

experts united by their interests in the relationship between reproductive behaviors and demographic trends, and the expert community, once formalized, advised the government on population matters. These developments, brought together by technical bureaucrats like Shinozaki, characterized additional ways population science resonated with the governing of the population in Japan in the specific context of post-WWII reconstruction.

Shinozaki's stories are particularly insightful because they confirm how, even after the fascist regime crumbled with the defeat of the war, technical bureaucrats continued to exert influence over the interplay between science making and the state's effort to govern the population. Shinozaki's case is also illuminating because it shows how technical bureaucrats' personal motivations supported post-WWII reconstruction as a national project. Specifically, it illustrates how Shinozaki's enthusiasm for policy-relevant birth control work was fueled by both his sense of duty as bureaucrat and his own aspirations to establish himself in the government office and population scientists' community. By addressing these personal factors that shaped the government's attempt to control the population size for the purpose of postwar reconstruction, this chapter presents a nuanced history of state population governance.

Discourse of Overpopulation and Birth Control Policy during and after the Occupation

For the Japanese government, the August 1945 surrender signified the coming six and a half years of occupation by the victor nations initially headed by Douglas MacArthur, the commander of the US Army Forces in the Far East and the Supreme Commander for Allied Powers (SCAP).[8] For the population experts and technical bureaucrats, it meant a balancing act between their Japanese colleagues and relevant personnel at the occupation's Supreme Commander for Allied Powers General Headquarters (SCAP-GHQ) over the problem of "overpopulation."

Population growth was not a new problem (see Chapter 3). However, the post-WWII discourse of "overpopulation" emerged out of the specific situation postsurrender. This time, commentators stressed that population growth – the addition of eight million people to the total population in the first years after the war – was caused by rising birth

---

[8] Deborah Oakley, "American-Japanese Interaction in the Development of Population Policy in Japan, 1945–52," *Population and Development Review* 4, no. 4 (1978): 617–43.

rates triggered by the repatriation of soldiers and an influx of repatriates from their former colonies.[9]

While agreeing with the "overpopulation" argument in principle, policy advisors summoned by the government to deliberate on post-WWII population matters disagreed on what exactly constituted the problem of overpopulation, in ways that reflected their own disciplinary backgrounds. Those with an inclination toward Malthusianism focused on already evident food shortages and future land erosion.[10] Economists argued that "overpopulation" would cause mass unemployment and distort the distribution of populations to the various industrial sectors, thereby impeding state efforts to reindustrialize.[11] Medical experts tended to claim overpopulation was in part caused by effective "death control": the decline in mortality rates due to improved public health.[12] Those with eugenic tendencies raised concerns that overpopulation, if unchecked, would trigger "reverse selection" (see Chapter 3).[13] Despite these diverse interpretations, one issue gained unanimous support: "overpopulation" would impede state efforts to "reconstruct" their war-obliterated nation.

Beyond government circles, birth control activists were fervently discussing population dynamics. Like in the prewar period, neo-Malthusians claimed overpopulation would exacerbate poverty, while socialists argued it would benefit exploitative capitalists by generating surplus labor.[14] From among these figures, Katō (Ishimoto) Shizue, a leading figure in popular birth control activism since the 1920s (see Chapter 3), articulated overpopulation in terms of the challenges Japanese women confronted in their everyday lives.[15] For activists struggling to rebuild their movement after government suppression, the narrative of overpopulation provided an impetus to relate their raison d'être to post-WWII reconstruction.

---

[9] Miho Ogino, "Jinkō seisaku no sutorateji 'umeyo fuyaseyo' kara kazoku keikaku e," in *Tekuno/baio poritikkusu: Kagaku iryō gijutsu no ima*, ed. Kaoru Tachi (Sakuhinsha, 2008), 145–59; Ogino, *"Kazoku keikaku" eno michi*; Norgren, *Abortion before Birth Control*; Oakley, "American-Japanese Interaction," 619.

[10] Dinmore, "A Small Island Nation Poor in Resources," 111–36.

[11] Minoru Tachi, "Japan's Population To-Day," *Japan Planned Parenthood Quarterly* 1, no. 1 (1950): 3–5.

[12] Homei, "The Science of Population and Birth Control"; Crawford F. Sams and Zabelle Zakarian, *"Medic": The Mission of an American Military Doctor in Occupied Japan and Wartorn Korea* (Armonk, NY: M. E. Sharpe, 1998).

[13] Yoko Matsubara, "Nihon ni okeru yūsei seisaku no keisei" (PhD diss., Ochanomizu University, 1998), 91–92.

[14] Ogino, *"Kazoku keikaku" eno michi*, 152–54.

[15] Hopper, *A New Woman of Japan*, 175–250. In 1944, Shizue remarried and adopted her new husband's surname, Katō.

Under the occupation, "overpopulation" also rapidly became a priority within the SCAP-GHQ.[16] Officials within SCAP-GHQ were concerned about a potential negative impact on the occupation's most important mission: transforming Japan into an independent sovereign state.[17] Due to food shortages, Japan had relied on US food aid since the occupation's onset,[18] and more mouths to feed would hinder progress toward national independence.[19] The SCAP-GHQ personnel were also concerned that overpopulation would hamper their efforts to revitalize the capitalist economy through reindustrialization. Finally, the idea that "overpopulation" would lower living standards and lead to political instability was a source of concern within SCAP-GHQ in light of the escalating Cold War. As overpopulation was understood to intersect with wide-ranging issues, such as food, land, labor, health, and security, various officers within SCAP-GHQ engaged with the issue.

The narrative of a population crisis prompted the Japanese government and SCAP-GHQ to build an institutional infrastructure to tackle the problem.[20] Within SCAP-GHQ, officers and consultants in the Public Health and Welfare Section (PH&W) and Natural Resources Section (NRS) held various conferences on the issue.[21] Within the Japanese government, IPP was reinstalled as early as May 1946 to assist the MHW, which continued to take charge of population issues.[22] In April 1949, the government founded the Advisory Council on Population Problems (ACPP) within the cabinet. The government also advised that the IRPP – dormant during a brief period after the end of the war – resume its policymaking activities. In April 1946, the IRPP set up a Committee on Population Measures (IRPP-CPM) dedicated solely to the deliberation of population policies.[23] From that time forward, these three organizations – IPP, IRPP, and ACPP – established a collaborative relationship so tight-knit that population policy expert Nagai Tōru (see Chapter 3) described it as a "trinity for the deliberation of population issues" in post-WWII

[16] Oakley, "American-Japanese Interaction."
[17] Dinmore, "A Small Island Nation Poor in Resources," 111–59.
[18] Stephen J. Fuchs, "Feeding the Japanese: Food Policy, Land Reform, and Japan's Economic Recovery," in *Democracy in Occupied Japan: The U.S. Occupation and Japanese Politics and Society*, eds. Mark E. Caprio and Sugita Yoneyuki (London: Routledge, 2007), 26–47.
[19] Dinmore, "A Small Island Nation Poor in Resources," 133.
[20] Naho Sugita, "1950-Nendai no nihon ni okeru jinkōgaku no kenkyū kyōiku taisei kakuritsu ni muketa ugoki ni tsuite," *Jinkōgaku kenkyū* (May 2019): 4–7.
[21] Oakley, "American-Japanese Interaction."
[22] "Manual of the Institute of Population Problem" 1951. Rockefeller Foundations Archive, Record Group 1.2, Series 609s, Box 55, Folder 586. Rockefeller Archive Center, Sleepy Hollow, NY. (Hereafter referred to as RAC.)
[23] Sugita, "1950-nendai no nihon ni okeru jinkōgaku," 6.

Japan.[24] According to this model, the IPP would conduct research and generate data on population, the IRPP would deliberate over population issues using IPP data and submit draft recommendations to the ACPP, and finally, the ACPP would deliberate and submit resolutions to the government.[25] Over the 1950s, the "trinity" system became so firmly established that recommendations made by the three organizations were readily taken up by the government, and they shaped key population policies.[26]

Contraceptive birth control – used synonymously with "family planning" (*kazoku keikaku*) – was a key population policy the "trinity" enthusiastically pursued during the 1950s.[27] However, the path to making birth control policy was far from smooth. Before the war, official discourse on birth control was largely negative, reflecting the fact that the majority of policy intellectuals and policymakers were concerned with "reverse selection" and socialist, feminist, and labor movements (see Chapter 3). During the war, the government's pronatalist policy made birth control unpopular among policymakers (see Chapter 4). This tendency continued in the period immediately after the war. Minister of Health and Welfare, and a member of the conservative Liberal Party, Ashida Hitoshi publicly expressed his reservations about birth control, arguing that an "uncontrolled practice of birth control" would lead to a "decrease in the population."[28] MHW technical bureaucrat Yokota Toshi acknowledged the potential benefit of birth control for slowing down the rate of population growth, but he hesitated to openly endorse birth control because he believed unchecked birth control use would trigger "reverse selection."[29]

The social environment immediately after the war fueled the officials' anxieties about birth control even more. Print media caricatured disabled, vagrant repatriate soldiers, orphaned street children, and promiscuous *panpan* street girls courting American GIs as embodying a eugenic

---

[24] Sugita, *Jinkō, kazoku, seimei*, 207.

[25] Ibid., 204.

[26] Ibid., 207.

[27] Terms referring to fertility regulation changed throughout Japan's modern history. Roughly speaking, *jutai chōsetsu, sanji chōsetsu*, and *sanji seigen* correspond to "birth control" and had been in use since the 1910s, whereas *kazoku keikaku* emerged in the postwar period as a direct translation of "family planning." For the semantics of this, see Michiko Obayashi, "Sengo nihon no kazoku keikaku fukyū katei nikansuru kenkyū" (PhD diss., Ochanomizu University, 2006).

[28] "Birth Control for Japan Opposed by Welfare Head," *New York Times* (December 16, 1945), 3.

[29] Toshi Yokota, "Yūseugaku kara mita sanji seigen," in *Sanji seigen no kenkyū*, ed. Kakuitsu Andō (Nihon Rinshosha Tokyo Shikyoku, 1948), 164–79.

crisis that was emerging from the postwar rubble.[30] Responding to this kind of press coverage, in 1946, Ashida proclaimed Japan was hit by a racial crisis; therefore, "the revival of race" (*minzoku fukkō*) should be a priority for his ministry.[31] Ashida's claim facilitated the Eugenic Protection Law (EPL) that was issued in 1948. The EPL enforced compulsory sterilization, legalized abortion for patients with hereditary diseases, and simplified the procedures for eugenic and voluntary sterilization stipulated in the wartime National Eugenic Law.[32] Under these circumstances, many in the government thought unchecked birth control would catalyze "reverse selection" and be a eugenic hazard, slowing the pace of "reviving" the Japanese race.

The occupation government also maintained a cautious stance toward birth control. There were chiefly two reasons for this attitude: First, with the advent of the Cold War, SCAP-GHQ officials feared that the Soviets might purposefully portray the SCAP's endorsement of birth control in Japan as an act of genocide by a conqueror race, compare it to the medical crimes committed by Germans on the Jews, and use these arguments to topple US hegemony in world politics.[33] Second, the SCAP-GHQ's hesitant attitude toward birth control reflected the personal situation surrounding Douglas MacArthur. MacArthur avoided the matter of birth control altogether because he feared it would negatively impact his own political career back home if the SCAP-GHQ officially endorsed birth control – in particular, it could negatively impact his plan to run for the Republican presidential nomination based on support from Catholics.[34]

Yet, some within the SCAP-GHQ, especially in the NRS, were of the opinion that the occupation authorities should help diffuse birth control in Japan. This opinion became pronounced after the delegate from the Far Eastern Commission visited Japan in January 1946 and suggested the SCAP-GHQ consider ordering the Japanese to use birth control.[35] The sense of urgency was intensified when the delegates of the mission group Reconnaissance in Public Health and Demography in the Far

---

[30] See Robert Kramm, *Sanitizing Sex: Regulating Prostitution, Venereal Disease, and Intimacy in Occupied Japan, 1945–1952* (Berkeley: University of California Press, 2017); Sarah Kovner, *Occupying Power: Sex Workers and Servicemen in Postwar Japan* (Stanford: Stanford University Press, 2012); Mire Koikari, *Pedagogy of Democracy: Feminism and the Cold War in the U.S. Occupation of Japan* (Philadelphia: Temple University Press, 2008).

[31] Yoko Matsubara, "Nihon ni okeru yūsei seisaku no keisei," 6–7.

[32] Ibid., 194–97.

[33] Takeuchi-Demirci, *Contraceptive Diplomacy*, 118–19; Oakley, "American-Japanese Interaction," 625.

[34] Oakley, "American-Japanese Interaction," 628.

[35] Ibid., 624.

East, dispatched by the Rockefeller Foundation (RF) in 1948, met high-rank officials within the SCAP-GHQ – most notably Crawford F. Sams, chief of the PH&W. Sams, originally wary of birth control, changed his opinion after the delegates' visit.[36] Eventually, the SCAP-GHQ adopted a policy that Deborah Oakley called "protective neutralism."[37] With this policy, the SCAP-GHQ publicly proclaimed neutrality on the issue of birth control but actively supported Japanese initiatives to popularize birth control behind the scenes.

Within the Japanese government, too, a critical watershed moment for birth control came in the late 1940s. Around this time, medical and demographic researchers began to publicly endorse contraception.[38] Behind these endorsements were general concerns regarding "overpopulation," but more immediately, the rise in the abortion rate that they witnessed during the period played a critical part. Abortion, illegal since the Criminal Code in 1880 (see Chapter 2), became de facto legalized under the 1949 amendment of the EPL, which permitted women to have abortions for economic reasons. After the amendment, abortion rates surged.[39] Medical researchers attributed the expanding availability of induced abortions, which was made possible by the EPL, to the rise in the abortion rates, and from the perspective of maternal and infant health, they began to lobby for contraceptive birth control as a viable alternative to induced abortion.[40]

Policy intellectuals around this time also felt an urgent need to wade into popular birth control practice from a eugenic point of view. As some population experts in the ACPP saw it, birth control thus far, because it was unchecked, "now spread ... among those who have intellectual professions in cities." It was virtually absent in "special areas" where "a large number of people with undesirable hereditary qualities live together" and in the places that acted as "hotbeds of all sorts of social evils," "plagued

[36] Takeuchi-Demirci, *Contraceptive Diplomacy*, 97–103, 123–26; Aiko Takeuchi, "The Transnational Politics of Public Health and Population Control: The Rockefeller Foundation's Role in Japan, 1920s–1950s," Rockefeller Archive Center Research Reports Online (RAC, 2009); Yu-ling Huang, "The Population Council and Population Control in Postwar East Asia," Rockefeller Archive Center Research Reports Online (RAC, 2009).

[37] Oakley, "American-Japanese Interaction," 624–34.

[38] H. Kubo, "Kōshū eisei yori mitaru jinkō mondai kaiketsu eno ichi shian," in *Jinkō mondai no igakuteki kenkyū*, "Showa 23-nendo monbushō gakujutsu kenkyū kaigi dai 9-bu igaku dai 21-pan hōkokusho (1949)," 9–23. In the English-language publications, Kubo's first name appears either as Hidebumi or Hideshi.

[39] Councillors Room, Prime Minister's Office, "Recommendations of the Population Problem Council in the Cabinet," November 29, 1949, Rockefeller Family Archive, Record Group 5, Series 1, Subseries 5, Box 80, Folder 672, RAC, 11.

[40] Councillors Room, Prime Minister's Office, 12.

with venereal diseases, alcoholism, and narcotic poisoning.... If conception control is not propagated in these areas," they continued, "the so-called reverse selection is sure to appear, and the future of the nation will present a gloomy prospect."[41] Fueled by this doomsday picture, the ACPP members came to think that it would be better off if the government had a heavy-handed approach to birth control rather than leaving its current laissez-faire state alone.

Consequently, from around the spring of 1949, there were moves to endorse a birth control policy. To start with, in April 1949, taking up the pro–birth control opinion of American sociologist Warren S. Thompson, now temporarily serving as consultant to the NRS of the SCAP-GHQ, Prime Minister Yoshida Shigeru himself stated in the Diet that birth control was necessary for solving the population problem.[42] Following Yoshida's statement, in the same month, the cabinet formed the ACPP. In May 1949, the group within the MHW in charge of deliberating on the actual terms for the popularization of birth control began to consider the contraceptive methods that would be recommended for use in community health centers.[43] Throughout the month, the MHW also issued an approval for the distribution of twenty-six birth control medicines, including the contraceptive jelly brand Contra.[44] In June 1949, the ACPP subdivision in charge of matters concerning population control reached an agreement that "birth control is necessary as a check on population control" and began to work on the actual recommendations.[45]

Eventually, in November 1949, the ACPP submitted a policy proposal to the government.[46] The proposal characterized abortion as an invasive surgical procedure harming mothers' health and recommended birth control to replace induced abortion. It also portrayed the current state of unchecked birth control as a risk factor for "lowering" the quality of the Japanese race and also recommended the popularization of *guided* birth control on the grounds of eugenics.[47] With the proposal, the ACPP responded to the government need to tame the surplus population *and* to protect the quality of the Japanese population.

---

[41] Councillors Room, Prime Minister's Office, 13.
[42] Oakley, "American-Japanese Interaction," 633–34.
[43] Diary of Oliver R. McCoy, May 27, 1949.
[44] Yasuko Tama, "Jutai chōsetsu (bāsu kontorōru) to botai hogohō," in *Umi sodate to josan no rekishi*, ed. Shirai, 114; Kubo, *Nihon no kazoku keikaku shi*, 97.
[45] Diary of Oliver R. McCoy, June 25, 1949.
[46] Councillors Room, Prime Minister's Office.
[47] Jinkō Mondai Shingikai, "Jinkō mondai shingikai kengi" (November 1949), 21.

Following these moves by the occupation and Japanese authorities, on October 26, 1951, the cabinet decided to popularize birth control throughout the country, with the MHW responsible for its implementation. In 1952, the EPL was amended to accommodate this new policy, which assigned the Eugenic Marriage Consultation Offices based within local health offices the task of providing guidance on birth control to married couples.[48] The amendment also created a new category of healthcare professionals: "birth control field instructors" (*jutai chōsetsu jicchi shidōin*).[49] Recruited primarily from the pool of midwives and public health nurses, "birth control field instructors" were expected to hold seminars on birth control and distribute contraceptives to married couples in their communities at wholesale prices. Recommendations by the "trinity" mobilized central and local government health agencies to organize activities to disseminate the knowledge and practice of birth control across the country.[50]

Birth control in the 1951 policy was defined officially as a means to protect the "life and health of mothers." However, it was clear to the people in the know that it was a measure for population control.[51] A change came in July 1954, when the IRPP-CPM submitted the "Resolution Concerning the Popularization of Family Planning as Population Measure" to urge the government to clarify its position on the relationship between birth control and population control. Based on the IRPP-CPM proposal, in August 1954, the ACPP further submitted the "Resolution Concerning the Quantitative Adjustment of the Population" and demanded the government define birth control explicitly as a population policy.[52] Responding to the ACPP proposal, in the same year, Minister of Health and Welfare Kusaba Ryūen publicly announced the government's commitment to birth control as a population policy.[53] From then on, the term "population control" (*jinkō yokusei*) was used openly to characterize the government-endorsed activities propagating the knowledge and practice of birth control. Throughout the process, until 1952, SCAP-GHQ officers and

---

[48] Yasuko Tama, *Boseiai toiu seido: Kogoroshi to chūzetsu no poritikkusu* (Keiso Shobo, 2001); Norgren, *Abortion before Birth Control*; Matsubara, "Nihon ni okeru yūsei seisaku no keisei."
[49] Kimura, *Shussan to seishoku*; Obayashi, *Josanpu no sengo*.
[50] Homei, "The Science of Population and Birth Control."
[51] Ministry of Welfare, Japan, "Item Concerning the Promotion of Conception Control, Comprehended at Cabinet," October 26, 1951, Rockefeller Foundations Archive, Record Group 2-1951, Series 609, Box 543, Folder 3629, RAC.
[52] Ogino, *"Kazoku keikaku" eno michi*, 193–94.
[53] Ogino, "Jinkō seisaku no sutorateji," 194.

consultants were behind the scenes, offering tangible help for the Japanese to make progress in policymaking.[54]

The process of establishing a government birth control policy illustrates how various assertions about "overpopulation," fueled by eagerness for national reconstruction, prompted the state to attempt to discipline reproductive bodies through policy. Yet, the "state," striving to control its population size via birth control, was not a faceless administrative machine. It was supported by the daily activities of technical bureaucrats who were trusted as state actors.

Role of Technical Bureaucrats in Birth Control Policy

From the 1920s onward, research and technical bureaucrats became increasingly important for the creation of population policies (see Chapters 3 and 4).[55] Originally, their contributions were centered on policy-oriented population research and policymaking, and only a few took part in the actual implementation of measures. However, with a post-WWII birth control policy that required concrete initiatives, some technical bureaucrats eagerly undertook more practical roles.

Among them, Shinozaki Nobuo stood out.[56] He was among the first anthropology graduates at the University of Tokyo after Hasebe Kotondo (1882–1969) founded the Department of Anthropology in 1939. Following graduation, he remained in the department as an assistant professor until June 1943, when he joined the MHW-RI Department of Population and Race. After the war, as a midcareer bureaucrat affiliated with the reinstalled IPP, Shinozaki assumed an important position within the "trinity" as secretary for the IRPP-CPM's Second Sectional Meeting and actively participated in the policy debate on population control. And, in the 1950s, Shinozaki made efforts to implement birth control programs as part of government policy.

Shinozaki became involved in official birth control work immediately after the war, when groups of up-and-coming bureaucrats and population experts were pursuing the possibility of adopting fertility regulation as a viable technique of population control.[57] However, he nurtured an

[54] Takeuchi-Demirci, *Contraceptive Diplomacy*; Ogino, "Jinkō seisaku no sutorateji";
Norgren, *Abortion before Birth Control*; Oakley, "American-Japanese Interaction."
[55] Sugita, "*Yūsei*," "*yūkyō*"; Sugita, *Jinkō, kazoku, seimei*.
[56] Yoichi Okazaki, "Tsuitō Shinozaki Nobuo hakase (tsuitōbun)," *Jinkōgaku kenkyū* 24 (June 1999): 74.
[57] Ogino, "*Kazoku keikaku*" *eno michi*, 154–55.

interest in fertility during the war. In the context of wartime race and population research, when an investigation of the eugenic implications of a group's ability to reproduce was a priority, Shinozaki conducted a survey of research on miscegenation across the world and studied how these works indicated the influence of miscegenation on birth rates.[58] Even after the war, this theme – the correlation between a race's biological profile and fertility pattern – continued to shape Shinozaki's work, but this time, Shinozaki carried out a study that examined, from the eugenic perspective, the effect of consanguineous marriage on fertility within Japan.[59] These eugenic and demographic studies on race and fertility certainly acted as a foundation for Shinozaki's involvement in the official campaign to spread birth control after the war.

In addition to this, Shinozaki's birth control work was also driven by the conviction that Japanese women of the new, post-WWII era should adopt what he called "modern birth control," or rationally planned, proactive contraceptive practices based on scientific principles.[60] In a small survey he conducted immediately after the war, Shinozaki observed that women were resorting to abortion to terminate unplanned pregnancies, and this was impacting population trends. Even worse, he found that many women seemed to be embracing the erroneous assumption that the "abortion and infanticide of the Tokugawa era" was "modern birth control." In Shinozaki's view, this tendency symbolized how the backward past was lingering in the modern era, causing friction with official efforts to build a modern, rational, and democratic society. In contrast, "modern birth control," or the rational application of contraceptive practices, would directly assist the government's endeavor by creating a suitably disciplined family. Based on this view, Shinozaki claimed that education about "modern birth control," embedded in an "active effort to rationalize and improve life," was urgently needed.[61]

In the 1950s, Shinozaki was committed to realizing this through the post-WWII New Life Movement, a half-state, half-private initiative

---

[58] Nobuo Shinozaki, "Minzoku konketsu no kenkyū," *Jinkō mondai kenkyū* 4, no. 9 (1942): 12–31.

[59] Nobuo Shinozaki and Hisao Aoki, "Ketsuzoku kekkon buraku no yūseigakuteki chōsa gaihō (dai ippō)," *Jinkō mondai kenkyū* 7, no. 1 (1951): 105–14; Nobuo Shinozaki, Keiko Yoshida and Hisao Aoki, "Ketsuzoku kekkon buraku no yūseigakuteki chōsa gaihō (dai nihō)," *Jinkō mondai kenkyū* 7, no. 2 (1951): 52–66.

[60] Nobuo Shinozaki and Hisao Aoki, *Anatano kazoku keikaku* (Nihon Kazoku Keikaku Fukyūkai, 1959).

[61] Kōseishō Jinkō Mondai Kenkyūsho, "Tokyo-to wo chūshin tosuru sanji seigen no jittai ni kansuru shiryō" (Kōseishō Jinkō Mondai Kenkyūsho), 23 in Irene B. Taeuber (1906–1974), Papers, 1912–1981, C2158, Folder f.2224, The State Historical Society of Missouri, Columbia, MO, USA. Hereafter Taeuber Papers.

intended to promote a democratic, efficient, and cultured life by rationalizing everyday activities.[62] He became a member of the New Life Guidance Committee, a committee within the IRPP to deliberate on introducing family planning into the New Life Movement. The movement's ethos resonated perfectly with his concept of "modern birth control." Furthermore, Shinozaki believed the movement's existing structure, as well as fostering the private-government partnership, would make it easier for him to entice the birth control program's major stakeholders, namely, private corporations, into joining his cause. Finally, Shinozaki was convinced a program within the movement promoting "modern birth control" would be a win-win for both government and private corporations. Disciplined families with fewer children, attained through "modern birth control," would produce financial benefits for both – the government would save on child welfare costs while companies would pay less for benefit packages. Shinozaki used this argument to garner support for his birth control program within the New Life Movement.[63]

Between December 1954 and September 1957, Shinozaki, together with the IRPP head Nagai Tōru and Aoki Hisao (1922–80), another IRPP committee member and Shinozaki's colleague at the IPP, championed these plans to over 140 companies and chambers of commerce.[64] Once the New Life Movement family planning program had been launched, Shinozaki organized a national council of private corporations to facilitate communication between company representatives.[65] He also liaised between the MHW's Department of Welfare and private corporations to ensure the program's smooth operation. By 1958, in part due to Shinozaki's activities, eighty-three public and private corporations had launched or begun to prepare in-house family planning programs.[66]

Through his birth control campaign for the New Life Movement, Shinozaki became instrumental in the initiatives underpinning the official birth control policy. He took advantage of his position within government to influence policymaking. He implemented policy by liaising

---

[62] Ogino, *"Kazoku keikaku" eno michi*, 208–13; Yasuko Tama, *"Kindai kazoku" to bodī poritikkusu* (Kyoto: Sekai Shisōsha, 2006), 100–61; Andrew Gordon, "Managing the Japanese Household: The New Life Movement in Postwar Japan," in *Gendering Modern Japanese History*, eds. Barbara Molony and Kathleen Uno (Cambridge, MA: Harvard University Asia Center, 2005), 423–51; Takeda, *The Political Economy of Reproduction in Japan*, 127–52.
[63] Gordon, "Managing the Japanese Household"; Obayashi, *Josanpu no sengo*, 210.
[64] Tama, *"Kindai kazoku" to bodī poritikkusu*, 107–8.
[65] Ibid., 118–19.
[66] Ibid., 110–11.

between government officials and other stakeholders, thereby reducing the distance between the government and the governed, and facilitating state efforts to manage the population.

While organizing the family planning initiative for the New Life Movement, Shinozaki led a series of birth control survey research. Shinozaki's research highlights the unique contributions technical bureaucrats made to the government's efforts to govern the population at a specific moment in postwar history when the boundaries of the Japanese population once again were in flux.

Shinozaki and Birth Control Research

Immediately following the war, as "overpopulation" became a topic of policy discussion, research institutes in Japan – both state and nonstate funded – began to survey opinions on birth control. As early as August 1946, the IRPP published an internal document based on a research survey titled "A Trend in the Public Opinion of Birth Control."[67] Jiji Press and *Asahi Newspaper* conducted similar surveys in 1949.[68] Within the government, the National Public Opinion Research Institute of the Prime Minister's Office conducted a survey of the public opinion on birth control; the report came out in 1952.[69] In 1947, the IPP assumed responsibility for birth control survey research from the IRPP. Since then, the IPP technical bureaucrats have published key surveys on birth control practices. Between 1946 and around 1960, government agencies and media organizations generated a seemingly inexhaustible supply of birth control surveys.

Within the IPP, Shinozaki was made responsible for the majority of its birth control surveys. From 1947 on, his survey team conducted pilot surveys. Between 1951 and 1952, to construct a "national" picture of birth control practices, the team visited seventeen prefectures and collected data from a total of 44,509 individuals.[70] During this period, Shinozaki was certainly dedicated to this survey work.

[67] Zaidan Hōjin Jinkō Mondai Kenkyukai, ed., "Sanji seigen ni kansuru yoron no dōkō August 1946," in *Sei to seishoku no jinken mondai shiryō shūsei*, vol. 8 (Fuji Shuppan, 2002), 94–98.

[68] Lt Col Thomas, "Birth Control, Public Opinion Survey," Box No. 9344, Sheet no: PHW 02611, Class no. 710, 751. National Diet Library GHQ/SCAP Records, 1949.

[69] Sōrifu Kokuritsu Yoron Chōsajo, "Jutai chōsa nikansuru yoron chōsa," March 24, 1952 in *Sei to seishoku no jinken mondai shiryō shūsei*, vol. 11 (Fuji Shuppan, 2002), 227–28.

[70] Kōseishō Jinkō Mondai Kenkyūsho, "Kenbetsu oyobi toshi chōson betsu sanji chōsetsu jittai chōsa shūkei kekkahyō: Showa 24–25-nendo zenkoku 17-ken ni okeru chōsa," 1952, 76.

What motivated Shinozaki to conduct this birth control survey research? I argue that we need to pay attention to his multiple attributes as a government bureaucrat, birth control campaigner, and scientific expert, because this triple identity fueled his enthusiasm for the state-endorsed birth control work, including the survey research. First, and most obviously, he was involved in the work because it was part of his job. But, his dedication to the surveys was not merely the result of his role as a state bureaucrat.[71] Shinozaki's enthusiasm for this work also stemmed from his belief that the survey research would validate the corporate-based New Life Movement campaign.[72] Because of Shinozaki's conviction, the IPP research was often targeted at workers. A pilot survey recruited participants from the staff at the University of Tokyo's School of Medicine, the Japan Steel Pipe Company (Nihon Kōkan , a.k.a. NKK), Fuji Electronics Appliances Company, the MHW and the Tokyo Metropolitan Government, in addition to the Ajinomoto Company.[73] By organizing this type of survey, Shinozaki was able justify and even improve his family planning initiative in the New Life Movement. In other words, surveys helped to establish the postwar officially endorsed birth control campaign Shinozaki had vested interests in.

Finally, his attributes as a race and population scientist were also critical and motivated his leadership in the survey research. Shinozaki was an active member of the Population Association of Japan (PAJ), founded on November 11, 1948, as the first professional organization dedicated to the advancement of demographic studies in Japan.[74] From its inception, at almost every annual meeting, Shinozaki supported the PAJ's activities. In the early period, he actively presented his research on birth control, fertility, miscegenation, population quality, the mental ability of children of lower-class parents, etc.[75] Shinozaki continued to occupy a central position in the organization later in his career. In 1974, he headed the organizing committee for the annual meeting, and in 1983–84, he served

---

[71] Nobuo Shinozaki, "The Actual State of Spread of Birth-Control in Suburbs of Tokyo: To Analyze the Conditions of It" (Institute of Population Problems, 1952), 2, Taeuber Papers, 1912–1981, C2158, Folder f.2226. It must be noted that not all technocrats were driven to influence policies through their scientific research. Tatsuma Honda, "Sanji chōsetsu no fukyū jōkyō nikansuru chōsa" (Kōseishō Jinkō Mondai Kenkyūsho, 1953), 1.

[72] Shinozaki, "The Actual State of Spread of Birth-Control in Suburbs of Tokyo."

[73] Nobuo Shinozaki, "Sanji seigen jittai chōsa nikansuru gaikyō," in *Jinkō mondai kenkyusho Shōwa 22-nen 3-gatsu 19-nichi kenkyū hōkokukai hōkoku gaiyō* (Jinkō Mondai Kenkyūsho, 1947).

[74] Nihon Jinkō Gakkai Sōritsu 50-shūnen Kinen Jigyō Iinkai, *Nihon jinkōgakkai 50-nenshi* (Nihon Jinkō Gakkai, 2002), 9.

[75] Ibid., 163–77.

as the president. In this context, the birth control survey research in the early 1950s carried a special meaning for Shinozaki as a midcareer population expert: It helped advance his status as a scientist and PAJ member. Shinozaki's birth control work in bureaucracy, birth control activism, and his science-building activities were all tightly enmeshed, and the manner in which the science-building activities were coproduced with policymaking explains Shinozaki's commitment to the birth control work.

In the birth control survey research, Shinozaki and his colleagues collected data through questionnaires and interviews. Questionnaires were mainly multiple choice, inviting respondents to tick the one or more boxes they deemed closest to their answers. For the question, "Why have you not practiced birth control?," for example, the questionnaires provided the following options: "(1) Because I do not know about contraception (2) Because I feel it is a burden (3) Because I do not like it (4) Because either a husband or a wife is infertile."[76] The interview that typically followed was intended to fill gaps in the knowledge gained from these questionnaires. Often conducted by a fieldworker at a respondent's home, these included intimate details about sexual behavior as well as opinions on contraception.[77]

For the most part, Shinozaki's team reported the survey results in the form of numerical data, categorized by the respondents' personal attributes. For instance, in the survey of birth control practices in the Tokyo suburbs of Musashino City, Abiko Town, Tanaka Village, Tomise Village, and Kobari Village, the team would first calculate the actual number and percentage of "practitioners" against the total populations of the five administrative units – 361 (43.1%), 218 (15.3%), 52 (6.9%), 76 (12.3%), 9 (3.7%) – then catalog the data according to husband's occupation and level of education, wife's age and occupation, duration of marriage, number of children, and, finally, the amount of cultivated land.[78] The assumption was that this way of displaying numerical data would accurately reveal not only the opinions of the respondents but also the "actual state of the spread of birth control," as expressed in the title of the research survey.

The survey was conducted on the premise that collecting respondents' opinions on the "actual state" (*jittai*) of their reproductive and sexual

---

[76] Nobuo Shinozaki, Akira Kaneko, and Kazumasa Kobayashi, n.d. "Summary of The Investigation of the Actual State of the Practice of Contraception (First Report)," 4, Taeuber Papers, 1912–1981, C2158, Folder f.2235.

[77] Shinozaki, "Sanji seigen to seiseikatsu no jittaiteki chōsa."

[78] Shinozaki, "The Actual State of Spread of Birth-Control."

lives would facilitate the creation of successful policies.[79] This assumption needs further analysis: How could the opinions about birth control contribute to the government's efforts to govern Japan's population? Koyama Eizō, first director of the National Public Opinion Research Institute and a former colleague of Shinozaki at the IPP during the war, held ideas that help us address this question.[80] According to Koyama, opinion research never simply mirrored the mood of the general public but was rather a force in itself, shaping the current of mainstream opinion. Koyama further contended that the role of the government in this situation was that of a doctor: coordinating opinions based on research results and intervening if attitudes revealed "maladies."[81] Faith in the corrective power of the knowledge produced by their surveys was widely shared by IPP officials, who believed their research into public opinions on family size and ideal contraception practices would be utilized by the government to influence reproductive behaviors as required. Birth control surveys thus enjoyed a special status within the state, in part because of the certainty that public opinion could be utilized as a tool of governance.

Despite this confidence, Shinozaki's fieldwork at times faltered, especially when research subjects refused to cooperate. For instance, in 1949–50, when Shinozaki's survey team conducted fieldwork in Aomori, Iwate, and Miyagi Prefectures, as many as 2,073 married couples either did not return the questionnaire or returned it incomplete.[82] Further investigation revealed that 455 of those couples had found the questionnaire too difficult to understand, 85 could not be bothered to fill it in, and a small number stated they were simply "not interested" or "did not like to be asked such questions."[83] The scientific knowledge gleaned from such fieldwork could potentially detail only a partial picture of the "actual state."

Nevertheless, throughout the 1950s, Shinozaki tirelessly led survey research with the conviction that cumulative data on opinions about birth control on a regional level would eventually form a big picture, capturing what he termed the "actual state of the spread of birth control" throughout Japan, and that this big picture would enable the government

---

[79] Ibid.
[80] Morris-Suzuki, "Ethnic Engineering."
[81] Ibid., 515.
[82] Kōseishō Jinkō Mondai Kenkyūsho, "Tohoku sanken ni okeru sanji chōsetsu jittai chōsahyō miteishutsusha no miteishutsu riyū oyobi chōsa ni taisuru iken no jitsujō ni tsuite," (1951).
[83] Ibid. 2–3.

to effectively implement population policies.[84] His trust that scientific data could facilitate the governing of Japan's population reinforced his passion for the research.

Reimagining the Japanese Population

In the context of postwar Japan, when the idea of "Japan's population" itself was in flux, Shinozaki's research did not merely create knowledge about reproductive bodies for the state. Since it clarified patterns of reproductive behaviors that would directly inform the future profile of population dynamics, Shinozaki's surveys participated in the broader bureaucratic activity of compiling demographic data to establish a new interpretation of the Japanese population.

Immediately after the war, scientific investigations to collect numerical facts about the population emerged as a major bureaucratic objective in Japan. Following surrender, the Japanese Empire collapsed almost overnight, triggering territorial disputes and mass migration on a scale not witnessed in the previous era.[85] Migrations and shifting national borders challenged the existing notion of the "Japanese population" that had held currency under colonial rule.[86] For both the Japanese state and Occupation governments, this was highly problematic: Various factors shaping sovereignty, such as citizenship and land ownership, relied on this destabilized category.[87] Under these circumstances, examination of the Japanese population became a priority.

Thus, government offices swiftly began to compile population statistics. Censuses were carried out by the SCAP-GHQ six times between September 1945 and October 1950,[88] and the Japanese government's IPP and Cabinet Bureau of Statistics were assigned similar tasks.[89] The

---

[84] Shinozaki, "The Actual State of Spread of Birth-Control."

[85] Lu, *The Making of Japanese Settler Colonialism*; Iacobelli, *Postwar Emigration to South America*; Yoshikuni Igarashi, *Homecomings: the Belated Return of Japan's Lost Soldiers* (New York: Columbia University Press, 2016); Shinzo Araragi, ed., *Teikoku igo no hito no idō: Posutokoroniarizumu to gurōbarizumu no kōsakuten* (Bensei Shuppan, 2013); Lori Watt, *When Empire Comes Home: Repatriation and Reintegration in Postwar Japan* (Cambridge, MA: Harvard University Press, 2009).

[86] Shiode, *Ekkyōsha no seijishi*.

[87] Tessa Morris-Suzuki, "Beyond Racism: Semi-Citizenship and Marginality in Modern Japan." *Japanese Studies* 35, no. 1 (2015): 67–84; Chapman, "Geographies of Self and Other".

[88] Oakley, "American-Japanese Interaction," 622.

[89] "Chikaki shōrai naichi (Hokkaido, Honshu, Shikoku oyobi Kyushu) ni oite fuyō subeki jinkō no suikei," PDFY090805020, 1945, Tachi Bunko.

actual processes of collecting and presenting demographic data were complex, indicating how much effort population bureaucrats made to stabilize knowledge of the Japanese population.[90] Constant adjustments were required when calculating demographic data to accommodate ongoing political changes.[91]

For technical bureaucrats in charge of compiling population figures, this kind of adjustment was a standard administrative task. Yet, in the specific context of postwar Japan, it was simultaneously more than just routine work: These endeavors intimately interacted with the process of redrawing the boundaries of Japanese citizenship.[92] To conform to this legal practice, compiling population statistics required constant negotiations over who should be included in, or expunged from, the category of Japanese. Through the adjustment work of technical bureaucrats, the population of Japan would soon be repackaged as a historically consistent, ethnically homogenous, national group in accord with the new political outlines of the Japanese state.

Similarly, Shinozaki's research portraying the individuals who would produce the future Japanese population also contributed to the image of ethnically homogenous Japanese nationals. However, in contrast to the census work, which categorized various constituents of the former empire along ethnic and territorial lines, the birth control research confirmed this image by presenting ethnicity as a nonissue: Generally silent on the racial identity of research participants, the surveys suggested this was self-evident. On the odd occasion that race was mentioned, it was depicted as a foreign phenomenon. For instance, one of Shinozaki's surveys introduced a table showing ethnicity – "black" and "white" – as a factor in the correlation between pregnancy rates and socioeconomic class, but this was simply a citation from research conducted in the United States.[93] As the table was for reference only, the impression was given that ethnicity was tangential to reflections on Japanese demographic phenomena. By presenting race in this manner, the research projected a message that was then flourishing in official discourse: Only reproduction by ethnic Japanese people should count in the reconstruction of Japan as a nation.

In reality, the boundaries of the Japanese population during this period was far more contested than Shinozaki's research suggests, mirroring the

[90] Kōseishō Kenkyūsho Jinkō Minzokubu, "Showa 20-nen ikō Showa 22-nen ni itaru zaigai heiryoku no fukuin oyobi zaigai naichijin no hikiage ni yoru naichi jinkō no suikei (zanteikō) (hi)," 1945, PDFY090805026, Tachi Bunko.
[91] Sōmushō Tōkeikyoku, "Danjobetsu jinkō."
[92] Morris-Suzuki, "Beyond Racism."
[93] Shinozaki, "Sanji seigen jittai chōsa ni kansuru gaikyō."

reconfiguration of postwar Japan that Lori Watt once characterized as "the uneven and incomplete process of absorbing and re-categorizing the fragments of empire within Japan."[94] Following the collapse of the empire, Japanese citizens in the former colonies and soldiers at the front were redefined as "people of Japan Proper placed externally" (zaigai naichijin). Some repatriated to Japan, others stayed away. Among those who did not return were young women marrying into Chinese families, who became known as "remaining women" (zanryū fujin), as well as adopted Japanese orphans, the "children left behind in China" (chūgoku zanryū koji). While the majority of the 700,000 Koreans forcibly migrated to Japan during the war were repatriated to the Korean Peninsula after 1945, those who stayed in Japan became known as zainichi Koreans.[95] Furthermore, with the advent of the US Occupation, the people of Okinawa were now legally called Ryukyuans and declared "foreigners," along with former Korean, Chinese, and Taiwanese colonial subjects.[96] Finally, immediately after the occupation, the Japanese press declared a national crisis over the existence of orphaned "mixed-blood children" (konketsuji).[97] Whether these groups belonged to the Japanese population and what criteria should be used to determine eligibility were thorny issues for policymakers and technical bureaucrats specializing in population.

It was against this backdrop that Shinozaki's policy-oriented survey research attempted to uncover the "actual state" of the Japanese people's birth control practices. In a context in which the definition of the Japanese population itself remained uncertain, the quantification of reproductive experience was not simply a mathematical practice; it also intersected with the question of how to recognize the Japanese population in the face of the shifting geopolitical landscape of East Asia. The birth control research engaged with this issue primarily by maintaining silence on the issue of race. This act of silence, I argue, ultimately served to stabilize increasing official claims of Japan's ethnonational identity.

[94] Watt, When Empire Comes Home, 5.
[95] Shinzo Araragi, "The Collapse of the Japanese Empire and the Great Migrations: Repatriation, Assimilation, and Remaining Behind," in The Dismantling of Japan's Empire in East Asia: Deimperialization, Postwar Legitimation and Imperial Afterlife, eds. Barak Kushner and Sherzod Muminov (London: Routledge, 2017), 66–84; Toyomi Asano, "Zentai no shikaku: Hikiage no tenkai to zaisan wo meguru teikoku no butsuriteki kaitai to chiikiteki saihen," in Sengo nihon no baishō mondai to higashi ajia chiiki saihen, ed. Toyomi Asano (Jigakusha Shuppan, 2013), 1–27.
[96] Shiode, Ekkyōsha no seijishi, 351–411.
[97] Shimoji, "Konketsu" to 'nihonjin,' 61–133; Seiji Kamita, "Konketsuji" no sengoshi (Seikyusha, 2018); Roebuck, "Orphans by Design."

In addition, birth control surveys adopted an extra framework that corresponded to the domestic goal of reconstruction. As I will explain in the next section, this was an introspective perspective that compelled viewers to focus on the Japanese as a productive unit contributing to the reconstruction effort through economic means.

Designing a Productive Population for the Nation's Bright Economic Future

If race was not a primary category for classifying participants in the birth control research, then what was? As suggested above, Shinozaki's survey employed sociological classifications and internal regional differences to explore demographic variations. The report, submitted on the authority of the IPP on February 1, 1952, was based on the survey research led by Shinozaki and also used these classifications. First, the report – consisting mainly of numerical data presented in tables – classified research participants into two categories, "those practicing birth control" (*jikkōsha*) and "those who are not" (*fujikkōsha*), and then further classified them according to social and geographical categories. Social categories included the husband's educational level and occupation; geographical categories included prefectures, then the subcategories of city, town, farming village, mountain village, and fishing village.[98]

Why did the IPP adopt these categories? There is no doubt that disciplinary conventions played a role; it was a long-established standard in social scientific studies to categorize data according to region and socioeconomic status. However, the fact that the IPP surveys prioritized these specific categories over other possibilities does merit attention.

To account for the inclusion of occupation and education level, Michelle Murphy's concept of the "economization of life" is useful.[99] Murphy uses this term to refer to a "historically specific and polyvalent mode for knitting living-being to economy" and to describe "practices that differently value and govern life in terms of their ability to foster the macroeconomy of the nation-state."[100] In Japan after the WWII, the "economization of life" acted as a guiding principle, especially in programs like the New Life Movement, in which efforts to discipline reproductive bodies were articulated in relation to reconstructing the national economy. However,

---

[98] Kōseishō Jinkō Mondai Kenkyūsho, "Kenbetsu oyobi toshi chōson betsu sanji chōsetsu jittai chōsa shūkei kekkahyō" (1952), in *Sei to seishoku no jinken mondai shiryō shūsei*, vol. 11 (Fuji Shuppan, 2009), 235–42.

[99] Michelle Murphy, *The Economization of Life* (Durham: Duke University Press, 2017).

[100] Ibid., 13, 6.

Murphy also claimed that this "economization of life" had reestablished race as a category in order to determine which lives were worth reproducing. In the case of postwar Japan, at least in the domain engaging with population policies, race was associated less with the fostering of economy than with nationalism, in part due to the aforementioned assumption of ethnic homogeneity. I argue that, in this context, other kinds of social attributes, such as education and occupation, were regarded as more appropriately informative of an individual's economic value. Birth control research embodied this logic within postwar population management.

In parallel with this, the predominant demographic discourse emerging at the time, which incorporated a progressivist narrative, acted as a crucial background for the presence of regional categories in the IPP's research. This discourse, embodied in the so-called demographic transition theory, maintained that a correlation existed between fertility patterns and socioeconomic developments on a linear time scale.[101] The model, which also embraced the modernization theory later associated with the economist Walt W. Rostow, claimed that demographic patterns universally shifted from a "high-birth," "high-death" to a "low-birth," "low-death" model as a society progressed from the "pre-industrial" to "post-industrial" stage. In the Cold War, this discourse was used to justify transnational family planning aid programs in "underdeveloped" nations to establish a "free-world" alliance revolving around the United States.[102] In post-WWII Japan, the same discourse reinforced a deep-seated stereotype that cast rural areas and lower socioeconomic classes as the source of the nation's "overpopulation" problem and cities as enlightened spaces where the educated classes voluntarily practiced birth control. It simultaneously sanctioned the diffusionist view inscribed in the state campaign: The idea and practice of birth control would necessarily flow from "modern" urban centers to peripheral backwaters.

The IPP birth control survey internalized this developmentalist narrative and opted for regional analytic categories. A focus on regions went hand in hand with the diffusionist perspective, which was even integrated into research questions. For instance, a survey conducted by Shinozaki in the suburbs of Tokyo asked, "How much is 'birth control' diffused

---

[101] Carole R. McCann, *Figuring the Population Bomb: Gender and Demography in the Mid-Twentieth Century* (Seattle: University of Washington Press, 2017); Simon Szreter, "The Idea of Demographic Transition and the Study of Fertility Change: A Critical Intellectual History," *Population and Development Review* 19, no. 4 (1993): 659–701.

[102] John Sharpless, "World Population Growth, Family Planning, and American Foreign Policy," *Journal of Policy History* 7 (1995): 72–102; Peter J. Donaldson, *Nature Against Us: The United States and the World Population Crisis, 1965–1980* (Chapel Hill: University of North Carolina Press, 1990).

as one travels from the center of Tokyo to its neighboring towns and villages?" and compared data collected from three regions: Tokyo, "cities and towns in the suburbs of Tokyo," and "villages in the suburbs of Tokyo."[103] The survey's results revealed a higher degree of "indifference" to birth control among people in the rural district compared to cities, confirming not only assumptions about lower socioeconomic development in rural areas but also the argument dominating policymaking at that time: State birth control initiatives should target the countryside.[104] This approach to data thus enabled researchers to craft their research findings in ways that were comprehensible for the policy agenda.

On the surface, this focus on regional categories appears disconnected from economic rationale. However, in reality, consideration of the national economy was an omnipresent backdrop. For instance, a mid-1950s policy discussion on the rural population was dominated by the issue of how the economy could absorb the expanding labor force to prevent them from becoming "the complete unemployed" (*kanzen shitsugyōsha*).[105] Even after the Japanese economy experienced high economic growth in the late 1950s and concerns about unemployment had dissipated, economic considerations formed the core of policy discussions on peripheral populations. In the early 1960s, when members of the IRPP-CPM Second Special Committee brought up the issues of "population quality" as a policy agenda, family planning was linked to issues of "regional development" (*chiiki kaihatsu*). A type of social policy was emerging as a response to Japan's post-WWII reconstruction efforts, which were by that time being criticized as too weighted toward economic development.[106]

Through sorting data by socioeconomic and regional characteristics, the birth control research inscribed the economic rationale underpinning the state's objective to reconstruct the nation. In so doing, it simultaneously categorized respondents' sexual lives in terms of their reproductive *and* productive capacities. The image of the Japanese population that emerged as a result was that of an aggregate of individuals whose ability to produce labor and Japanese offspring would contribute to the reproduction of the national economy and the nation's population. This portrayal of the Japanese population consolidated the official standpoint

---

[103] Kōseishō Jinkō Mondai Kenkyūsho, "Tokyo-to wo chūshin tosuru," 21.

[104] Ibid., 14.

[105] Scott O'Bryan, *The Growth Idea: Purpose and Prosperity in Postwar Japan* (Honolulu: University of Hawai'i Press, 2009).

[106] Naho Sugita, "Nihon ni okeru jinkō shishitsu gainen no tenkai to shakai seisaku: senzen kara sengo e," *Keizaigaku zasshi* 116, no. 2 (2015): 59–81; Zaidan Hōjin Jinkō Mondai Kenkyūkai, "Jinkō shishitsu kōjō ni kansuru taisaku yōkō ketsugi" (Zaidan Hōjin Jinkō Mondai Kenkyūkai, 1962), 39.

and enabled the prioritization of supposedly stable internal subcategories over the haphazard movements of people breaching Japan's newly formed territorial borders. It certainly left little space for any reflection on racial politics, but rather it provided an opportunity to reinforce postwar Japan's officially sanctioned identity, which was based on a narrative of ethnic homogeneity and amnesia about the country's colonial past.

Conclusion

The story of the birth control survey boom during post-WWII reconstruction is significant for the following three points. First, it details how the survey research was uniquely and directly embedded in postwar Japan's search for a new identity – a specific historical juncture when the dismantling of the Japanese Empire fundamentally reconfigured politics, the economy, and society.[107] Second, it highlights how knowledge produced through the official birth control survey research deftly paralleled state efforts to govern the Japanese via birth control policy, both mobilized for the grand mission of postwar reconstruction. Finally, state-led population management did not simply happen because a policy acted as an embodiment of diffused power; rather, it was shaped by the everyday activities of technical bureaucrats. Birth control survey research and the state's population control effort realized through the birth control policy maintained a coproductive but fundamentally complex relationship. Yet, ultimately, the outcomes of this relationship reinforced the process of post-WWII reconstruction by providing an epistemological framework with which to imagine Japan's population in terms of ethnic homogeneity and economic rationale. The influence of this social imaginary was expansive: The resulting narrative of the Japanese population had a profound impact on the contours of population science and the mode of state population management for many years to come.

This chapter gives the impression that postwar birth control work was strictly a "national" project: It has illustrated how the work was directly accountable for the domestic population policy. The main focus of the chapter was a technical bureaucrat who served the national government. The chapter also depicted how the research subjects in the birth control surveys were the citizens who made up the postwar Japanese nation-state. However, postwar birth control research was never solely a domestic endeavor. It was firmly embedded in the transnational population control movement.

---

[107] Kushner and Muminov, *The Dismantling of Japan's Empire.*

6 Public Health Demography
Local, National, and Transnational Efforts to Govern Lower-Class Populations

Shinozaki had an effective colleague at the Institute of Public Health (IPH), just six kilometers south of central Tokyo where the government offices were clustered, to help with his birth control campaign and the initiative to launch the first association for population science: Koya Yoshio (1890–1974), the IPH director-general, also a renowned scholar, government advisor, and social reformer advocated to eugenics and racial hygiene.[1] During World War II (WWII), Koya was involved in developing reproductive policies.[2] He continued to do so after the war, but this time, going along with the government line, he promoted birth control instead of pronatalism. Koya spearheaded the move within the "trinity" to promote a birth control policy on the grounds of maternal health and eugenics. For the same reasons, he joined force with birth control activists to form the Family Planning Federation of Japan (FPFJ) in 1954. Koya, even more so than Shinozaki, was at the center of the movement to popularize birth control in 1950s Japan.

Like Shinozaki, Koya also integrated his scientific activities with policy-relevant birth control advocacy work. However, Koya did it on a larger scale. Far more advanced than Shinozaki in terms of his career, Koya was able to influence the government birth control policy and popular birth control movement on a more fundamental level. An outcome of Koya's policy-oriented birth control work was the formation of the Department of Public Health Demography (DPHD), the academic department Koya created within the IPH. Reflecting Koya's position, the DPHD fostered a lively community that consisted of collaborations among health officials, research specialists in reproductive medicine and

[1] For Koya's biography, see Izumi Takahide, ed., *Nihon kingendai jinmei jiten 1868–2011* (Igaku Shoin, 2012), 265; "The Rockefeller Foundation, Division of Medicine and Public Health, Personal History Record and Application for Travel Grant: Koya Yoshio," n.d. c.1954, Rockefeller Foundations Archive, Record Group 1.2, Series 609, Box 6, Folder 46, RAC.
[2] Koya was also closely working with Tachi during the war. Koya Yoshio and Tachi Minoru, *Kindaisen to tairyoku, jinkō* (Osaka: Sogensha, 1944).

population statistics, and birth control activists. Through these collaborations, the DPHD supported the post-WWII government's effort to govern the population via the promotion of birth control. Like Shinozaki's survey research, stories about the DPHD certainly point to the intimate relationship between the development of population science and the postwar domestic population policy, which unfolded within the movement to popularize birth control across the nation.

What was particularly remarkable about Koya's stories, which was less obvious in those about Shinozaki, was that this seemingly "national" endeavor was buttressed both by "transnational" and "local" elements.[3] Specifically, through the birth control pilot studies based at the DPHD, Japanese population science and the state's efforts to govern the population became woven into the transnational endeavor to manage the world's population. At the same time, fieldwork for the DPHD pilot studies fundamentally relied on various local governing infrastructures in order to actualize the intended goal. This chapter thus elaborates the history of the science–state collaboration on population management by showing how local, national, and transnational population governance efforts were entangled in the process to develop science – which, in this case, was the field of public health demography.

Koya's Birth Control Advocacy and Public Health Demography

Stories about the Japanese government's engagement with the issues of reproductive bodies in the period between the 1930s and 1960s are incomplete if they do not mention Koya Yoshio. Koya, a graduate of Japan's top medical program, the Department of Medicine at the Imperial University of Tokyo, cultivated his career as a specialist and promoter of racial hygiene. After studying in Berlin – as many elite medical students had done since the Meiji period – Koya taught racial hygiene at Kanazawa Medical University. In 1939, he joined the MHW as a technical bureaucrat (*gikan*), and from December 1940 on, he directed the MHW-RI Department of Welfare Science. After the war, appointed by the SCAP, Koya assumed directorship of the IPH, which last until September 1956, when he moved to Nihon University. Throughout his career, Koya maintained close contact with population scientists in

---

[3] On the "transnational" aspect of Shinozaki's birth control work, see Nobuo Shinozaki, Akira Kaneko, and Kazumasa Kobayashi, "Summary of The Investigation of the Actual State of the Practice of Contraception (First Report)," n.d., Taeuber Papers, C2158, Folder f.2235.

various fields, including those introduced in this book, and he was a central figure at the institutions that critically shaped – and were shaped by – official reproductive policies.

Yet, in part due to his proximity to official reproductive policies, Koya had a rather complicated history with birth control advocacy.[4] In the early 1930s, as a conservative racial hygiene scholar, Koya firmly opposed the popular birth control movement. Like Nagai Hisomu (see Chapter 3), Koya feared the current state birth control – highly popular among the urban educated class but not among the less-educated lower socioeconomic class – would jumpstart "reverse selection" and consequently "lower" the quality of the Japanese race. In 1930, together with Nagai Hisomu, Koya established the Japanese Association of Racial Hygiene (*Nihon Minzoku Eisei Gakkai*) and lobbied for eugenic policies that would prevent the use of birth control and induced abortion from spreading further. During the war, and now a technical bureaucrat at the MHW, Koya helped draft key wartime eugenic and population policies that banned contraceptives now deemed "harmful."[5]

After the war, Koya completely flipped his position on birth control – as did the government. Witnessing the surge in abortions after the amendment of the Eugenic Protection Law (EPL) in 1949 (see Chapter 5), Koya persuaded Minister of Health and Welfare Hashimoto Ryōgo to consider promoting birth control as an alternative to induced abortion.[6] He was also a leading figure in the "trinity" structure, actively involved in drafting the proposal that culminated in the birth control policy of 1951.[7] In the 1950s, he led birth control advocacy by acting as a go-between for the government and existing birth control activists. By the end of the decade, Koya was known as the most eminent health official involved in the popular birth control movement.

The above story suggests that Koya's attitude toward birth control shifted dramatically after the war. Yet, in reality, his philosophy supporting the attitude changed very little. Beginning in the prewar era, the fear of laissez-faire birth control facilitating "reverse selection" continued to dominate his understanding. Thus, when the government was inclined to adopt birth control as a national policy, Koya problematized *unchecked* birth control practice, while fervently endorsing a *guided* birth control

---

[4] Miho Ogino, *"Kazoku keikaku" eno michi*, 179–81.

[5] Hiroyuki Matsumura, "'Kokubō kokka' no yūseigaku: Koya Yoshio wo chūshin ni," *Shirin* 83 (2000): 102–32.

[6] Yoshio Koya, "The Program for Family Planning in Japan," *Journal of Japanese Medical Association* 24, no. 9 (September 1950): 2.

[7] Yoshio Koya, "Whither the Population Problem in Japan," *Bulletin of the Institute of Public Health* 1 (1951): 2.

program. A *guided* program would specifically target the lower socioeconomic classes. It would be carefully coordinated by local public health authorities, supervised by medical practitioners who were familiar with eugenics, and assisted by a trained midwife or a nurse.[8] Koya believed such a program could successfully lower abortion and pregnancy rates without triggering "reverse selection." Thus, what appeared on the surface as a drastic transformation was in fact not so drastic, and even after the war, Koya's attitude was fueled by his fervor for racial hygiene.[9]

Another aspect of Koya's work that changed very little was his willingness to use scientific expertise for his cause. In wartime, Koya applied his expert knowledge in racial hygiene to mold reproductive policies. After WWII, he insisted that the birth control movement should be informed by science. In 1957, from his position as the president of the FPFJ, Koya argued "the correct way for spreading F.P." was one predicated on "a scientific basis" and insisted the FPFJ advocate for such "scientific" birth control.[10] As a technical bureaucrat who specialized in racial hygiene, Koya firmly believed that knowledge and techniques buttressed by eugenics and population science were essential for birth control advocacy.

In the latter half of the 1940s, using his position as the IPH director-general, Koya strove to create an academic department that would provide "a scientific basis" for his birth control advocacy. He used the widespread anxiety about "overpopulation" to negotiate with the SCAP-GHQ and the Japanese government and managed to persuade the MHW to authorize the formation of a department that would respond to the population problem from the viewpoint of public health. Consequently, on May 17, 1949, the DPHD was established within the IPH, with Koya as its head.

Because Koya was a driving force behind the department's creation, the DPHD unabashedly reflected Koya's ambition to build a scientific institution that would directly contribute to his birth control campaign. The official document defined the area of expertise for each IPH department

---

[8] Yoshio Koya, "The Family Planning Program Should Be Promoted in Cooperation of Physicians More Intensively," n.d. c.1958, Series III, Box 95, Folder 1565, Gamble, Clarence James Papers, 1920–1970s, Center for the History of Medicine. Francis A. Countway Library of Medicine, Boston, MA. (Hereafter referred to as Gamble Papers).

[9] Maho Toyoda, "Sengo nihon no bāsu kontorōru undō," 58.

[10] Edna McKinnon, "Report for the Far East and Austrasian Region of I.P.P.F. Tokyo, Japan, September 14 to 26, 1960," 3, 1960, Series III, Box 97, Folder 1582, Gamble Papers; "FPFJ Is Planning to Conduct Important Projects Shortly," *Family Planning News*, no. 2 (March 1957): 1.

that the DPHD oversaw: "(1) Matters concerning public health linked to the population phenomenon; (2) Matters concerning family planning; and (3) Matters concerning eugenics and physical quality."[11] This focus directly corresponded to the objectives Koya frequently articulated in his post-WWII birth control activism: to reduce a population size through contraception and to improve the biological quality of the Japanese population through eugenics. They would be realized through a scientifically informed *guided* family planning program tied to government policy.

The recruitment process in the early days also illustrate Koya's ambition. The IPH recruited staff from within the organization who would be able to instantaneously assist Koya's birth control research. Ultimately, Kubo Hidebumi (Hideshi), Kumazawa Kiyoshi, and Muramatsu Minoru were transferred to the DPHD. Along with Koya, they constituted the founding members of the new department.[12] In June 1950, epidemiologist Yuasa Shū replaced Kumazawa. In May 1952, Ogino Hiroshi, son of the prominent gynecologist Ogino Kyūsaku, became the fifth member of the DPHD, and in 1957, biostatistician Kimura Masabumi joined the department.[13] In addition to this permanent staff, in the 1950s, external colleagues either taught courses for the DPHD or joined the department for specific research projects. They included eminent Professor of Obstetrics-Gynecology Moriyama Yutaka from the Yokohama Medical School and Koya's son, the gynecologist Koya Tomohiko from the Department of Obstetrics and Gynecology at Tokyo Teishin Hospital. Staff members at the DPHD were well versed in either biostatistics or obstetrics-gynecology, the two medical fields Koya deemed were key to sustaining his policy-relevant birth control research.

Once established, the DPHD became a hub for the Japanese birth control movement. Notable figures such as Katō Shizue and Kitaoka Juitsu were among the frequent visitors to the department.[14] They met with Koya at the DPHD to discuss a wide range of topics, such as the future of activism, methods of contraception, and the eligibility of doctors permitted to perform induced abortions under the EPL.[15] In the mid-1950s, after Koya became the FPFJ president, the organization

---

[11] Kokuritsu Kōshū Eisei In, *Kokuritsu kōshū eisei in sōritsu jūgo shūnen kinenshi* (1953), 22.

[12] Kokuritsu Kōshū Eisei In, *Kokuritsu kōshū eisei in sōritsu jūgo shūnen kinenshi*, 29.

[13] Ogino Kyūsaku was internationally renowned for his Ogino contraceptive rhythm method.

[14] Diary of Oliver R. McCoy, August 20, 1949; December 13, 1949; December 30, 1949, Rockefeller Foundations Archive, Record Group 12.1, Box 83, Folder 266–7 at RAC.

[15] Diary of Oliver R. McCoy, August 20, 1949; December 13, 1949.

moved its institutional base to the DPHD. It was the DPHD that administered the budget for the FPFJ and published its English-Japanese bilingual newsletter, *Family Planning News*. The DPHD provided Koya with a physical space in which he could develop his birth control activism.

As well as offering him a venue for his activism, the DPHD was doubly valuable for Koya because its core activities also directly fed into his birth control advocacy. Like the other departments within the IPH, the DPHD's core work consisted of professional teaching and scientific research in public health. With regard to teaching, family planning dominated the DPHD teaching curriculum shortly after the government turned birth control into a national policy in 1951. The training program was a direct outcome of Koya's birth control activism. As the FPFJ president, he once commented that both doctors and healthcare practitioners should learn contraception systematically so they could effectively run a guided birth control initiative.[16] With this idea in mind, in September 1951, Koya met Minister of Health and Welfare Hashimoto and recommended that the government should train local doctors, health practitioners, and health officials in the "methods of regulating contraception."[17] After the government amended the EPL in 1952 in accordance with the cabinet decision and created the role of "birth control field instructor" (see Chapter 5), the government entrusted the DPHD with offering a week-long course on contraception to prefectural health officials and medical professionals. After completing the course, the attendants were expected to return home and train and certify local midwives and nurses to act as "birth control field instructors." In 1952 alone – the first year the course ran – 111 individuals completed the course.[18] For the rest of the 1950s, the course continued, acting as an integral part of the state birth control initiative Koya fervently promoted.

The research conducted at the DPHD was significant for Koya's birth control advocacy. Between the early 1950s and mid-1960s, DPHD researchers carried out a stream of policy-relevant, applied, semi-longitudinal public health and demographic studies on people's birth control practices that primarily aimed to assess the efficacy of a contraceptive method or the impact of a birth control initiative on vital statistics, in particular on birth, pregnancy, and abortion rates.[19] Starting with the

---

[16] Koya, "The Family Planning Program Should Be Promoted."
[17] Diary of Oliver R. McCoy, September 24, 1952.
[18] Kokuritsu Kōshū Eisei In, *Kokuritsu kōshū eisei in sōritsu jūgo shūnen kinenshi*, 76.
[19] For the list of publications coming out of the aforementioned projects at DPHD, see Yoshio Koya, "Papers of Dr. Koya and his associates, Department of Public Health Demography, The Institute of Public Health, Tokyo, Japan. 1950–1959," January, 1959 (PC, Box 19 Folder 302).

abortion survey in 1950, DPHD staff and associates launched the internationally celebrated seven-year birth control pilot project, the "three-village study" (1950), "Research Concerning Induced Abortions" (February 1952), a sterilization study (May 1953), the five-year birth control pilot scheme involving coal miner's communities in Ibaraki Prefecture that were employed by the Joban Coal Mine Co. Ltd. (October 1, 1953), the Kajiya Village study experimenting with the recently developed foam tablet (October 1, 1953), and another pilot study involving the employees of the National Railways (1961). Finally, when the neighboring countries of Taiwan and South Korea began to use modern intrauterine devices (IUDs) for their state-led family planning activities in the mid-1960s, the DPHD also launched a collaborative project in 1966 that first tested the Japanese intrauterine contraceptive device Ota Ring and later American-made IUDs, with researchers affiliated with ob-gyn departments at the Iwate, Gunma, Tokyo, Niigata, Osaka Medical, Okayama, and Kyushu universities.[20] These DPHD projects directly supported Koya's birth control advocacy by producing medico-scientific evidence that would undergird his campaign to promote "scientific" family planning.

Public health demography, however, was not merely the title of an academic department. In the 1950s, it also represented a research strand that dominated the field of population science in Japan. This was apparent in the administrative structure and research activities of the PAJ. First, the secretariat of the PAJ was located at the DPHD in the early years. Second, the first seven PAJ annual meetings took place at the IPH. Third, the presentations given at the PAJ were mainly on topics exploring the relationship between public health and demographic issues, similar to the kind of research conducted at the DPHD. For instance, the very first meeting on March 19, 1949 had presentations that included ones by: Shinozaki on birth control social surveys, Tachi on the relationship between social and population dynamics, Tachikawa Kiyoshi on the current trend in induced abortion, and Kubo and two other colleagues on the link between occupation and fertility in Tokyo's suburbs.[21] Needless to say, Koya was behind this profile of the PAJ. While assuming the positions as the IPH director-general and the DPHD director, Koya was also the vice president of the PAJ and the editor of the association's flagship journal, *Journal of the Population Association of Japan*. Under Koya, the

---

[20] See the annual reports of Kokuritsu Kōshū Eisei In (*Kokuritsu kōshū eisei in nenpō*) published 1966–73.
[21] Nihon Jinkō Gakkai Sōritsu 50-shūnen Kinen Jigyō Iinkai, *Nihon jinkōgakkai 50-nenshi*, 163.

research agendas of the DPHD and PAJ became blurry. Consequently, the kind of fertility research conducted under the name of public health demography became mainstream in population science in the 1950s.

As the subsequent sections elaborate, Koya actively used the research in public health demography to influence population policies. At the same time, Koya's policy agenda directly shaped the contours of the scientific research undertaken in this field of inquiry.

Policy-Relevant Abortion Studies and Birth Control Pilot Projects

Among the DPHD research, studies on induced abortion were among the first to tangibly influence Koya's policymaking activities. The impetus for abortion studies came with the rising abortion rate after the amendment of the EPL in 1949.[22] Some in or affiliated with the DPHD, such as Moriyama, were concerned it could become a medical ethics issue, especially since many doctors seemed to be performing abortions repeatedly for financial benefit and at the cost of a mother's health.[23] Moreover, officers within the SCAP-GHQ, not least Sams, feared the phenomenon might turn into a political controversy if the department was implicated in promoting abortion.[24] Witnessing this trend, Koya, too, was alarmed that the rising abortion rate would be a serious problem for public health in the near future.

In this context, in the early 1950s, the DPHD organized a fact-finding mission on induced abortions. The team first conducted a preliminary survey, asking doctors who were registered to perform abortions under the EPL to collect questionnaires from their patients. After the preliminary survey, the team, managed by Muramatsu, conducted a pilot study in January 1952 in the working-class Kawasaki Health Center district in Kanagawa Prefecture. In the meantime, DPHD staff and affiliates, Dr. Agata, Dr. Teramura, and Koya Tomohiko, were assigned to conduct interviews with ninety-nine women who had an abortion between August 1949 and August 1950 under Article 13 of the EPL.[25] Based on these test studies, starting in mid-February 1952, the team conducted interviews

---

[22] Yoshio Koya, "Preliminary Report of a Survey of Health and Demographic Aspects of Induced Abortion in Japan," December 1951, Rockefeller Foundations Archive, Record Group 1.2, Series 609, Box 6, Folder 45, RAC; "Wither the Population Problem in Japan: Report at the 7th Annual Meeting of the Association of Public Health and Welfare," n.d. c.1950, Rockefeller Foundations Archive, Record Group 2-1950, Series 609S, Box 501, Folder 3354, RAC.

[23] Diary of Oliver R. McCoy, January 12, 1950.

[24] Diary of Oliver R. McCoy, June 2, 1950.

[25] Diary of Oliver R. McCoy, January 21, 1952.

in the three health center districts in Shizuoka Prefecture, which is to the west of Kanagawa Prefecture.[26] In March, the investigation expanded to include interviews with residents of health center districts in Yokosuka, Odawara, Mizaki, and Atsugi under the auspices of the Kanagawa Prefectural Health Department.[27] The team managed to cover cities, towns, and villages. By the winter of 1953, the team had completed a total of 1,487 interviews, but only 1,382 fit the specifications for the study (462 in "large cities," 464 in "small cities," and 456 in "rural areas").[28]

The preliminary survey, examining the women's motivations for opting for induced abortions instead of contraception, gave Koya an indication of how to organize a policy-relevant birth control program effectively. From the questionnaires returned by 337 doctors, the team found that the majority of respondents answered that either a lack of information on contraceptive methods (37%) or it being "too troublesome" to use contraceptives (35%) were their primary reasons for their decision.[29] The survey convinced Koya of the importance of education when it came to replacing abortion with contraception.

Meanwhile, the full-scale abortion study presented two points that could affect policy. First, it showed that abortions were practiced widely, cutting across geographical and social boundaries.[30] This point was indeed problematic from the perspective of public health, especially because it confirmed women, irrespective of their social status and areas of domicile, were exposed to the health risk of abortion.[31] Second, it revealed women who had induced abortions tended to have shorter interval between subsequent pregnancies than women who went through "normal" childbirth. This finding was disturbing from a demographic point of view. It indicated that women who had induced abortions would contribute to rising fertility rates in the long run. In other words, as Koya and Muramatsu saw it, "induced abortion cannot be regarded as an efficient means of family limitation."[32] Based on these findings, Koya

---

[26] Diary of Oliver R. McCoy, March 10, 1952.

[27] Diary of Oliver R. McCoy, March 24, 1952.

[28] Diary of Oliver R. McCoy, March 26, 1953.

[29] Diary of Oliver R. McCoy, June 1, 1950.

[30] Yoshio Koya and Minoru Muramatsu, "A Survey of Health and Demographic Aspects of Induced Abortion in Japan-Special Report No. 2," *Bulletin of the Institute of Public Health* (December 1953): 18–24.

[31] Yoshio Koya and Minoru Muramatsu, "A Survey of Health and Demographic Aspects of Induced Abortion in Japan-Special Report No. 3," *Bulletin of the Institute of Public Health* (1954): 1–9.

[32] Yoshio Koya and Minoru Muramatsu, "A Survey of Health and Demographic Aspects of Induced Abortion in Japan-Special Report No. 4," *Bulletin of the Institute of Public Health* (June 1955): 7.

concluded that induced abortions were not only a public health risk but also a liability for the government, which was tackling the problem of surplus population by fertility regulation.

The results of these abortion studies substantiated Koya's campaign to persuade the government to establish a birth control policy from the point of public health. Koya's wish came true with the cabinet's decision to popularize birth control in October 1951 (see Chapter 5). However, Koya was simultaneously confronted with another problem: There was a dearth of information on the kind of guided birth control initiative he wanted to promote in Japan. Then, "the idea occurred to [Koya] of setting up test-projects for family planning in order to get much needed information which would assist the Government when it started its campaign."[33] This was the idea behind the series of family planning pilot projects Koya ran with DPHD staff and associates in the 1950s.

The first pilot project Koya headed was the aforementioned "three-village study," which began in September 1950. Reflecting Koya's objective, the project was designed to test the effectiveness of a guided birth control program for lowering the birth *and* abortion rates. In cooperation with local health authorities and other community-based organizations, DPHD researchers set up three "family planning model villages" (*kazoku keikaku moderu mura*) and recruited a total of 7,133 persons from 1,325 households.[34]

The "model villages" selected for the pilot study were Kamifunaka and Fukuura Villages in Kanagawa Prefecture and Minamoto Village in Yamanashi Prefecture. According to the project, Kamifunaka represented a "rice-cultivating" village, Fukuura a "fishing" one, and Minamoto a "farming" one.[35] Together, the three villages epitomized "typical" Japanese rural communities.[36] The choice of the rural area for the pilot project was obvious for Koya. The birth rate in the countryside was higher than in cities. Recognizing this fact, Koya thought rural communities should be a target of the government's guided birth control program, thus he selected these "typical" villages for the pilot project.

To test the efficacy of medical guidance, Koya made sure medical practitioners played a central role in the pilot scheme. He assigned his medically qualified subordinates at the DPHD to the roles of local

---

[33] Yoshio Koya, "The Experience of Seven Years Guidance in Family Planning in Farming Areas in Japan," October 1957, 2, Series III, Box 96, Folder 1570, Gamble Papers.

[34] Koya et al., "Seven Years of a Family Planning Program," 364.

[35] Ibid., 363–64.

[36] Koya et al., "Test Studies of Family Planning," 3.

doctors, in addition to being medical researchers. Thus, to start with, the DPHD researchers visited the villages at least once a month. At one visit quite early on, the principle staff in charge of the village gathered some women and gave them a lecture on the benefit of contraception. After the lecture, he ran group sessions to discuss the techniques for using contraceptives in more detail. At another visit, he stayed for several days and conducted one-on-one interviews with the women.

After the initial stage, a local midwife employed by the study ensured the day-to-day running of the initiative. For instance, in Minamoto Village, the local nurse and midwife Mrs. Amari Tatsuyo checked whether or not the women had correctly recorded their menstrual cycles and use of contraceptives.[37] She then collected the data from the women and hand them over to the local intermediaries, who passed them on to the DPHD staff. The DPHD team compiled a two-page summary sheet every six months based on the data; it contained charts on the birth rates, pregnancy rates, contraceptive use, reasons for the contraceptive choice, and so on.[38] This data became the foundation for the pilot project's analysis. Throughout the process, the principle DPHD staff fulfilled the roles of local doctors and midwives to assess the feasibility of birth control guidance work.

The pilot project, which in the end lasted for seven years, was evaluated as a success, at least by Koya. First, it showed that the crude birth rate dramatically decreased over the period. From 1949–50, before the beginning of the project, the crude birth rate per 1,000 persons was 26.7. The rate began to decrease soon after the pilot study commenced. The decline was so steep that it went down to 13.6 in 1956–57, the final years of the project. Second, the rate of induced abortions per 1,000 persons in all three villages began to decrease from the third year of the project, while the national average kept rising throughout the period. Toward the end of the project, the rate was kept as low as 1.4 in the model villages, whereas the nationwide figure rose to a phenomenal 13.0. Based on these figures, Koya concluded that medical guidance work was effective for persuading women to opt for contraceptives instead of induced abortions.

---

[37] Homei, "Midwife and Public Health Nurse Tatsuyo Amari."

[38] Kokuritsu Kōshū Eisei In Eisei Jinkō Gakubu, "San moderumura ni okeru kazoku keikaku jisshi jōkyō (Shōwa 30-nen 6-gatsu 1-nichi genzai)," June 1955, Unit 398-E1016, Slip Number 0116, Folder 180, Files 0182-0186, 0189-0190, 0201, Minamoto Aiikukaikan Hozon Shiryō, Minamiarupusu-City, Yamanashi-Prefecture, Japan [thereafter Minamoto Papers].

The pilot study gave a boost of confidence to Koya as a policy advisor. Now, inspired by the study's results, Koya tried to exert further influence within the government. He negotiated with a succession of health ministers to increase the government's budget for the birth control guidance work. Koya's lobbying activities were effective. In 1952, the national appropriation of birth control education increased from the equivalent of $75,000 to approximately $110,000 in 1953 and 1954.[39] Furthermore, when the government raised the budget for family planning again in the fiscal year of 1957, the government earmarked an increase in paying allowances to "case-workers (midwives and nurses) to encourage their work."[40] The government clearly responded to Koya's requests he made based on the three-village study.

As such, the three -illage study illustrates how Koya used the pilot project to carve out his role within the government. At the same time, Koya's perception of the project's value can be explained by his position within the broader popular birth control movement. In the mid-1950s, when the study was in full swing, Japan witnessed a convergence of the until then fragmented birth control campaigns. In April 1954, the FPFJ was launched as a nationwide birth control advocacy organization, and the activists united to host The Fifth International Conference on Planned Parenthood in Tokyo in October 1955. However, beneath the surface, schisms continued to exist. Some doctors participating in the activism were dissatisfied with the fact that they were seen in the same light as "lay" activists under the united front.[41] As a medically trained technical bureaucrat, Koya was similarly unhappy with the loosening boundary between medical and lay activists. What bothered him most was that "lay" activists, who he saw did not have any knowledge of statistics, were gaining public attention by stressing the effectiveness of their campaigns using numbers. Koya's opinion of The Research Institute for Better Living – led by Ishikawa Fumiko and Tanabe Hiroko and consisting of twenty-two wives of prominent industrialists and businessmen – perfectly epitomized his views on such "lay" activists. When Edna McKinnon visited Japan as a member and representative of the

[39] R. B. Gamle and C. J. Gamble, "Summary of Japan," 1, March 1953, Series III, Box 94, Folder 1539, Gamble Papers; "Yearly Budget for Family Planning," n.d. c.1959, Series III, Box 95, Folder 1568, Gamble Papers.
[40] "The Budget for Family Planning in the Fiscal Year of 1957," *Family Planning News*, March 1957, Series III, Box 95, Folder 1561, Gamble Papers.
[41] However, the tension between "lay" and "medical" birth control activists was nothing new. According to Tenrei Ōta, a socialist birth control activist and the doctor credited with the invention of an intrauterine device, it already existed in the 1920s. Tenrei Ota, "Ota ringu no hanseiki," *Gendai no me* 20, no. 9 (1979): 236–45.

International Planned Parenthood Federation (IPPF) Far East and Australasian Region to observe the birth control situation and asked Koya to comment on the organization, Koya dismissively told McKinnon: "I … can't understand what they are doing, though they are always talking about how they have got effective results … they have never shown them clearly scientifically."[42] Koya thought lay activists who were untrained in biostatistics could never show the effectiveness of their campaigns "scientifically," unlike doctor-activists such as himself. In this context, Koya's birth control pilot projects could be read as his attempt to draw a clearer line between "medical" and "lay" activists.

Koya's birth control pilot projects also depict how public health demography and the state-led population management endeavor influenced each other. Based on Koya's campaign to promote a scientifically proven birth control program, public health demography supplied useful knowledge to policymaking, while the agenda in the national policy further molded the research content in public health demography. Through Koya, birth control research in public health demography and the government birth control policy had a coproductive relationship. As the next section shows, the coproductive relationship also worked to consolidate the preexisting social order.

The Coproductive Effort to Govern Lower- and Working-Class Populations

Though the three-village study was going smoothly, in the first half of the 1950s, Koya increasingly saw the need for further pilot studies. The three-village study was demonstrating how effective birth control education work was at lowering the birth rate, but Koya thought it did not sufficiently engage with the issues of population quality. Specifically, the study was not responding well to the question of whether or not the education work could stop, or at least slow, the process of "reverse selection."[43] During this period, he felt a sense of urgency, particularly when he found out that the birth control endorsed by the government had "become widespread only among the intellectual class and is not easily disseminated among extremely poor people of the lower class of

---

[42] McKinnon, "Report for the Far East and Austrasian Region," 1.

[43] Another reason was because the three-village study had also received criticism that it had purposefully chosen the villages in order to succeed. Oliver R. McCoy, "Oliver R. McCoy to Donald H. McLean, Jr.," April 15, 1953, Population Council Collection, Record Group 1, Accession 1, Series 1, Box 18, Folder 300, RAC.

society."[44] As in wartime, Koya feared the trend would ultimately "lead to a lowering of the average quality of the nation."[45] He began to think that the situation would need to be investigated thoroughly, and if necessary, there would need to be government intervention.

Government officials shared Koya's sentiments, not least represented in Minister of Health and Welfare Ashida Hitoshi's statement about Japan's being confronted with the racial crisis (see Chapter 5). Following Ashida's remark, the Round-Table Conference on Population Problems (*Jinkō Mondai Kondankai*), organized by the MHW in January 1946, recommended that the postwar Japanese government should prioritize measures that would assist the "improvement of the hereditary and acquired quality of the population."[46] Following this recommendation, after the enactment of the EPL in 1948, the MHW promoted "eugenic marriage" and entrusted the local Eugenic Marriage Consultation Offices to propagate a practice of birth control that would follow eugenic principles.[47] Yet, in the early 1950s, as Koya pointed out, a rift in the practice of birth control emerged among social classes, heralding "reverse selection." Under these circumstances, MHW officials were seeking countermeasures to this trend.

In this context, in April 1953, Koya launched pilot projects aiming to assess the feasibility of a birth control "education program" for the two distinct populations he deemed as making up "the lower class of society," "a most miserable group because of their poverty and excessive family size."[48] One was the Katsushika Ward study, targeting the residents of Katsushika Ward in Tokyo who received public relief under the Livelihood Protection Law.[49] Another was the Joban study, which involved coalminers employed by the Joban Coal Mine Co. Ltd. in Fukushima

---

[44] Yoshio Koya, "Family Planning in the Population on the Public Relief Program," n.d. c.1956, 3, Rockefeller Family Archive, Record Group 5, Series 1, Subseries 5, Box 80, Folder 671, RAC, 3, Gamble Papers.

[45] Yoshio Koya, "Review of Past Achievements and Planning of Future Studies," n.d. c.1961, Series III, Box 97, Folder 1585, Gamble Papers.

[46] Cited in Naho Sugita, *Jinkō, kazoku, seimei*, 209.

[47] Takeuchi-Demirci, *Contraceptive Diplomacy*, 164–67.

[48] Yoshio Koya, "Family Planning Practice in Households on Public Relief of Katsushika Ward, Tokyo," April 1956, Population Council Collection, Record Group 1, Accession 1, Series 1, Box 18, Folder 300, RAC; and Yoshio Koya, "Present Situation of Family Planning among Farmers and Coal Mine Workers in Japan," *Archives of the Population Association of Japan* 3 (March 1955): 6.

[49] Ibid. For a social history of poverty and social benefits in post-WWII Japan, see Yoshiya Soeda, *Seikatsu hogo seido no shakaishi [zōhoban]* (Tokyo Daigaku Shuppankai, 2014); Masami Iwata and Akihiko Nishizawa, *Poverty and Social Welfare in Japan*, Japanese Society Series (Melbourne: Trans Pacific Press, 2008).

Prefecture, approximately 160 kilometers northeast of Tokyo.[50] For the Katsushika Ward study, researchers initially recruited 418 women but in the end could only analyze 277.[51] For the Joban study, the study initially selected Joban's Iwasaki District, with a population of 3,672 forming 716 households.[52] On October 1, 1953, the site expanded to include Nakago District, and Kaminayama District was included on October 1, 1955.[53] For both studies, Koya headed the project and Kubo, Ogino, and Yuasa from the DPHD did the fieldwork.[54] The Katsushika Ward study lasted for three years and the Joban study for five.[55]

By assuming that their subjects were fertile because they had no idea about birth control, researchers in both studies were initially convinced that their projects would encounter challenges. For the Katsushika Ward study, Kubo had anticipated that participants would reject their pilot project because of the "fixed opinion that poor people are more ignorant and uncooperative toward conception control.. as attested by the proverb, 'poor people have more children [binbō kodakusan].'"[56] Likewise, researchers in the Joban study started the fieldwork without doubting the "reputation" that coalminers "produc[e] many children, and [these children are] attendant hardships."[57] According to them, coalminers' quality, "known colloquially in Japanese as Tanko Binbo Ko-dakusan ['coalmine poverty and fecundity'']," originated from their low level of education, lack of skills to organize life and fatalistic attitude to life

[50] Yoshio Koya, "Five-Year Experiment on Family Planning among Coal Miners in Joban, Japan," *Population Studies* 13, no. 2 (November 1959): 157.
[51] Yoshio Koya, *Pioneering in Family Planning* (New York: Population Council, 1963), 59.
[52] Yoshio Koya, "A Study of Family-Planning in Coal-Miners," n.d., 2, C2158, Folder f.2240, Taeuber Papers.
[53] Yoshio Koya, "The Progressive Reduction of Pregnancy Rates and Birth Rates during Five Years of Family Planning Programs in Japan," n.d. c.1958, 2, Series III, Box 96, Folder 1570, Gamble Papers.
[54] Yoshio Koya, "Good Result of Conception Control in Down Town. Poor Families Welcome Contraception. Deliveries Decreased to One Third," April 13, 1954, Series III, Box 94, Folder 1548, Gamble Papers.
[55] Koya, "Five-Year Experiment on Family Planning among Coal Miners in Joban."
[56] Koya, "Family Planning Practice in Households on Public Relief of Katsushika Ward."
[57] Koya, "Five-Year Experiment on Family Planning among Coal Miners in Joban," 157. However, Koya also acknowledged the status of coal miners as "an economically important group because this industry furnishes a valuable source of energy to Japanese industrial development." Koya, "Present Situation of Family Planning among Farmers and Coal Mine Workers," 6. The coal mine industry in Joban would see a decline as Japan's source of energy shifted to petroleum. To survive, Joban Coal Mine Ltd. diversified its operations to tourism and opened a spa resort called Joban Hawaiian Center in 1966. Joban was also an area affected by three disasters – an earthquake, tsunami, and the nuclear explosion that occurred in northeastern Japan – in March 2011. Mire Koikari, *Gender, Culture, and Disaster in Post-3.11 Japan* (London: Bloomsbury Publishing, 2020), 108.

originating from the exposure to risk at work.[58] This way of portraying the research subjects was based on widely held assumptions about the lower socioeconomic classes. Phrases such as *binbō kodakusan*, which linked poverty, ignorance, and fecundity, was a common trope describing this social group, and it certainly informed the medical researchers' views of their research subjects at the beginning of the pilot studies.

But, as the studies progressed, researchers were compelled to revise their views. In the Katsushika Ward study, Koya and his colleagues were pleasantly "surprised" that 63.1% of the research participants "actually wanted teaching [on birth control]."[59] On one occasion, a woman desperately expressed her wish for their guidance by confiding in the researchers that she intentionally stayed up and did extra manual work "just to avoid the chance of pregnancy."[60] In view of this, they came to think that the majority of the research participants in Katsushika had a large family size, not because they were inherently fecund but because various external circumstances had prevented them from accessing contraceptives thus far.

Likewise, in Joban, 352 out of the 716 coalminers' wives in Iwasaki District who were interviewed for the study expressed their willingness to practice birth control. Of the 352, 204 even stated they wanted to practice contraception *because of* their economic condition.[61] Furthermore, they also learned that 106 already had previous experience with contraceptives[62] and 172 with induced abortions.[63] Also surprising to them was the fact that 45 of the 106 women who had practiced birth control opted for the "safe period," or the rhythm method, which Koya had deemed unsuitable for individuals with a low level of education, allegedly because it required calculation skills.[64] These findings compelled the medical researchers to conclude that "coalmine poverty and fecundity" was a myth, at least in Iwasaki.

---

[58] Yoshio Koya, "Preliminary Report: Study of Family Planning Guidance in a Coal Mining Village in Fukushima Prefecture, Japan," July 15, 1953, Population Council Collection, Record Group 1, Accession 1, Series 1, Box 18, Folder 300, RAC, 1; "A Study of Family-Planning in Coal-Miners," 5.

[59] Koya, *Pioneering in Family Planning*, 57.

[60] Yoshio Koya, "Good Result of Conception Control in Down Town."

[61] Koya, "Preliminary Report: Study of Family Planning Guidance in a Coal Mining Village in Fukushima Prefecture, Japan," 4.

[62] "The methods used by the 106 women ... in the past were as follows: Condom 41; Tablet 5; Jelly 2; Withdrawal 3; Safe period 22; Safe period with condom 22; Safe period and withdrawal 1; Condom or withdrawal 7; Condom or tablet 2; Douche and tablet 1." Ibid., 6.

[63] Koya, "Five-Year Experiment on Family Planning among Coal Miners in Joban," 158.

[64] Koya, "Preliminary Report: Study of Family Planning Guidance in a Coal Mining Village in Fukushima Prefecture, Japan," 5–6.

Faced with these findings, researchers in both sites conducted further interviews to ascertain the reasons behind the high birth rate. In both cases, they found many couples were indeed not practicing birth control and thought this was certainly a factor for high fertility. But they also discovered not all nonpractice was culpable. For instance, some reasons women provided for their nonpractice, such as they "were sterile or had been sterilized," "passed childbearing age," lactating, or just married and wishing to have the first baby, had no bearings on the fertility rate.[65] At the same time, other reasons, in particular that couples "did not know [about] the methods of contraception" and that they "could not buy the contraceptive chemicals or instruments," could have a direct impact.[66] When analyzing the interviews, researchers therefore focused on these latter statements and concluded that they would stress the guidance work and distribute contraceptives for free in their studies. Consequently, both in Katsushika and Joban, the research teams tested how much impact these specific measures had for lowering the birth rate.

In the end, both the Katsushika Ward and Joban studies produced impressive results, at least in numbers. In Katsushika Ward, among the 277 women who stayed in the project until the end, pregnancies dropped from 92 to 33 cases between the start of the project in April 1953 and its completion in April 1956. In tandem with this, the number of live births and abortions was more than halved during the period.[67] Like the three-village study, the decline in fertility corresponded to the fall in the number of induced abortions, from 45 to 22 cases. Similarly, in Joban, the pregnancy rate calculated using the Stix-Notestein method declined from 41.0 between 1952 and 1953 before the project began to 15.9 between 1957 and 1958 at the end of the project, which marked a reduction of 61 percent. During the period, the number of live births in Iwasaki dropped from 130 cases out of 716 families participating in the project to 17 out of 590.[68] Furthermore, in the third year of its guidance, the decline in pregnancy and birth rates occurred simultaneously with a reduction in abortions.[69] With the pilot studies, the DPHD team showed that targeted birth control guidance work, combined with the free distribution of contraceptives, could reduce the number of induced abortions and pregnancies.

[65] Koya, *Pioneering in Family Planning*, 57.
[66] Koya, "Good Result of Conception Control in Down Town."
[67] Koya, *Pioneering in Family Planning*, 59.
[68] Koya, "Five-Year Experiment on Family Planning among Coal Miners in Joban," 160.
[69] Ibid.

Based on this experience in Katsushika Ward, Koya expanded his campaign to include education work specifically targeted at populations whose fertility practices he reckoned would exacerbate differential fertility. One element of this was what Koya called an "enlightenment" or "educational" activity, namely, a lecture tour he made in rural areas to preach the benefits of birth control. In the 1950s, Koya traveled across Japan, from Aomori, the northern tip of the main Honshu Island, to as far south as Ibusuki, located in the southern shores of the southern island of Kyushu.[70] Everywhere he went, he seemed to attract a crowd. In 1959, when he made a tour in the north, Koya found himself with "more than 1,000 people" who wanted to try the contraceptives he introduced while there.[71] As Koya saw it, these "enlightenment" activities were particularly effective precisely because they enabled him to reach out to the populations that he considered should practice birth control.

The Katsushika Ward study also encouraged Koya to persuade the government to support a family planning program targeting the poor.[72] The study's results gave "a good reference to the Government," and consequently, the government "establish[ed] a budget" for this purpose.[73] In 1955, the government enforced a birth control program among "the extremely poor."[74] The government scheme aimed to cover approximately 276 thousand couples, which consisted of 155 thousand who were on the public relief scheme and the remaining 121 thousand were "so-called borderline groups – not so indigent as the former groups but … living at the subsistence level." By 1957, the amount the government allocated to the project amounted to 32,375,000 yen (90,000 US dollars). Within the scheme, contraceptives were distributed free

[70] Yoshio Koya, "Statement of Expenditure of the Fund Provided by Pathfinder Foundation from 1962," January 1963, Series III, Box 97, Folder 1596, Gamble Papers; Yoshio Koya and Tomohiko Koya, "Statement of Expenditure from the Grant in Aid by Pathfinder Foundation (January 1–December 31, 1961)," February 1, 1962, Series III, Box 97, Folder 1588, Gamble Papers; Koya to Gamble, September 4, 1959, Series III, Box 95, Folder 1567, Gamble Papers.

[71] Koya to Gamble, September 4, 1959.

[72] Koya, "Family Planning in the Population on the Public Relief Program," 4.

[73] Yoshio Koya, "Supplementary Note on the Items Discussed with Mr. Rockefeller 3rd on February 27, 1957," 1957, Population Council Collection, Record Group 1, Accession 1, Series 1, Box 19, Folder 301, RAC.

[74] Koya, *Pioneering in Family Planning*, 53; "Family Planning for the Poor," *Family Planning News*, June 1961, Series III, Box 97, Folder 1587, Gamble Papers; "Fertility Control among Indigent People," *Family Planning News*, July 1957, Series III, Box 95, Folder 1558, Gamble Papers; Frederick Osborn, "Memorandum of Conversation with Dr. McCoy of the Rockefeller Foundation, September 22, 1955 Re Japan," September 1955, Population Council Collection, Record Group 1, Accession 1, Series 1, Box 18, Folder 300, RAC.

of charge, or at one-half the normal price, to the groups, and the participants were offered opportunities for "practical consultations about conception control." As of March 1957, reportedly over 217 thousand couples were covered by this program.[75] The Katsushika Ward study led to the government program to popularize birth control among the lower socioeconomic classes.

In turn, the Joban study maintained close ties with state-endorsed corporate family planning campaigns, not least the one that became the New Life Movement. At quite an early stage of the Joban study, Nagai Tōru, as the New Life Movement's IRPP representative, and Fujiwara Kanji, a consultant for *Mainichi Newspaper*, visited the president of Joban Coal Mine Ltd. to take part in the New Life Movement.[76] The company was initially reluctant, but after meeting with the president, it responded positively to their invitation and joined the movement. Eventually, Joban Coal Mine Ltd. received an award from the MHW for its pioneering role in the corporate-led birth control program and "for the striking interest and co-operation with the [state-initiated] family planning program."[77] The Joban study certainly triggered the private-government cooperation in regulating workers' fertility for the sake of economic prosperity of the nation.

The Katsushika Ward and Joban studies show how the coproductive relationship between the scientific practice in public health demography and the government's policy to control size and quality produced a set of concerted efforts to regulate lower- and working-class fertility. Underlying these efforts was the anxiety, fueled by a classist attitude, about the possibility of the nation's decline brought about by the unhinged sexual activities of poor and blue-collar workers. Spurred by this anxiety, the coproductive relationship produced knowledge about how to reduce fertility among the target groups as well as concrete policy actions that aimed to intervene in their reproductive lives. From the perspective of policymakers and population experts concerned about "overpopulation" and the decline in the quality of the Japanese population, the science–policy nexus observed in this episode yielded the intended outcome. However, by not seriously addressing the prevailing class ideology inscribed in the ways knowledge about the target groups' fertility was produced, the coproductive relationship ultimately failed to lead to a conceptual shift in the existing social order.

[75] Koya, "Fertility Control among Indigent People."
[76] Yasuko Tama, *"Kindai kazoku" to bodī poritikkusu*, 108–9.
[77] Koya, "Five-Year Experiment on Family Planning among Coal Miners in Joban," 163.

Struggling to Produce Relevant Knowledge: Local Conditions and Infrastructures

What made the DPHD pilot studies particularly successful in terms of policy terms – what made the coproductive relationship actually "productive," as described above – was that the studies produced relevant knowledge in the eyes of the policymakers. The policymakers wished to see evidence that a public health program promoting contraception would sway the behaviors of women, in particular wives from poor families and married to blue-collar workers, away from induced abortions and toward contraceptives. The DPHD studies impeccably demonstrated that the women taking part in the pilot schemes took up contraceptives and turned away from abortion and, consequently, fertility rates fell. The DPHD delivered knowledge almost tailor-made for the policymakers.

Yet, the process of making this tailor-made knowledge was not easy. Unlike the three-village study, both the Katsushika Ward and Joban studies were rife with problems from the beginning. To start with, in Katsushika, the staff found it difficult to keep track of research participants because of the high turnover rate. For the study, the three staff managed to recruit 418 women from the community. However, by April 1, 1955, they had to shed the data collected from 60 participants because they no longer were relief recipients, had moved elsewhere, or for some other reason had dropped out of the project. By April 1, 1956, a further 133 had left the project for the same reasons.[78] In the meantime, a few new women joined the project. In the end, only data from 277 households was valid for analysis.[79]

Next, the routine guidance work established by the three-village study did not work in Katsushika.[80] Women simply did not attend the group sessions, which were a core component of the guidance work. The trend became most conspicuous when the participants were asked to pay for the transportation to attend the sessions.[81] Home visits, another crucial activity for the guidance work, turned out to be futile as well. Typically, a home visit included conducting interviews that required a space where the research participants felt comfortable enough to share their personal information with the interviewers. However, the DPHD staff found that the participants' houses were typically crowded and thus not conducive to the interviews.[82]

[78] Koya, "Family Planning Practice in Households on Public Relief of Katsushika Ward."
[79] Koya, *Pioneering in Family Planning*, 59.
[80] Ibid., 58–59; Koya, "Family Planning in the Population on the Public Relief Program," 15–16.
[81] Koya, "Family Planning in the Population on the Public Relief Program," 15.
[82] Ibid.

Finally, in an entirely opposite manner from the three-village study, the DPHD team failed to offer a variety of contraceptive options to the Katsushika Ward study participants. The challenge stemmed from the lack of an appropriate physical setup for proper guidance. In Katsushika, the research team could not offer diaphragms, for example, because this would require a room with at least basic healthcare equipment.[83] This potentially had a negative impact on the study's outcome. In the three-village study, one crucial factor leading to the study's success was the wide range of contraceptive choices given to the research participants. The limited contraceptive choices in the Katsushika Ward study therefore meant the pilot study itself might fail.

In Joban, the biggest challenges were the lack of company support in the initial stage and the recalcitrantly high abortion rates.[84] With regard to induced abortion, to the researchers' dismay, they found that the abortion rate in Iwasaki was already higher than the national average when the project began, and it became even higher as the project progressed.[85] Even worse, in the first year, it was induced abortion, not contraception, that brought a sharp reduction in crude birth rates, from 33.5 to 20.8 per 1,000 persons.[86] The research team speculated that the high abortion rate was ironically facilitated by the medical subsidy offered as part of the company welfare package.[87] As Koya once pointed out, "facilities of the Company-operated hospital were made readily available to those who wanted to [have] induced abortion.... For wives of employees the charge for this operation was only 300 yen (80 cents)."[88] Because of the locally specific health infrastructure, women found it easy to have abortions, which contributed to the rising abortion figures.

Faced with these challenges, the DPHD research team was compelled to be flexible when adapting to local conditions. In Katsushika, after

---

[83] Koya, "Family Planning Practice in Households on Public Relief of Katsushika Ward."

[84] With regard to the lack of company support, see Yoshio Koya, "Estimates of Economic Effect to Be Obtained by Birth Control at Joban Coal Mine," April 9, 1955, Series III, Box 95, Folder 1553, Gamble Papers; "Birth Rate Has Fallen Half – Results of 2-Year's Guidance of Family Planning Revealed," May 16, 1955, Rockefeller Family Archive, Record Group 5, Series 1, Subseries 5, Folder 671, RAC. With regard to the high abortion rates, see Koya, "Present Situation of Family Planning Among Farmers and Coal Mine Workers in Japan," 6.

[85] Koya, "Five-Year Experiment on Family Planning among Coal Miners in Joban," 160.

[86] The figures look even more remarkable when compared to the nationwide average, which were 23.4 before the project and 21.5 in 1954. Ibid., 160; Koya, "Yoshio Koya to Clarence J. Gamble," n.d. c.1954, Series III, Box 94, Folder 1548, Gamble Papers.

[87] Koya, "Present Situation of Family Planning Among Farmers and Coal Mine Workers in Japan," 4–5.

[88] Koya, "Preliminary Report: Study of Family Planning Guidance in a Coal Mining Village in Fukushima Prefecture, Japan," 7.

exploring various possibilities, the team eventually decided to liaise with the local welfare agency.[89] They thought that waiting at the local welfare office would be a sure way to "get in contact with [the research participants] satisfactorily," because relief recipients came to the office once a month to receive funds. Taking advantage of these circumstances, the team managed to run interviews and individual guidance sessions once a month in a separate room within the welfare office.[90] In turn, to solve the issue of the limited range of contraceptives, the team in the end decided to change the framework of the Katsushika Ward study. They now argued they were focusing on analyzing the increased use of "simple contraceptive methods" (e.g., condoms and foam tablets) women could use without prior medical knowledge or supervision.[91]

In Joban, to tackle the problem of noncooperation, Koya used a financial argument to persuade the company.[92] He claimed:

At present, the Company provides its employees with the childbirth allowance of ¥1,000 nursing allowance of ¥300 per month for six months, as well as with the family allowance of ¥400 per month per child up until the child gets [to be] 18 years old. If it was assumed that the birth rate ten years later would be 13.7 per 1,000 population [instead of 33.5 currently], it is estimated that a huge sum, 266 million yen, will be saved by the practice of family planning for the Joban Coal Mining Company alone ... covering 8,700 households in total.[93]

The financial incentive presented by Koya seemed to shift the company's attitude. By 1955, there had been a clear sign that the company was endorsing the program fully.[94] By then, the preliminary results "pleased the company staff greatly," so much so that "another coal mine village, which [is] situate[d at a] 10 mile distance from Nakago, also is hoping to have our guidance very earnestly."[95] In addition, "[c]o-operation reached a point in 1956 where the company offered to pay part of the cost of contraceptives and salaries of midwives."[96]

[89] Koya, "Family Planning in the Population on the Public Relief Program," 15.
[90] Ibid., 16.
[91] Among the 277 participants, figures for their contraceptive use were as follows: "Condom alone 103; tablet alone 40; condoms or tablet 8; withdrawal 5; sponge 5; safe period 3; condom or withdrawal 2; tablet, safe period 1; condom or tablet, safe period 1; intrauterine ring 1; sterilization 6." Koya, "Family Planning Practice in Households on Public Relief of Katsushika Ward."
[92] Koya, "Estimates of Economic Effect to Be Obtained by Birth Control at Joban Coal Mine"; "Birth Rate Has Fallen Half."
[93] Koya, "Birth Rate Has Fallen Half."
[94] Yoshio Koya to Clarence J. Gamble, March 4, 1955. Series III, Box 95, Folder 1553, Gamble Papers.
[95] Ibid.
[96] Koya, "Five-Year Experiment on Family Planning among Coal Miners in Joban," 163.

In conjunction with this, to rectify the issue of the high abortion rate, the team took two strategies: First, in the guidance work, the DPHD team stressed the health hazard of repeated induced abortions.[97] Second, they aligned themselves with the company doctor Kimura to persuade the women to opt for contraceptives instead of induced abortions. The two tactics seemed to work.[98] From the second year on, the abortion rates began to decline.[99]

Ultimately, the challenges in both studies made the DPHD staff acutely aware that birth control education work was so embedded in locally unique situations that it would be impossible to implement a standardized program across the country. Time and again, the DPHD team observed how the methods established in the three-village study could not be replicated in Katsushika or Joban, and each time, they were compelled to come up with creative solutions to adapt to the specific needs of the local community in order to tease out reliable data. Behind what seemed, at least on the surface, as an impeccably policy-relevant pilot program on the national level was researchers' efforts to flexibly adapt to locally specific issues on the ground level.

In addition to the ability to flexibly adapt to unexpected local situations, the support of existing local organizations, especially local women's groups, mitigated researchers' struggles with fieldwork. For the Joban study, a housewives' association in the Iwasaki District helped the midwife in charge, Mrs. Sagawa, and made sure that its members participating in the study would gather to watch the birth control promotion video and receive the pamphlet.[100] A similar phenomenon was observed outside of Joban and Katsushika. In Minamoto Village – from the three-village study – the local branch of the Imperial Gift Foundation Aiiku-kai, which had already been providing maternity and infant care for the villagers since before the war, provided the venue for the study and recruited women on behalf of the researchers.[101] In Kajiya, another test site, leaders of the women's organization affiliated with the local branch of the Japan Agriculture Cooperation offered reliable help.[102]

---

[97] Ibid., 160; Koya, "Present Situation of Family Planning Among Farmers and Coal Mine Workers in Japan," 5.

[98] Koya, "Five-Year Experiment on Family Planning among Coal Miners in Joban," 159.

[99] Yoshio Koya to Dudley Kirk, July 30, 1955, RAC Population Council Collection, Record Group 1, Accession 1, Series 1, Box 18, Folder 300.

[100] "Joban tankō no genchi wo yuku," *Kazoku keikaku*, May 20, 1956, Population Council Collection, Record Group 1, Accession 1, Series 1, Box 18, Folder 300, RAC.

[101] Yoshinaga, "The Modernization of Childbirth."

[102] "Jōzai ga unda mikan mura no kazoku keikaku," *Kazoku keikaku dayori*, July 15, 1960, 2–3, Series III, Box 97, Folder 1580, Gamble Papers. The Kajiya pilot study will be introduced in more detail below.

As Sheldon Garon's canonical work on prewar social management illustrates, in Japan, since the prewar period, women's organizations mobilized local women to reform everyday lives for the sake of the nation.[103] The DPHD researchers took advantage of this existing infrastructure to ensure the smooth operation of the pilot studies.

Finally, local midwives also helped the DPHD researchers tackle issues arising in the pilot projects. In many test villages, the biggest obstacle for the study was recruiting local women in the first place. Especially in small villages, mothers-in-law posed as an obstacle for the recruitment. In Kajiya, for instance, an elderly woman refused to let her daughter-in-law join the study, saying "the god of the child would punish them if you let women use medicines to stop a child from being born."[104] In this hostile environment, local midwives, working alongside female leaders in the local communities, helped the researchers recruit a sufficient number of women for the study.[105] In Kajiya, the midwife Mrs. Kimura did a round of home visits with Mrs. Kashiwagi, the head of the local branch of Japan Agriculture's women's group, to persuade the women directly. For the distribution of contraceptives, Mrs. Kimura talked with the head of the local neighborhood group (hanchō). Finally, in the actual field-work, Mrs. Kimura made sure to visit a woman at home when she knew the mother-in-law was absent. If the mother-in-law was in, she tried to pass the contraceptives to the wife discretely behind her back.[106] These efforts by the local midwives in their everyday conduct were certainly behind the success of the DPHD pilot projects.

However, what was particularly remarkable about the midwife in the DPHD pilot projects was that she also determined the quality of the pilot project's outcome as a *scientific investigation*.[107] For the pilot project to work as scientific research, it was vital that the project organizers obtained information about the sexual lives of the couples participating in the study. However, as Muramatsu recalled, this was a challenge. Out of "shyness," wives in the test villages often hesitated to talk about a subject as intimate as sex directly with the DPHD researchers, who were male and outsiders.[108] Under these circumstances, the midwife surfaced

---

[103] Sheldon Garon, *Molding Japanese Minds: The State in Everyday Life* (Princeton: Princeton University Press, 1997), 115–45.

[104] "Jōzai ga unda mikan mura."

[105] Homei, "Midwife and Public Health Nurse"; "Shidō rokunengo wo miru," *Kazoku keikaku*, August 20, 1956, Series III, Box 95, Folder 1557, Gamble Papers; "Joban tankō no genchi wo yuku."

[106] "Jōzai ga unda mikan mura no kazoku keikaku," Population Council Collection, Record Group 1, Accession 1, Series 1, Box 18, Folder 300, RAC.

[107] This did not mean that the DPHD researchers did not play any role in this task.

[108] Minoru Muramatsu, "Some Field Studies on Family Planning Practice among the Japanese and Their Overall Implications," 6, April 1959, Series III, Box 95, Folder 1569, Gamble Papers.

as the most reliable research assistant. She was not only a female health practitioner well connected with the local women but also well versed in their reproductive histories through her usual work. Furthermore, in the context of childbirth culture in 1950s Japan, where home visits and home births were still dominant, a midwife was used to visiting women at their homes, and in turn, many women shared the intimate details of their lives with the midwife. In other words, the midwife was a particularly effective data collector for the DPHD studies because she possessed a unique set of social attributes that allowed research participants to willingly disclose the information necessary for the study. For this reason, wherever possible, the DPHD assigned the local midwife taking part in the pilot scheme to gather raw data on behalf of the researchers.

In Minamoto Village, midwife/public health nurse Mrs. Amari was singled out for the role, and she diligently obtained data.[109] She checked whether or not the women taking part in the study in the village correctly filled in the postcard-sized calendar sheet they were given to record their menstrual cycles. She also interviewed the woman and noted relevant details, such as their choice of contraceptives, number of children, children's approximate age, and future family plans. These were cursory memos but contained personal information vital for the study, for instance: "Name: Mr. and Mrs. A; Number of Children: 3; Current Methods of Contraceptives: diaphragm and jelly; Notes: the youngest child in school this year. Planning to stop with three children."[110] Finally, if she found a woman became pregnant, Mrs. Amari paid an extra home visit and through an interview ascertained whether or not the pregnancy was due to a failure in the use of contraceptives. For instance, she noted that Mrs. B, who chose to use condoms for the study, became pregnant between January and June 1930, and she scribbled "sometimes [Mrs. B and her husband] didn't use [a condom] because they couldn't be bothered."[111]

The data collected by the midwife then became the basis for the study. Back in Minamoto Village, once she was done with home visits, Mrs. Amari passed the memos and calendars on to the local Aiiku-kai branch office, which collated her data and made fortnightly reports to the IPH with lists and charts.[112] Based on the reports, the IPH team did a more

---

[109] However, this did not mean that the midwife was a docile follower who simply obeyed the researcher's request. Sometimes she gave advice to women based on her own decisions. Homei, "Midwife and Public Health Nurse."

[110] "Kazoku keikaku," n.d., Unit E1298, Folder number 0200, 0601-0673, Files 0200-607, Minamoto Papers. In the original, the memos noted their full names.

[111] "Kazoku keikaku," n.d., Unit E1298, Folder number 0200, 0601-0673, Files 0200-620, Minamoto Papers.

[112] Ibid.

sophisticated calculation on contraceptive usage, as well as birth, fertility, and abortion rates, and produced their own reports.[113] Finally, based on their reports, Koya and his colleagues at the DPHD authored scientific papers.[114] Local midwives such as Mrs. Amari were therefore the initial point at which knowledge about the test villagers' reproductive habits and contraceptive use was entered and turned into scientific knowledge. The midwife's contribution was immeasurable for the production of what policymakers deemed relevant knowledge.[115]

Behind what appeared to be a smooth interplay between the research in public health demography and the national birth control policy was a number of challenges in the fieldwork that were rooted in locally specific conditions. The DPHD researchers tried to overcome the problems by flexibly adapting to the conditions and by gaining support from the local women and midwives. The midwife's support was invaluable. She not only helped the study by managing the day-to-day running of the pilot initiative but also by acting as a reliable fieldworker collecting data for the study. Because of the efforts of various groups participating in the pilot studies, the studies managed to produce the kind of knowledge that was highly desirable for domestic policy.

However, the reach of public health demography was not confined to domestic policy. During the 1950s, it became widely recognized within the international scientific communities that specialized in family planning and population studies. The discipline's globally celebrated status was due to cooperation with US-based philanthropists, in particular the Rockefeller Foundation, the Population Council, and Clarence J. Gamble played pivotal roles.

Integrating Public Health Demography into the Transnational Population Network: Rockefeller Foundation, Population Council, and Clarence J. Gamble

In the 1950s, in parallel with establishing public health demography as an applied research field primarily accountable for the domestic birth control policy, Koya also looked beyond the national government to expand the field. Fortunate for him, at the time, the transnational

---

[113] Kokuritsu Kōshū Eisei In Eisei Jinkō Gakubu, "San moderumura niokeru kazoku keikaku jisshi jōkyō (Shōwa 30-nen 6-gatsu 1-nichi genzai)," June 1955.

[114] Koya et al., "Seven Years of a Family Planning Program"; "Kazoku keikaku moderu mura no kenkyū," *Nihon iji shinpō*, no. 1475 (August 2, 1952): 3–11.

[115] "Kazoku keikaku shō ni kagayaku hitobito," *Kazoku keikaku*, 4–5, November 20, 1956, Series III, Box 95, Folder 1557, Gamble Papers.

movement aiming to contain the growth of world's population through fertility reductions in "developing countries" was quickly expanding, giving a great impetus to policy-oriented demographic studies. In particular, noted US-based philanthropic organizations pushed the movement forward by offering generous financial support to population research on a global scale. Koya saw the trend as an opportunity. Over the decade, he approached these American organizations to receive assistance and also urged his colleagues in the field to do the same. Thanks to this American support, public health demography became an internationally recognized field. Population experts outside of Japan called the DPHD a "unique existence in the world" and acknowledged its "pioneering" role in the field of population and family planning.[116]

The American philanthropists and organizations involved in the expansion of public health demography were the Rockefeller Foundation (RF), the Population Council (PC), and Clarence J. Gamble. Throughout the 1950s, they helped to integrate the scientific field into the transnational population network by financing the kind of research activities in Japan – in particular, those based in the DPHD – that they deemed would satisfy the needs of the international communities specifically dedicated to demographic science and population control.

The involvement of the RF in DPHD population research stemmed from its long-standing interest in the issues of population and race biology in Asia.[117] After WWII, the conversation between RF officer Roger F. Evans at the Social Sciences Division and SCAP-GHQ advisor Warren S. Thompson, as well as the reports by a number of RF officers who visited Japan, additionally raised John D. Rockefeller III's interest in Asia's population growth. In 1948, through the urging of Rockefeller III, the RF scientific directors authorized a survey trip to East Asia called the "Reconnaissance in Public Health and Demography in the Far East," or the Far East Mission as it was more commonly known. The delegates consisted of Frank W. Notestein and Irene B. Taeuber from the Office of Population Research at Princeton University (OPR) and Evans and Marshall C. Balfour from the RF, and the Far East Mission was to advise the RF on its future policy concerning population issues and studies in East Asia. The delegates toured Japan, China, Taiwan, Korea, Indonesia, and the Philippines from September to December 1948.[118]

---

[116] Kokuritsu Kōshū Eisei In, ed., *Kokuritsu Kōshū Eisei In sōritsu gojusshūnen kinenshi* (1988), 103.
[117] Takeuchi-Demirci, *Contraceptive Diplomacy*, 99–103.
[118] Ibid., 123–26.

The report the Far East Mission produced after the tour explicitly recommended a US-Japan interaction within population studies. The report claimed that, despite the imminent population crisis, the "research necessary for policy is neither available nor in process" in Japan. Based on this observation, it recommended the RF concentrate on assisting activities that would foster "cooperation" between Americans and the Japanese in "demographic research untrammelled by restrictions." In Japan, this specifically meant the "strengthening of the Institute of Population Research [most likely it referred to the IPP] and the Institute of Public Health for population study."[119] Such assistance, the report concluded, would "contribute indirectly to the development of feasible solutions" for the Japanese population problem.[120]

Between the two Japanese institutions that the Far East Mission suggested to be the recipients of the RF assistance, the IPH had a special historical link with the RF. In the first place, it was the RF's International Health Division (IHD) that gave financial assistance for the establishment of the IPH in 1938. After the war, in June 1948, the IHD dispatched its officer, Oliver R. McCoy, a Johns Hopkins graduate and specialist in parasitology, to Tokyo. McCoy was officially affiliated with the IPH as an advisor, while also serving for the SCAP-GHQ as an informal consultant. Through McCoy, the RF established its special status within Japan as an unofficial liaison between American and Japanese state and nonstate actors who had a say in the country's public health administration. The IPH offered a venue for the RF to establish this status.

Partly due to the IPH's special role with the RF, after the Far East Mission, the RF began financially supporting population research within the IPH.[121] After the establishment of the DPHD in 1949, the RF concentrated its funding stream on the DPHD. Recognizing that the Far East Mission produced a long report on the Japanese response to the EPL, the RF was willing to support research on the effects of abortion and on demography and public health. Thus, the RF funded the aforementioned abortion study.[122]

[119] Frank W. Notestein et al., "The Rockefeller Foundation Reconnaissance in Public Health and Demography in Far East," November 1949, 44c, Rockefeller Foundations Archive, Record Group 1.2, Series 600, Box 2, Folder 12, RAC.

[120] Ibid., 40.

[121] "Grant In Aid to Japan – Institute of Public Health, Tokyo, to Provide Additional Funds for the Purchase of Teaching Materials, Including Foreign Books, Periodicals, Motion Pictures, Etc.," November 1950, Rockefeller Foundations Archive, Record Group 1.2, Series 609, Box 6, Folder 44, RAC.

[122] "Grant in Aid to the Institute of Public Health, Tokyo, Japan for the Department of Public Health Demography for Continued Investigation of the Health and Demographic Aspects of Induced Abortion and, in Addition, Sterilization, as

Behind the scenes, McCoy made every effort to bring RF funding to the DPHD. For instance, it was McCoy that suggested Koya apply for RF funding for the abortion study.[123] Once Koya decided to apply for RF funding, McCoy helped him with the application process. Prior to the application, McCoy circulated application drafts among other RF staff for "comments" and lobbied for the application within the RF once it was submitted.[124] Thanks to the tireless efforts of McCoy, Koya's application was accepted. During the term of his office at the IPH, McCoy offered the DPHD staff generous help, and consequently, the DPHD as an organization managed to enjoy quite a sum of RF funding.

Despite its involvement in population research, the RF played a smaller role when it came to the birth control pilot studies. In the early 1950s, the RF as institution was reluctant to be seen openly supporting birth control research because of the politically charged nature of the subject.[125] For this reason, while in principle the RF would happily fund DPHD demographic studies, it avoided supporting its birth control pilot studies. Under these circumstances, the PC filled the void created by the RF, and it supported the birth control pilot project that presented the DPHD to an international audience in the 1950s.

The PC was founded on November 7, 1952 as a result of John D. Rockefeller III's arduous campaign to boost the transnational population control movement. Since the late 1940s, John D. Rockefeller III had been making efforts to persuade the RF to get involved in global population control, but he often met with resistance. A turning point for John D. Rockefeller III came in December 1951, when he met with Detlev Bronk, Lewis Strauss, and Warren Weaver.[126] Spurred by Harry

---

Authorized by the Japanese Eugenic Protection Law," December 10, 1952, Rockefeller Foundations Archive, Record Group 1.2, Series 609, Box 6, Folder 44, RAC; "Grant in Aid I the Amount of $2,500 to the Department of Public Health Demography, Institute of Public Health, for the Investigation of the Influences of Induced Abortion and Guidance in Conception Control upon the Recent Downward Trend in the Birth Rate in Japan," November 29, 1951, Rockefeller Foundations Archive, Record Group 1.2, Series 609, Box 6, Folder 43, RAC.

[123] See, for example, Diary of Oliver R. McCoy, July 12 1951.

[124] Oliver R. McCoy to Marshall C. Balfour, July 13, 1951, Rockefeller Foundations Archive, Record Group 1.2, Series 609, Box 6, Folder 44, RAC; Oliver R. McCoy, "Oliver R. McCoy to Andrew J. Warren," October 16, 1952, Rockefeller Foundations Archive, Record Group 1.2, Series 609, Box 6, Folder 44, RAC.

[125] Rebecca Williams, "Rockefeller Foundation Support to the Khanna Study: Population Policy and the Construction of Demographic Knowledge, 1945–1953," Rockefeller Archive Center Research Reports Online (Rockefeller Archive Center, 2011), www.issuelab.org/resources/28011/28011.pdf, p. 3.

[126] Bronk, at the time, was the director of the National Academy of Sciences but was soon to become the president of Rockefeller University. Strauss was the chairman of

S. Truman's famous inaugural address in 1949, which called for a US foreign policy that would offer technical assistance to developing countries in the context of the Cold War, the four agreed to plan a conference discussing the issues of global population and socioeconomic development.[127] The plan came to fruition with the Conference on Population Problems at Colonial Williamsburg, Virginia on June 20–22, 1952. The conference gathered notable demographers and public health specialists in the United States. At the end of the conference, the participants passed a resolution to establish an organization dedicated to issues related to global population control. The resolution gave birth to the PC.

Since its inception, Japan played an important role in the PC. As mentioned above, it was the Japanese population problem that ignited John D. Rockefeller III's interest in global population control in the first place. Moreover, among the attendants of the Williamsburg conference were demographers such as Thompson, Notestein, and Taeuber, who had recently visited Japan and observed the various governmental and nongovernmental responses to the problem of the surplus population. These demographers highlighted the importance of the Japanese case: It created a precedent, especially in Asia – arguably the problem area in terms of world population growth – for the kind of birth control program that should be endorsed as part of the transnational population control movement.[128]

The interest in Japan was maintained by those inside the country as well. For that, McCoy played a key role. McCoy saw the establishment of the PC as an opportunity for the DPHD, giving it an extra source of income. Thus, within six months of its establishment, McCoy contacted Donald H. McLean, Jr. of the PC and explored the possibility of the organization "mak[ing] a grant of assistance" to the "Department of Public Health Demography." In so doing, McCoy promoted DPHD birth control research with fervor, arguing that it was "a worthy project which should contribute to the solution of the population problem in Japan" and that the "implications [of the research] have had considerable influence on government policy."[129] By lobbying for the DPHD

the U.S. Atomic Energy Commission but also was a consultant for the Rockefeller brothers.

127 Michael E. Latham, *The Right Kind of Revolution: Modernization, Development, and U.S. Foreign Policy from the Cold War to the Present* (Ithaca: Cornell University Press, 2011), 93–122.

128 Yu-ling Huang, "The Population Council and Population Control in Postwar East Asia," Rockefeller Archive Center Research Reports Online: Rockefeller Archive Center, 2009, accessed June 28, 2011. www.rockarch.org/publications/resrep/huang .pdf.

129 McCoy, "Oliver R. McCoy to Donald H. McLean, Jr.," April 15, 1953.

study, McCoy not only informed the PC of the population situation in Japan but also kept the PC's interest in the country high.

In part due to McCoy's campaign, starting in the mid-1950s, the PC began to bankroll the DPHD's birth control studies. To start with, in 1953, it granted the DPHD $3,200 "for a three-year study of the feasibility of guidance in family planning among the lower economic classes in Japan," which was used to run the Katsushika and Joban studies.[130] In 1954, Koya requested further assistance on the projects, to which the PC gave approximately $2,730 to the DPHD.[131] In the face of the RF's noncommittal attitude to applied population research, public health demography nevertheless thrived in the early 1950s, in part because of the PC's financial support.

Along with the PC fund, equally significant to the expansion of public health demography was the funding offered by the medically trained American philanthropist Clarence J. Gamble (1894–1966).[132] Gamble, an heir of the famed Proctor and Gamble fortune and a graduate of Princeton University and Harvard Medical School, stumbled across the birth control movement after his academic research career with the pharmacologist Alfred N. Richards at the University of Pennsylvania became compromised. Using the financial independence granted by his family fortune, Gamble first joined in the cause by offering the Pennsylvania-based Committee for Maternal Health Betterment services for assessing contraceptives, which included jelly, suppositories, foam powder, and condoms.[133] Based on this experience, in 1934, Gamble established a research program at the National Committee on Maternal Health (NCMH), which aimed to discover better, more affordable, and "simple" contraceptives for the "unprivileged masses."[134] Between the 1930s and 1957, when he established the Pathfinder Fund, NCMH was the institutional base for Gamble to develop collaborative relationships

[130] McCoy, "Oliver R. McCoy to Yoshio Koya," April 23, 1954, Population Council Collection, Record Group 1, Accession 1, Series 1, Box 18, Folder 300, RAC; Donald H. McLean, "Donald H. McLean Jr. to Yoshio Koya," May 5, 1953, Population Council Collection, Record Group 1, Accession 1, Series 1, Box 18, Folder 300, RAC.

[131] Frederick Osborn to Yoshio Koya, July 1, 1954; Yoshio Koya to Frederick Osborn, June 11, 1954, Population Council Collection, Record Group 1, Accession 1, Series 1, Box 18, Folder 300, RAC; Oliver R. McCoy to Frederick Osborn, June 19, 1954, Population Council Collection, Record Group 1, Accession 1, Series 1, Box 18, Folder 300, RAC.

[132] Toyoda, "Sengo nihon no bāsu kontorōru undō."

[133] James Reed, *The Birth Control Movement and American Society: From Private Vice to Public Virtue* Princeton Legacy Library (Princeton: Princeton University Press, 1978), 233.

[134] Ibid., 241.

with researchers and public health officials who were striving to ascertain the best contraceptives and birth control initiatives for their locales. His activities were initially confined to the US South, but soon he also worked in Puerto Rico.[135] After WWII, Gamble joined the transnational movement to control the population in Asia and Latin America through his quest for "simple" contraceptives.[136]

Japan was the first country in Asia that received Gamble's support for birth control work. In 1949, Gamble learned about the population debate and birth control movement in Japan when he stopped to see Notestein at his Princeton reunion.[137] At the urging of Notestein, Gamble wrote to McCoy and Thompson to ask if it was possible for him to "give a contraceptive program there a push" by giving out "a few hundred dollars, a thousand or less" for each cause.[138] This initial contact led to Gamble's involvement in the Japanese birth control movement, enabling him to become connected to prominent figures such as Katō Shizue, Kitaoka Juitsu, Majima Yutaka (Kan), Tanabe Hiroko, and the Amano couple. Throughout the 1950s and until shortly before his death in 1966, Gamble, behind the scenes, bankrolled some of their birth control work and acted as their foreign mentor.

Among the Japanese birth control activists, Koya transpired to be by far the closest to Gamble. Koya and Gamble quickly forged an amicable relationship due to their shared intellectual profile and interests (Figure 6.1). To start with, Koya was medically trained, like Gamble. Furthermore, for both Koya and Gamble, eugenics was the driving force behind their birth control advocacy. As has been explained above, eugenics was the backbone of Koya's professional activities.[139] In turn, Gamble became a proponent of birth control when he became concerned that the post–Depression era policies, such as the New Deal, were leading to an expansion of the welfare-dependent population. Gamble believed birth control could cut

---

[135] Johanna Schoen, *Choice & Coercion: Birth Control, Sterilization, and Abortion in Public Health and Welfare* (Chapel Hill: University of North Carolina Press, 2005); Laura Briggs, *Reproducing Empire: Race, Sex, Science, and U.S. Imperialism in Puerto Rico* (Berkeley: University of California Press, 2002).

[136] Raúl Necochea López, "Gambling on the Protestants: The Pathfinder Fund and Birth Control in Peru, 1958–1965," *Bulletin of the History of Medicine* 88, no. 2 (2014): 344–71.

[137] Clarence J. Gamble to Warren S. Thompson, June 17, 1949, Series III, Box 94, Folder 1530, Gamble Papers.

[138] Ibid.; Clarence J. Gamble to Oliver R. McCoy, September 10, 1949, Series III, Box 94, Folder 1530, Gamble Papers.

[139] Yoshio Koya, "Statement of The Expenditures For The Work on Family Planning in Japan. From Jan. 1959 to Jan. 1960." Koya to Gamble, January 22, 1960, Series III, Box 97, Folder 1578, Gamble Papers.

Figure 6.1 A snapshot of Gamble and Koya, 1963.
Source: Private collection of Clarence J. Gamble's family member.

the welfare cost down, as well as the socially undesirable population's numbers.[140] Their common professional background and shared sense of values about population management facilitated the collaboration between
Gamble and Koya.

Throughout the 1950s, Gamble tirelessly supported a wide range of
Koya's birth control work. Among the activities he funded, some, such as
the aforementioned "enlightenment" or "educational" activity, contributed to Koya's work as a birth control activist. Still, as medical researchers, both Koya and Gamble were convinced of the importance of birth
control pilot studies. For Koya, they were indispensable for the growth
of the DPHD. For Gamble, they provided a valuable opportunity to test
the efficacy of a recently developed contraceptive in an actual village in
Asia, the target of global population control. For this reason, Gamble
ceaselessly funded the DPHD's birth control studies. It was Gamble who

---

[140] Toyoda, "Sengo nihon no bāsu kontorōru undō," 61–62; Reed, *The Birth Control
Movement and American Society*, 233–35.

bankrolled the three-village study.[141] He also supported others, namely, the part of the Joban study targeting the Nakago District, the Kajiya Village foam tablet experiment, and the National Railways pilot study.[142] Following his usual style, Gamble authorized the payment of between $1,000 and $3,000 annually, and an additional $500–$700 each time for extra research projects that he thought were worth supporting, and sent the funding to Koya via the NCMH and later the Pathfinder Fund.[143] Although the amount sent each time was small, it eventually accumulated into a large sum. Between 1950 and 1957 alone – the period during which the three-village study took place – the amount Gamble funded to Koya's pilot projects totaled $23,500.[144] In each of these studies, Koya duly produced and supplied Gamble with the data.

Over the 1950s, the news about DPHD birth control research reached noted demographers and advocates of global population control. Taeuber believed the "activities of Dr. Koya and the Institute of Public Health in the field of the diffusion of contraception are one of the most significant demographic experiments in the world today, if not the most significant, and those activities are very courageous ones" for the proponents of global population control.[145] Likewise, John D. Rockefeller III commended Koya after reading a number of his papers: "Each time I hear of the research work in the population field undertaken in Japan under your leadership, I am more impressed as to the magnitude and importance of the contribution which you have made. And what you have done is not only significant for Japan, but also on a much wider basis."[146] The financial support offered by the RF, PC, and Gamble not

[141] Koya et al., "Seven Years of a Family Planning Program in Three Typical Japanese Villages"; Koya et al., "Kazoku keikaku moderu mura no kenkyū"; "Review of Past Achievements and Planning of Future Studies," 1, Series III, Box 97 Folder 1585, Gamble Papers.

[142] Yoshio Koya, "A Family Planning Program in a Large Population Group: The Case of the Japanese National Railways," *The Milbank Memorial Fund Quarterly* 40, no. 3 (1962): 319–27; Yoshio Koya, "Economic Impact of Instruction in Family Planning," *Eugenics Quarterly* 7, no. 4 (December 1960): 212–16; Yoshio Koya and Tomohiko Koya, "The Prevention of Unwanted Pregnancies in a Japanese Village by Contraceptive Foam Tablets," *The Milbank Memorial Fund Quarterly* 38, no. 2 (1960): 167–70; Koya, "Five-Year Experiment on Family Planning among Coal Miners in Joban"; Koya, "Present Situation of Family Planning among Farmers and Coal Mine Workers."

[143] About Pathfinder International, see, e.g., James A. Miller, "Betting with Lives: Clarence Gamble and the Pathfinder International," Population Research Institute, July 1, 1996, www.pop.org/betting-with-lives-clarence-gamble-and-the-pathfinder-international/.

[144] Reed, *The Birth Control Movement and American Society*, 295.

[145] Irene Taeuber to Frederick Osborn, June 26, 1954, Population Council Collection, Record Group 1, Accession 1, Series 1, Box 18, Folder 300, RAC.

[146] John D. Rockefeller III to Yoshio Koya, December 7, 1959, Rockefeller Family Archive, Record Group 5, Series 1, Subseries 5, Box 80, Folder 671, RAC.

only stimulated population research in Japan, it also enabled the DPHD research to form an integral part of the transnational population control movement.

In addition to developing internationally recognized demographic research in Japan, the RF, PC, and Gamble supported various other activities that encouraged Japanese researchers in public health demography to forge links with international population studies communities. The first was a fellowship that allowed Japanese researchers to study at renowned academic institutions in the United States. The RF supported Muramatsu Minoru's study at the School of Public Health at Johns Hopkins University between 1950 and 1951 and then between 1958 and 1959 as a RF fellow.[147] Later, the PC funded Ogino Hiroshi, a DPHD staff, to study at the same institution as Muramatsu.[148] These fellowships gave the DPHD researchers a valuable opportunity to learn the latest knowledge and trends in the field and to connect with specialists through their studies.

Second, PC and RF travel grants enabled Japanese researchers to network with foreign colleagues by attending international meetings. The travel grants the PC awarded to Koya and Muramatsu in 1954 allowed them to present at the World Population Conference that took place in Rome in September of that year.[149] After the conference, Muramatsu proudly reported how his joint presentation with Koya was well received at the conference and was attended by "first-class demographers," even including "'iron curtain' delegates."[150] The PC gave another travel grant to Koya and one of his sons, Koya Giichi, in 1957.[151] This time, Koya not only participated in a population conference in Berlin in October, he also

[147] "Second Fellowship for Dr. Minoru Muramatsu – Japan," April 21, 1958, Rockefeller Foundations Archive, Record Group 10.1, Series 609, Subseries 609.E, Box 367, Folder 5417, RAC.
[148] See, e.g., Minoru Muramatsu to Marshall C. Balfour, June 19, 1959, Population Council Collection, Accession II, Foreign Correspondence Series, Box 15, Folder: Japan, Institute of Public Health, Muramatsu, Minoru, 1959–1964, RAC; Marshall C. Balfour, "Marshall C. Balfour to Paul A. Harper," September 8, 1959, Rockefeller Foundations Archive, Record Group 10.1, Series 609, Subseries 609.E, Box 367, Folder 5417, RAC.
[149] Frederick Osborn to Yoshio Koya, July 15, 1954 and Frederick Osborn to Minoru Muramatsu, July 21, 1954, Population Council Collection, Record Group 1, Accession 1, Series 1, Box 18, Folder 300, RAC; "Grant in Aid for Travel and Living Expenses of Dr. Yoshio Koya, Director of the Institute of Public Health, Tokyo, Japan," July 8, 1954, Rockefeller Foundations Archive, Record Group 1.2, Series 609, Box 6, Folder 46, RAC.
[150] Minoru Muramatsu, "World Population Conference," October 1954, 8–9, Population Council Collection, Record Group 1, Accession 1, Series 1, Box 18, Folder 300, RAC.
[151] Yoshio Koya to Clarence J. Gamble, August 12, 1957, Series III, Box 95, Folder 1559, Gamble Papers.

visited India, Ceylon, and Egypt – the focus countries of the PC population control initiatives at the time – and shared his past experiences with colleagues in the respective countries.[152] The RF made a similar arrangement when it gave a travel grant to Muramatsu in 1960. The grant permitted Muramatsu to make a world tour en route from his studies at Johns Hopkins University. With the grant, he was able to visit population experts in London, Copenhagen, Stockholm, Paris, Geneva, New Delhi, Calcutta, and Hong Kong.[153] Similar to the fellowships, the travel grants helped the Japanese researchers to connect with their international colleagues.

Among the many international links forged thanks to the travel grants, the one that left an impression on the DPHD researchers was with Indian colleagues. In the 1950s, India was one of the most active transnational birth control advocates.[154] The Family Planning Association of India (FPAI), established in 1949, was a founding member of the IPPF. Its founding president, Dhanvanthi Rama Rau, served as the president of the IPPF. The FPAI also hosted the Third International Conference on Planned Parenthood in Bombay between November 24 and 29, 1952. In parallel, the postcolonial Indian state, like the postwar Japanese government, considered birth control to be a solution to the problem of its surplus population. In 1951, the same year the Japanese Diet adopted birth control as a national policy, the Indian government set up a Ministry of Health and Family Planning as a sign of its state-level commitment to the popularization of birth control as part of its public health measures. Meanwhile, in the aforementioned Williamsburg conference, the problem of overpopulation in India attracted as much attention as

---

[152] Frederick Osborn to Yoshio Koya, June 26, 1957, Population Council Collection, Record Group 1, Accession 1, Series 1, Box 19, Folder 301, RAC.

[153] Minoru Muramatsu, "A Report of the Activities As Rockefeller Foundation Fellow," January 1960, Population Council Collection, Accession II, Foreign Correspondence Series, Box 23, Folder: Institute of Public Health Muramatsu, Minoru FC-O Japan 59–64, RAC.

[154] Mytheli Sreenivas, *Reproductive Politics and the Making of Modern India* (Washington, D.C.: University of Washington Press, 2021); Savina Balasubramanian, "Motivating Men: Social Science and the Regulation of Men's Reproduction in Postwar India," *Gender & Society* 32, no. 1 (February 2018): 34–58; Sanjam Ahluwalia and Daksha Parmar, "From Gandhi to Gandhi: Contraceptive Technologies and Sexual Politics in Postcolonial India, 1947–1977," in *Reproductive States: Global Perspectives on the Invention and Implementation of Population Policy*, eds. Rickie Solinger and Mie Nakachi (Oxford and New York: Oxford University Press, 2016), 124–55; Rebecca Jane Williams, "Storming the Citadels of Poverty: Family Planning under the Emergency in India, 1975–1977," *The Journal of Asian Studies* 73, no. 02 (May 2014): 471–92; Schoen, *Choice & Coercion*.

the Japanese actions.[155] Under the circumstances, Indian demographers and public health specialists participated in the transnational population network along with their counterparts in Japan.

In most instances, the institutions and individuals driving the transnational population network facilitated the encounters between the Japanese and Indian colleagues. Health Minister Rajkumari Amrit Kaur's visit to Japan was a result of the connection established through the PC. At Koya's invitation, Amrit Kaur arrived in Japan on July 4, 1956. During her stay in Japan, Koya took her to Fukuura Village, one of the villages in the three-village study.[156] When he visited India and Ceylon between October and November 1957, and later when Muramatsu visited in 1960, it was Marshall C. Balfour, a RF officer and staunch proponent of population control who was stationed in India, connected Koya with important birth control advocates, health officials, and population scientists in India, including Rama Rau.[157] As both Koya and Muramatsu testified after their trip, the visit to India was "one of the most interesting experiences."[158] The connections with the US-based foundations that were forged through their population research allowed them to network with their colleagues in Asia who were also participating in the transnational movement.

Financial support from the RF, PC, and Gamble for Japanese fertility research was not an isolated instance. During the twentieth century, US-based charity foundations gave financial assistance to a wide range of research programs across the world to tackle global issues such as poverty and development. Backed by primary mission, the foundations' assistance harnessed national and global networks of intellectuals, which consequently served what Inderjeet Parmar called an "American imperium," a global hegemony built around American cultural and intellectual

---

[155] Matthew Connelly, *Fatal Misconception: The Struggle to Control World Population* (Cambridge, MA and London: The Belknap Press of Harvard University Press, 2008), 166–73.

[156] Yoshio Koya to Clarence J. Gamble, July 7, 1956, Series III, Box 95, Folder 1556, Gamble Papers.

[157] Yoshio Koya to Dudley Kirk, November 17, 1957. RAC, Population Council Collection, Record Group 1, Accession 1, Series 1, Box 19, Folder 301; Marshall C. Balfour to Yoshio Koya, November 11, 1957. RAC, Population Council Collection, Record Group 1, Accession 1, Series 1, Box 19, Folder 301; Marshall C. Balfour to Minoru Muramatsu, June 25, 1959, Rockefeller Foundations Archive, Record Group 10.1, Series 609, Subseries 609.E, Box 367, Folder 5417, RAC.

[158] Yoshio Koya, "President Prof. Y. Koya Talks, Coming Back from His Trip," *Family Planning News*, no. 6 (February 1958): 1–2; Muramatsu, "A Report of the Activities As Rockefeller Foundation Fellow."

involvement.[159] By the same token, the financial assistance the RF, PC, and Gamble gave to public health demography integrated the Japanese researchers into the transnational population network and, in so doing, contributed to the global population control movement that was in many instances buttressed by American leadership. In the 1950s, American leadership within global population control carried a special meaning, as population control was discussed in relation to the American attempt to construct a new world order in the middle of the Cold War. In this context, American support for public health demography was part of the broader movement to create a Cold War world order that revolved around the "American imperium."

The Japanese, in turn, used the American assistance to their advantage. As mentioned, Koya had long desired to expand the field, and under the circumstance, the American financial help came at an opportune time. For younger scholars such as Muramatsu and Ogino, the American grants gave them a valuable opportunity to build their careers. However, by accepting the American assistance, the researchers inadvertently were co-opted into the transnational network that buttressed American hegemony in global population control.

The support from the American philanthropists did more than simply integrate Japanese birth control researchers into the transnational network. As the subsequent section shows, it also catalyzed negotiations that shaped the Japanese pilot studies in ways that were relevant for the transnational population control network.

Connecting Kajiya with Khanna: Technological Fix and Guidance Work

While the RF, PC, and Gamble all gave financial assistance to the population research in public health demography, their degree of involvement in the studies they funded were different. Generally speaking, the RF and PC kept a hands-off attitude toward the actual fieldwork in the pilot studies, although they would happily offer advice if the Japanese wished. In contrast, Gamble proactively shaped both the content and results of the pilot studies he helped. This difference in approach left a visible legacy among studies conducted under the name of public health demography.

One area that vividly depicts the impact of Gamble's hands-on attitude was a study testing the efficacy of foam tablet as a "simple"

---

[159] Inderjeet Parmar, *Foundations of the American Century: The Ford, Carnegie, and Rockefeller Foundations in the Rise of American Power* (New York: Columbia University Press, 2012), 2, 257.

contraceptive method. As mentioned above, since the late 1930s, Gamble had been at the forefront of the research on the development and assessment of "simple chemical contraceptives."[160] Gamble believed that the diaphragm – one of the few contraceptives available at the time – was too difficult for poor and less-educated women to use, thus he sought contraceptive chemicals as a "simpler" alternative.[161] In the 1940s, Gamble, together with NCMH statistician Christopher Tietze, conducted a clinical study to test the efficacy of contraceptive suppositories among women from impoverished rural communities in Alabama and Puerto Rico.[162] The Alabama study was not successful, but the project in Puerto Rico was more fruitful, in part because they used the data provided by local contacts.[163] After Puerto Rico, Gamble came to believe that birth control research required collaboration with locals. These local contacts should ideally be medically savvy and willingly produce reliable data on behalf of Gamble. In late-1940s Japan, Koya emerged as a perfect candidate for this purpose.

Meanwhile, in Japan, spermicidal chemicals were emerging as the contraceptive of the future. Among them was Eisai Pharmaceutical's foam tablet, Sampoon, the distribution of which the government approved in 1950.[164] By 1953, Sampoon had become one of the most popular birth control medicines in Japan.[165] The news about the popularity of Sampoon spread quickly among birth control activists.[166] Medical communities were eager to conduct trials of Sampoon, too. In the early 1950s, a small-scale investigation into Sampoon's efficacy was underway in the district covered by the Shibukawa Health Center in Gunma Prefecture, which was run by a female doctor, Aida Sakiko, and a public health nurse, Kobuchi Aiko, who were stationed at the health center.[167]

[160] Ilana Löwy, "'Sexual Chemistry' before the Pill: Science, Industry and Chemical Contraceptives, 1920–1960," *The British Journal for the History of Science* 44, no. 2 (June 2011): 257.

[161] Briggs, *Reproducing Empire*, 102–3.

[162] Clarence J. Gamble and Christopher Tietze, "A Field Study of Contraceptive Suppositories," *Human Fertility* 13 (n.d.): 33–36.

[163] Briggs, *Reproducing Empire*, 107, 227.

[164] Eisai Co., Ltd., "Ēzai no rekishi 80th," accessed March 11, 2018, www.eisai.co.jp/company/profile/history/index.html.

[165] "More 'Sampoon' Tablets Exported," *IPPF News* 4, no. 3–4 (December 1953): 70.

[166] T. Nasu, H. Nakai, and H. Kawakami, "Contraceptive Efficacy of 'Sampoon,'" *Japan Planned Parenthood Quarterly* 2, no. 3–4 (December 1951): 25–26.

[167] Sakiko Aida and Aiko Kobuchi, "Collective Experiment on Contraception with 'Sampoon' Tablets," *IPPF News* 5, no. 2 (June 1954): 15, 27.

Gamble latched onto the hype in Japan and tried to encourage Koya to study the foam tablet's acceptability. In June 1952, in response to Koya's letter to Gamble, in which Koya mentioned Sampoon in passing, Gamble told Koya he had "seen an advertisement of Sampoon Tablet" and urged Koya to do the "testing in 200 or 300 families."[168] Three months later, Gamble again asked if any scientific study had been done on Sampoon.[169] Indeed, Koya had been using foam tablets in his pilot studies, but it was presented merely as one of many contraceptive options. Having heard this, in May 1954, while visiting Japan with his son Richard, Gamble and Koya discussed plans to launch an independent "project where only foam tablets would be distributed. This would determine their effectiveness and acceptability." In order to persuade Koya, Gamble stressed the advantage of foam tablets over guidance work "for the budget-conscious mass birth control campaign in Japan" and pointed out how the foam tablet could lower the campaign's cost by "mak[ing] unnecessary the present cost of individual interviews with a physician."[170] Gamble offered to pay for such a test project.

Koya accommodated Gamble's needs. While committed to the guided birth control program, Koya also shared Gamble's sentiment that the contraceptives distributed to a target population should be affordable and "simple" because the population was generally regarded as poor and less educated. Partly for this reason, in the three-village study, Koya had been testing the efficacy of the salt-and-sponge method: a barrier method in which a woman inserted a sponge soaked in a homemade salt solution into her vagina immediately before intercourse.[171] He thought that the method was particularly suitable for a poor farmer because it was a budget method. As he said: "No Japanese farmer is so poor that he cannot buy one or two sponges a year. Salt is abundantly found in any farmer's kitchen."[172] However, the women taking in part in the three-village study did not particularly like the salt-and-sponge method, therefore, he was looking for an alternative. Koya thought, "it is quite essential for us to find out the best suited method for Japanese," and for "the Government," and

[168] Clarence J. Gamble to Yoshio Koya, June 18, 1952, Series III, Box 94, Folder 1538, Gamble Papers.

[169] Clarence J. Gamble to Yoshio Koya, September 3, 1952, Series III, Box 94, Folder 1538, Gamble Papers.

[170] R. B. Gamle and C. J. Gamble, "Summary of Japan," May 1954, Series III, Box 94, Folder 1547, Gamble Papers.

[171] Toyoda, "Sengo nihon no bāsu kontorōru undō," 62–63.

[172] Yoshio Koya, "Effectiveness of Contraception by the Use of Sponge (Used in Conjunction with 10% Saline Solution'," 1, n.d. c.1955, Series III, Box 95, Folder 1553, Gamble Papers.

wondered if the foam tablet could become such a method.[173] Koya agreed to set up a study specifically dedicated to Sampoon.

Based on Gamble's funding, which arrived via the NCMH, Koya set up a foam tablet pilot scheme in Hakusen Town and Hikawashita Town, with about 300 families.[174] Following on the pilot project, in 1955, a foam tablet study was formally launched in Kajiya Village, Kanagawa Prefecture, located about 100 kilometers west of Tokyo (Figures 6.2 and 6.3).[175] Koya was the project leader, but one of his sons, Koya Tomohiko (mentioned above), conducted the fieldwork.

At the beginning of the study, in 1955, Kajiya Village had a population of 1,789 in 298 households.[176] The village was once described as "uncultivated" in regard to birth control.[177] Researchers understood the village to be "of a feudalistic type in which the young people were under the strict control of the parents. Often this control was used to forbid contraception, though fortunately this opposition is now decreasing. As many of the young people had left the village to find employment, the number of families past childbearing age was surprisingly large."[178] In part due to the village's character, the researchers were able to select only 101 wives for the study whom they deemed as fertile.[179] These women were asked to take part in the project for exactly three years, from January 1, 1955 to December 31, 1958. In the end, the data gleaned from eighty-two women were entered for analysis.

In line with the "simple" nature of the contraceptive, the study's procedure was kept simple.[180] The wives agreeing to take part in the study were simply given the Sampoon tablet, told to place it with a finger as far into the vagina as possible just before intercourse, and record the usage. Mrs. Kimura, the aforementioned local midwife employed for the study, visited the wives approximately once a month to record its use, objections to the method if they stopped using it, and pregnancies.[181] In

---

[173] Apart from the sponge-and-salt method, he thought foam powder would be a "most promising ... for rural areas," Diary of Oliver R. McCoy, November 28, 1950.

[174] Yoshio Koya to Clarence J. Gamble, August 12, 1954, Series III, Box 94, Folder 1548, Gable Papers; Yoshio Koya and Tomohiko Koya, "The Effectiveness of Contraceptive Tablets in a Japanese Village," n.d. c.1956, Series III, Box 95, Folder 1557, Gamble Papers.

[175] "Jōzai ga unda mikan mura."

[176] Koya and Koya, "The Effectiveness of Contraceptive Tablets."

[177] "Jōzai ga unda mikan mura," 2.

[178] Koya and Koya, "The Effectiveness of Contraceptive Tablets."

[179] Ibid.

[180] Ibid.

[181] Yoshio Koya to Clarence J. Gamble, July 26, 1956, Series III, Box 95, Folder 1556, Gamble Papers.

Figure 6.2 A scene from the Kajiya village pilot study. A public health nurse is handing over "tablets" to a woman.
Source: "Mikan mura no kazoku keikaku," *Kensei nyūsu*, no. 399 (November 9, 1959): 1.

Figure 6.3 A scene of an interview between a doctor and local women in the Kajiya village pilot study.
Source: "Mikan mura no kazoku keikaku." *Kensei nyūsu*, no. 399 (November 9, 1959): 1.

addition, Koya Tomohiko visited the village fortnightly to interview the women. Based on the interviews and the records collected, Koya Tomohiko made what his father called a "family line," a horizontal line chart that represented the time each participant's reproductive life was marked (Figures 6.4 and 6.5). Based on the "family line," Tomohiko calculated the birth and pregnancy rates among the wives.

As in other pilot studies by Koya, the Kajiya Village study was characterized as a success. The study showed that the pregnancy rates of the 82 wives prior to the study between 1950 and 1954, during which the couples used no contraceptive, was 52.8, while the corresponding rate during the time of the study was 11.9 per 100 years of exposure.[182] With these figures, the Koyas concluded that Sampoon was "effective enough

[182] Koya and Koya, "The Prevention of Unwanted Pregnancies." The figures are original.

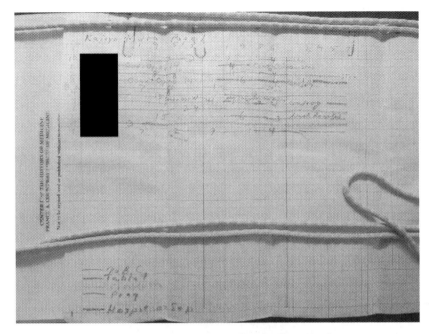

Figure 6.4 The "family line" in a handwritten draft note.
Source: Japan: two folders of 1957–1960 materials re Koya article, "The Prevention of Unwanted Pregnancies in a Japanese Village by Contraceptive Foam Tablets," Box 96, Folder 1577, Series: III. Countries Correspondence and Records, 1927–1965, Gamble, Clarence James, 1894-. Papers, 1920–1970s (inclusive), 1920–1966 (bulk): Finding Aid. (H MS c23) [Persistent ID: nrs.harvard.edu/urn-3:HMS.Count:med00082], Center for the History of Medicine. Francis A. Countway Library of Medicine.

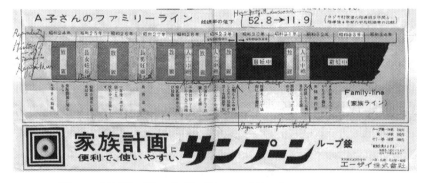

Figure 6.5 The "family line" in a publication.
Source: "Mikan no mura no kazoku keikaku," *Kensei nyūsu* no. 9 November 1959: 1.

to be recommended to the people at large."[183] The study demonstrated the benefit of Sampoon for a mass population control initiative.

Gamble became heavily involved in the study as soon as he learned Koya Tomohiko collated some data. From 1956 on, he began to ask Koya Yoshio about the "progress" in the "foam tablet project" and requested that Koya Tomohiko send him "a tabulation of the foam tablet cases" as well as "a new family line chart" periodically.[184] Once Koya Yoshio sent the requested documents, Gamble did some calculations for the study.[185] Later, when the study's results were taking shape, Gamble requested if Koya could approve to have a draft of the Kajiya paper mimeographed so it could be distributed it as widely as possible "to the people who are asking about how successful foam tablets can be."[186] Finally, Gamble steered the Koyas to consider having the article published in *Milbank Quarterly*, a journal highly reputed among proponents of global population control. Gamble even offered to help edit the English for the article.[187]

Why did Gamble have a heavy-handed approach to the foam tablet study in Kajiya Village? To start with, it was generally known that Gamble kept a close eye on the birth control projects for which he authorized funding. The Kajiya study was one of them, so it seemed evident that he provided the same oversight for the Kajiya study. However, for Gamble, who was closely watching the birth control initiatives in Asia, the Kajiya study had special meaning. Specifically, Gamble saw the Kajiya study as a parallel study to the Khanna study, another foam tablet study taking place in Asia.

The Khanna study was conducted between 1953 and 1960 in the Ludhiana district of Punjab in India as a collaborative project involving John Gordon's team from the Harvard School of Public Health, the government of India, and Ludhiana Christian Medical College. The primary aim of the study was to investigate population dynamics, specifically in relation to a birth control pilot initiative. Moreover, as in the Kajiya study, the assessment of the efficacy of foam tablets was an important mission in the Khanna study. The study was supported by the

---

[183] Ibid., 170.
[184] See Clarence J. Gamble to Yoshio Koya, December 3, 1956, Series III, Box 95, Folder 1556, Gable Papers; Gamble to Koya, July 26, 1957, Series III, Box 95, Folder 1559, Gamble Papers; Gamble to Koya, July 23, 1958, Series III, Box 95, Folder 1564, Gamble Papers.
[185] See Gamble to Koya, October 28, 1958, Series III, Box 95, Folder 1564, Gamble Papers.
[186] Gamble to Koya, August 24, 1959, Series III, Box 95, Folder 1566, Gamble Papers.
[187] Gamble to Koya, July 7, 1959, Series III, Box 95, Folder 1566, Gamble Papers.

RF grant-in-aid between 1953 and 1954, and Gamble also offered help behind the scenes.[188] The Khanna study, like the Japanese birth control pilot studies, occupied a special space within the history of population control in India and beyond. Specifically, the study became the first family planning pilot project that provided the government of India and international demographers with evidence that India's people wished to limit the size of their families.[189]

For Gamble, who had long been working on the development and assessment of "simple" contraceptives, the Khanna study was particularly interesting because it aimed to study a "simple" contraceptive that was quickly becoming popular in India but whose efficacy on population dynamics had yet to be established. In the 1950s, along with his fellow advocates in population control, Gamble began to pursue ways to further promote foam tablets in India. In India at the time, various foam tablets were already available due to the efforts of government officials, international public health and birth control campaigners, and pharmaceutical companies, but the majority of the tablets sold there was exported from outside Asia, so they were expensive due to the shipping charge.[190] An alternative was the Contab tablet, produced locally by Smith Stanistreet, Ltd. from Calcutta, but its quality was deemed questionable.

This was when Gamble discovered the existence of the Japanese Sampoon tablet. The Sampoon tablet was trialed in India during the early 1950s, and in 1952, 2,000 were imported from Japan via the IPH.[191] In the eyes of Gamble, the Sampoon tablet had advantages over the majority of foam tablets sold in India for two reasons: The first was the quality. Compared to other foam tablets that deteriorated after one to five months of storage, the Sampoon tablet could be active even after a year of storage.[192] Second, the price of the Japanese Sampoon tablet was competitive in comparison to other tablets exported from outside Asia.[193] Considering these points, Gamble must have concluded the Sampoon tablet could become a viable option in India or other regions in Asia confronted with overpopulation. However, in the 1950s, the efficacy of Sampoon for fertility reduction had yet to be established.

---

[188] Connelly, *Fatal Misconception*, 173.
[189] Williams, "Rockefeller Foundation Support," 5.
[190] Ilana Löwy, "Defusing the population bomb in the 1950s: Foam tablets in India." *Studies in History and Philosophy of Science Part C: Studies in History and Philosophy of Biological and Biomedical Sciences* 43, no. 3 (September 2012): 584–85, 589–600.
[191] "More 'Sampoon' Tablets Exported."
[192] Although it was later discovered that Sampoon produced too much gas to be effective. Löwy, "Defusing the Population Bomb," 585.
[193] Löwy, "Defusing the Population Bomb," 584–85.

Against this backdrop, Gamble urged Koya to carry out the Kajiya study, and once the study was established, he tried to make it known to the family planning officials and scholars in India via Koya.[194] Gamble contacted Frederick Osborn, vice president of the PC, to help Koya to travel to India so he could introduce the Kajiya study to Indian colleagues.[195] Gamble's lobbying culminated in the aforementioned trip to India in 1957 that was supported by the PC. During the tour of India, Koya tirelessly introduced the Kajiya study to the individuals and groups he met.[196] At the same time, with the help of Balfour, Koya traveled to Khanna. In Khanna, John B. Wyon, the field director of the Khanna study, took Koya around the test villages where he observed and even managed to interview the women participating in the study. After the trip, he excitedly reported to Gamble: "The most important experiences [I] got during this trip was about the foam-tablets. About every person for instance, Col. Raina, Mr. Karmakar, Dr. Balfour, Dr. Wyon, Dr. Mauldin … asked me to give them information on Sampoon."[197] What excited Koya the most was when Minister of Health Rajkumari Amrit Kaur asked him, "please send me every information on foam tablets which you have."[198] "I was much encouraged to have met such a [sic] great responses and welcomes."[199] It seemed like Gamble's attempt to connect those involved in the Kajiya study to the Khanna study ended in success.

For Gamble, it was evident that the Japanese involved in the Kajiya study and the Indians familiar with the Khanna study should be connected, because in his eyes, the two studies were related. Both the Khanna and Kajiya studies focused on the impact of a single, "simple" contraceptive on the dynamics of a targeted population. Both also took place in Asia, a target area of global population control. From Gamble's point of view, these studies, when combined, could provide him and his international colleagues with invaluable comparative data from Asia in order to consider birth control as a technological fix to the problem of global overpopulation. Thus, it only made sense that those involved in the studies should know each other's work so they could compare results.

---

[194] Koya to Gamble, August 12 1954, Series III, Box 94 Folder 1548, Gamble Papers.

[195] Clarence J. Gamble to Frederick Osborn, August 20, 1957, Population Council Collection, Record Group 1, Accession 1, Series 1, Box 19, Folder 301, RAC.

[196] Koya to Gamble, November 15, 1957 and November 21, 1957, Series III, Box 95, Folder 1559, Gamble Papers; Yoshio Koya to Dudley Kirk, November 17, 1957, Population Council Collection, Record Group 1, Accession 1, Series 1, Box 19, Folder 301, RAC.

[197] Koya to Gamble, November 21, 1957, Series III, Box 95, Folder 1559, Gamble Papers.

[198] Koya and Koya, "The Effectiveness of Contraceptive Tablets."

[199] Koya to Gamble, November 21, 1957, Series III, Box 95, Folder 1559, Gamble Papers.

In turn, Koya had a distinct perspective on the studies. While partially sharing Gamble's idea that the studies were comparable, Koya also believed they were different in another sense. First of all, the studies led to completely different outcomes: At the time of Koya's visit to India, the pregnancy rate in Kajiya was 13.4, whereas the equivalent in the Khanna village still remained 57.0.[200] Seeing the gap, Koya readily applied demographic transition theory (see Chapter 5) and claimed that the gap in pregnancy rates was rooted in the different cultural levels of the test villages: "In Japanese villages 98% can read newspapers, but in Indian villages only 2%. The age construction of wives in Kajiya-mura is much older than in Khanna village. Almost all villages in Japan, even though they are poor, must be said to be 'modern villages,' while the Indian village is to be said an 'ancient village.'" Based on this observation, Koya concluded that there was a "more than 100 years gap between" the Indian and Japanese villages, and this "gap" caused the studies' distinctive outcomes.[201] Related to this, Koya considered the gap to also stem from the distinctive methodology adopted by the respective pilot projects. He wrote to Gamble:

The way of work … in Khanna and Kajiya-mura are also different. The way in Khanna is rather a survey, not family planning work. 80% of it is concentrated how to get good records of this acceptability of foam-tablet and only 20% must be educational and leading work. On the contrary, in Kajiya-mura, 80% is educational and leading work … and only 20% is survey on the results. So, it is nonsense to compare the pregnancy rates of Khannas and Kajiyas.[202]

Koya's judgment stemmed from his conviction regarding the value of guidance in a birth control initiative. For as long as he was involved in the birth control campaign, Koya stressed education was effective at curbing the fertility rates and improving the population quality, therefore, it should be at the heart of a birth control program. Koya agreed that the application of a "simple" contraceptive method to the program was important, but only when it was combined with guidance work.[203] Based on this view, Koya reckoned the Kajiya study was successful precisely because it stressed the education work, while Khanna was doomed to fail because of the lack thereof. For Koya, the technological fix to the population problem should be tried in addition to long-term education work.

While holding onto this idea, in the late 1950s, Koya additionally strengthened the DPHD research on Sampoon primarily as a "simple"

---

[200] Koya to Gamble; November 15, 1957, Series III, Box 95, Folder 1559, Gamble Papers.
[201] McCann, *Figuring the Population Bomb*, 156–98.
[202] Koya to Gamble, November 15, 1957, Series III, Box 95, Folder 1559, Gamble Papers.
[203] E.g., "Some Findings in a Recent Survey Conducted by FPFJ," *Family Planning News*, June 1958, 1–2.

contraceptive method offering a technological fix to population issues. Koya managed to extend the period of study for Kajiya. Additionally, Koya Tomohiko, under the supervision of his father and his senior Dr. Furusawa Yoshio, launched a "clinical experiment" on the recently developed variant of Sampoon with female patients at the Sumida-Metropolitan Hospital.[204] Finally, in 1959, Koya received the research fund of $2400 from the PC to "undertake a study of the acceptability of the foam tablet in a Japanese Railway community."[205] What motivated Koya to shift his research emphasis?

Behind the change in the research direction was the combination of domestic and transnational factors. Within Japan, throughout the 1950s, the fertility rate plummeted. Under these circumstances, Koya was made acutely aware that his birth control campaign and pilot studies were quickly losing their raison d'être within the Japanese government.[206] By the early 1960s, the state-led birth control campaign that Koya had fervently supported ironically robbed him of a research opportunity, but the reception of the Kajiya study in India gave him a glimmer of hope. Koya thought that aligning the DPHD research with current trends in the transnational population control movement would grant the scientific field he had created further legitimacy. Thus, Koya "really recognized how important [the Sampoon] experiment is by the trip in India" and decided to "[strengthen] our work at the Sampoon village."[207] Yet, in practice, this meant that the trend in global population control strongly shaped the birth control research conducted in public health demography.

Conclusion

The birth control advocacy and research that developed in tandem with Koya's effort to expand the DPHD testified how Public Health Demography, a medico-scientific field studying reproductive practices and demographic patterns, was at once a national and transnational project and was strongly shaped by locally specific conditions. This story, on the one hand, confirms the point made in the previous chapters about

---

[204] Koya, "Review of Past Achievements and Planning of Future Studies," 5.
[205] Dudley Kirk to Yoshio Koya, December 23, 1959, RAC, Population Council Collection, Record Group 1, Accession 1, Series 1, Box 19, Folder 302; "Office Memorandum from WPM to DK Re Grant to Dr. Koya," December 14, 1959, Population Council Collection, Accession II, Foreign Correspondence Series, Box 24, Folder: Nippon Medical School, Koya, Yoshio FC-O Japan 59–62, RAC.
[206] Koya to Gamble, May 29, 1962, Series III, Box 97, Folder 1588, Gamble Papers.
[207] Koya and Gamble, December 14, 1957, Series III, Box 95, Folder 1559, Gamble Papers.

micropolitics' centrality in shaping the interplay between science making and policymaking. On the other, it also showed how transnational elements clearly participated in the interplay, shaping the scientific endeavor that was explicitly associated with domestic policy. The latter point suggests the fragility of a nation-focused perspective in the analysis of this interplay. It certainly urges us to come up with an alternative interpretation about the interactions between the "national," "local," and "transnational."[208]

If transnationalism was "a structure that sustains and gives shapes to the identities of nation-states ... and local institutions" as defined by cultural historian Patricia Clavin, I argue that transnationalism shaped the interplay between the DPHD birth control studies and the domestic population policy in two ways. First, by supporting the birth control studies, it helped perpetuate the view of the poor as fecund and uneducated. This view not only sustained the preexisting, locally embedded social hierarchy within Japan but also affected the ways in which Japanese actors caricatured the poor in the transnational population control movement in later years. Second, by supporting Japanese birth control work and making it visible to the relevant international communities, transnational elements that participated in the science-policy interplay aided the creation and consolidation of post-WWII Japan's identity as the nation that miraculously resurfaced from the rubble through economic development – which was achieved by highly organized fertility regulations.[209] This interplay was far more complex than a mere "national" endeavor. Instead, it was firmly embedded in, and mobilized by, the diverse networks and practices that entangled the national with the local and transnational.

---

[208] Garon, "Transnational History."
[209] Patricia Clavin, "Defining Transnationalism," *Contemporary European History* 14, no. 4 (November 2005): 421.

Conclusion

Science–Policy Nexus for Governing Shifting Demographic Realities, 1960s–2010s

Given the profound changes the Japanese population has undergone in the time since Shinozaki and Koya lobbied for birth control in the wake of WWII, one wonders how the two bureaucrats-cum-scientists would react to the population trend today. As opposed to the "high birth, high death" model presented in the 1947 vital statistics – the first official statistics published after the war – the latest statistics from 2016 clearly show the Japanese population following a "low birth, low death" model (Figure 7.1). Moreover, in contrast to 1947, when the population was growing by 19.7 per 1,000 population, in 2016, the population is contracting by 2.6 per 1,000 population. Life expectancy has also increased enormously, from 57.68 and 60.99 years for men and women in 1950 to 80.98 and 87.14 in 2016, respectively.[1] Due to drastic changes over the past seven and a half decades, the Japanese population, once vibrant and youthful, is now shrinking and aging (Figure 7.2).

Like in the periods covered in this book, after the 1960s, the shifting demographic realities, and the constantly changing population discourse that accompanied it, continued to act as a root of the story surrounding the interplay between the science of population and the state's effort to manage its population size and quality.[2] In the early 1960s, the dramatic fall in the Total Fertility Rate (TFR) from previous decades – from 4.54 children per woman in 1947 to 2.0 in 1960 – catalyzed this interplay.[3] While technical bureaucrats and policy intellectuals who were at the prime of their careers in the late 1940s might have celebrated the fertility decline as dissipating the problem of "overpopulation," their successors

---

[1] Futoshi Ishii, "Posuto jinkō tenkanki no shibō dōkō," 93.
[2] For the discussion that follows, I rely on the periodization presented by Kiyoshi Hiroshima, "Horon sengo nihon no jinkō," 301–13.
[3] Total fertility rate refers to the number of children a woman might have in her lifetime if the fertility rates observed at each age in a given year remained unchanged.

247

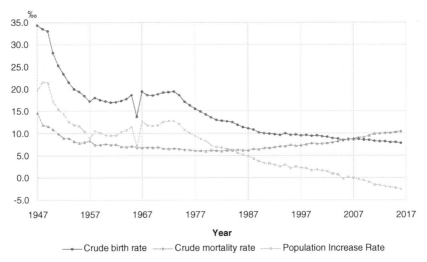

Figure 7.1 Crude birth rates, crude mortality rates, and population increase rates in Japan, 1947–2016.
Source: Ministry of Health, Labour and Welfare of the Government of Japan, "Wagakuni no jinkō dōtai. Tōkeihyō," www.mhlw.go.jp/toukei/list/81-1a.html, accessed July 6, 2021

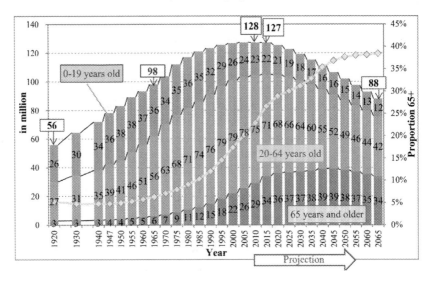

Figure 7.2 Population by age group and proportion of those aged 65 years and over, 1920–2065, cited from the National Institute of Population and Social Security Research, *Population and Social Security in Japan 2019*, IPSS Research Report No. 85, July 26, 2019, p. 3, www.ipss.go.jp/s-info/e/pssj/pssj2019.pdf, accessed July 16, 2021

in the early 1960s were worried that the same phenomenon might slow down the unprecedented pace of economic growth in Japan by causing a drastic shrink in the cohorts of working age persons in the near future.[4] Linked to this, they were also concerned that the child mortality rate, though declining in general, was still high compared to the United States and European countries, and infant mortality still posed a serious problem in rural areas.[5] In this context, they presented the idea of tackling the problems of an imminent labor shortage and child health through measures aiming to "improve population quality" by specifically targeted younger populations. The notion of "population quality" justified eugenic policies during WWII (see Chapter 4), but "population quality," as reformulated in the early 1960s, referred more broadly to the quality of "a fundamental, mental, and physical state of an individual determining the characteristics of a human group called population," as defined by the Foundation-Institute for Research of Population Problems (IRPP).[6] Arguably, it would be "improved" not only through eugenic policies like genetic screening but also through other wide-ranging social policies, such as child-support allowance and educational reform.[7]

Along with this, from the late 1950s onward, policymaking and policy-relevant research began to prioritize population quality. Sometime around 1958, population experts affiliated with the IRPP, including technical bureaucrats from relevant ministries, began to launch investigations into population quality.[8] In 1959, with a MHW research fund, a research team headed by Nagai Tōru, and also including Tachi and the economist Terao Takuma, launched a project to study the policy implications of the population quality problem.[9] Around the same time, the ACPP began to collect experts' opinions on population quality.[10] Based on this research, between 1959 and 1962, the IRPP Committee on Population Measures (IRPP-CPM) Second Special Committee deliberated

[4] Zaidan Hōjin Jinkō Mondai Kenkyūkai, "Jinkō shishitsu kōjō," 13–14.
[5] Population Problems Inquiry Councile [sic], "An Outline of Trends in Japan's Population," in *Activities in Japan for the World Population Year 1974*, n.d. c. 1974, 6, National Institute of Public Health Library archives, Wako-shi, Saitama Prefecture, Japan [thereafter NIPHL archives].
[6] Zaidan Hōjin Jinkō Mondai Kenkyūkai, "Jinkō shishitsu kōjō," 12.
[7] Jinkō Mondai Shingikai, "Jinkō shishitsu kōjō taisaku ni kansuru ketsugi," 3–14.
[8] Jinkō Mondai Shingikai, "Jinkō mondai shingikai dai 19 kai sōkai sokkiroku," 5.
[9] "Showa 34 nendo kōsei kagaku kenkyūhi niyoru kenkyū hōkokusho," in *Kōseishō daijin kanbō kikakushitsu kōsei kagaku kenkyū kadai 34-nendo, 35-nendo*, NIPHL archives.
[10] See Watanabe Jō, "Jinkō no shitsu no genjō to mondaiten," in *Jinkō mondai shingikai dai 20 kai sōkai giji sokkiroku*, ed. Jinkō Mondai Shingikai (March 25, 1960), 22–67. Also see Jinkō Mondai Shingikai, "Jinkō mondai shingikai dai 21-kai sōkai giji sokkiroku."

on measures to improve population quality.[11] In May 1962, the IRPP submitted the Proposal for the Outline of Measures on the Improvement of Population Quality to the minister of health and welfare.[12] The ACPP drafted a policy document based on the IRPP proposal, and in July 1962, it announced the Resolution on the Measures to Improve Population Quality.[13] Based on these recommendations, the 1962 *White Paper of Health and Welfare* had a whole chapter dedicated to the "improvement of population quality."[14] As a result, the MHW set up the Department of Population Quality within the Institute of Population Problems in 1963.[15] In the first half of the 1960s, population experts' awareness of a fertility decline quickly turned "population quality" into a buzzword in policy discussions and policy-relevant research on population.

As the new discourse of "population quality" was emerging, fertility in Japan kept falling and eventually went below what population experts called the net reproductive rate of 1.0 – the level at which a population could reproduce itself from one generation to the next without increasing or decreasing. Faced with this phenomenon, the 1964 *White Paper of Health and Welfare* problematized fertility decline as a cause of this phenomenon. The official recognition of further fertility decline led to two phenomena within policymaking: First, it consolidated the argument in favor of improving "population quality." Second, it gave rise to the narrative that the government needed to promote "social development" (*shakai kaihatsu*), which would indirectly boost fertility by removing various socioeconomic factors that were currently discouraging women from having babies.

Prior to this period, social development, a term originally coined by the United Nations in the 1950s to enable a more holistic approach to development, did not resonate with Japanese politicians and policymakers. Since the end of the occupation, the whole country was geared toward a reconstruction that was overly reliant on economic development.[16] But a decade later, political leaders in Japan became more susceptible to the concept as they were confronted with an increase in cases of serious

---

[11] The IRPP Committee on Population Measures was founded in April 1946.

[12] "Shishitsu no kōjō wo hakare jinkō kenkyūkai kōshō ni kengi e," *Asahi Shinbun* (May 21, 1962, morning edition): 1.

[13] "Jinkō mondai shingikai shishitsu kōjō no kengi'an naru," *Asahi Shinbun* (July 11, 1962, morning edition): 1.

[14] Kōseishō, *Kōsei hakusho (Shōwa 37-nendo ban)*.

[15] Jinkō Mondai Kenkyūsho, ed., *Jinkō mondai kenkyūsho sōritsu gojusshūnen*, 300–301.

[16] The United Nations General Assembly, resolution 1710 (XVI). United Nations Development Decade A programme for international economic co-operation (I)16 of December 19, 1961, 1084th plenary meeting.

pollution and damages to human health that suggested the adversarial effects of the economic-centric reconstruction effort.[17]

This is when Tachi and social policy specialist Ibe Hideo introduced "social development" in Japan, translating it into Japanese as *shakai kaihatsu* and defining it as the "development of the societal aspects of cities, villages, lodging, transportation, health, medicine, public health, social welfare, education, and so on" and "an attempt to directly improve human capacity and welfare."[18] Following this translation, for the rest of the 1960s, population experts began investigating various topics associated with education, town planning, public infrastructure, health and welfare, and other provisions associated with "social development" that were affecting people's lives. "Social development" came to underpin the government's commitment to social and welfare policy. In his inaugural speech in November 1964, Prime Minister Satō Eisaku proclaimed, "the promotion of social development will be the basis of [my government's] domestic policy ... with a long-term perspective ... to create a high-quality welfare state."[19] After the speech, on December 11, 1964, Economic Planning Secretary Takahashi Mamoru submitted the Basic Plan on Social Development to Satō.[20] Based on the plan, the guidelines for the government budget for the 1965 fiscal year stressed that it would be allocated to measures promoting social development and "regional development" (*chiiki kaihatsu*), another concept that overlapped with social development.[21] In the 1960s, social development surged as a catchphrase in policymaking, causing momentum for issuing major post-WWII health and welfare provisions, such as the Law on Social Welfare for the Elderly in 1963.

The emphasis on "population quality" and "social development" continued to center the policy discussions and research on population well into the early 1970s. However, during this time, the TFR suddenly increased because the cohort of baby boomers born in the late 1940s

---

[17] Walker, *Toxic Archipelago*, 137–210; Timothy S. George, *Minamata: Pollution and the Struggle for Democracy in Postwar Japan* (Cambridge, MA: Harvard University Asia Center, 2001).

[18] Jinko Mondai Shingikai, "'Chiiki kaihatsu ni kanshi, jinko mondai no kenchi kara tokuni ryūi subeki jikō' ni tsuite no iken" [August 17, 1963], in *Jūmin no seikatsu to shinsangyō toshi*, ed. Koseisho Daijin Kanbō Kikakushitsu (Okurasho Insatsukyoku, 1964), 165–67.

[19] "Sato shushō, hatsu no shoshin hyōmei," *Asahi Shinbun* (November 21, 1964, evening edition): 1.

[20] "Mazu jūtaku nado shichikōmoku shushō ryōshō shakai kaihatsu no kihon kōsō," *Asahi Shinbun* (December 12, 1964, morning edition): 1.

[21] "Sekkyokuteki ni shakai kaihatsu," *Asahi Shinbun* (December 18, 1964, evening edition): 1.

and early 1950s had babies. Although the surge in fertility was fleeting, policymakers and policy intellectuals became concerned that this phenomenon, in conjunction with sliding mortality rates, would push the population increase rate up over 12.0 per 1,000, whereas it had been kept in the single digit range in the 1960s. Under these circumstances, the narrative of "overpopulation" returned.

Against this backdrop, in July 1974, eminent policy intellectuals and family planning activists who had influenced the official discussion in the late 1940s – not least of whom was Shinozaki – organized the Japan Population Conference (*Nihon Jinkō Kaigi*), the largest population conference after WWII. Aiming to "study and discuss the various kinds of population questions which challenge the Japanese people to seek policies and actions appropriate for the future," the conference was not merely an abstract intellectual exercise but also clearly accountable for the government's population management effort, just like the Fourth National Conference on Population Problems held during the war (see Chapter 4). The conference therefore mirrored official concerns about the growing population. Partly in response to the environmental activism that arose during this period, and partly in response to the revitalization sentiments that dominated prewar discussions on resources within the NRS in the late 1940s, the conference problematized Japan's growing population in terms of its precarious status as a resource poor but densely populated nation, aiming to exchange dialog over "the resource-food-environment problems [which] recently have grown critical in many aspects."[22] At the conference, population experts involved in policymaking, including Muramatsu Minoru at the DPHD (see Chapter 6), discussed "Population, Resources and Food" and "Stabilization of Population and Strategy for Action."[23] At the end of the conference, participants "call[ed] for the Japanese Government's immediate action" on the recommendations they made, which, like in the early 1950s, included promoting new contraceptive methods and "population education through mass media."[24] Like the recommendations made by the IC-PFP in the early 1930s (see Chapter 3), the conference also requested the government to facilitate the "expansion of demographic research institutions."[25] The conference was designed to crystallize the science–policy nexus that was central to the state's population management exercise.

[22] "Programme for the Japan Population Conference," in *Activities in Japan for the World Population Year 1974*, 29.
[23] Ibid.
[24] "Declaration of the 1974 Japan Population Conference (July 4)," in *Activities in Japan for the World Population Year 1974*, 48.
[25] Ibid.

However, the return of the "overpopulation" discourse in the first half of the 1970s was short lived. In 1974, the TFR began to decrease, and this time, the fertility decline was gradual but persistent; despite minor fluctuations from year to year, the TFR kept falling and never rose significantly until 2005. Policymakers expressed concerns about the downward trend as early as 1977, as evident in the *White Paper of Health and Welfare* published that year.[26] They knew fertility decline would further accelerate the population aging that was already underway. In 1970, the ratio of the Japanese population of individuals sixty-five years old and older to the total population exceeded 7.0% for the first time since vital statistics were available. Throughout the 1970s, the figure kept rising, and the media extensively covered Japan becoming an aging society. Along with this, the government, fueled by an unprecedented high economic growth since the mid-1950s, was expanding social security provisions for older populations, which culminated in offering free healthcare to the elderly in 1973.[27] However, the economic downturn Japan experienced in 1974, a result of the previous year's global oil crisis, urged policymakers to revise the government's generous spending on social security. In this context, policymakers were worried that a prolonged fertility decline might put further pressure on the government budget by exacerbating problems associated with an aging population.

While officially acknowledging the need to tackle issues related to fertility decline and aging population, the government did little to fundamentally solve these problems until the 1990s.[28] In fact, faced with a tightened social security budget, for much of the 1970s and 1980s, the government established social policies that ended up farming out childcare and elderly care to individual families.[29] These policies worked out quite nicely for a while, especially among the urban middle class. Their nuclear family structure, buttressed by the gendered division of labor whereby men were expected to engage in waged work while women were looking after the families, filled the social welfare gap created by the government's stringent measures.[30] However, these policies, heavily relying on gender norms and the nuclear family model, ironically transpired to speed up the fertility decline. In addition to the consolidation of the "housewifization" of women, in the 1980s Japan also witnessed expanding work opportunities for them, symbolized by the enactment of the

---

[26] Koseishō, *Kōsei hakusho (shōwa 52-nendo ban)*.
[27] Campbell, "The Old People Boom," 353.
[28] Koichi Hiraoka, "1980-nendai ikō no nihon niokeru," 23–28.
[29] Hiroko Fujisaki, "Kea seisaku," 605–24.
[30] Emiko Ochiai, *Kindai kazoku to feminizumu*.

Equal Employment Opportunity Law in 1986. Furthermore, during this period the nuclear family model was dominating the public narrative of familyhood, but in reality, it was eroding, as young men and women were increasingly deciding to marry later, or not marry at all.

These situations eventually resulted in the 1989 phenomenon sensationalized by the media as the "1.57 shock"; the TFR that year was even lower than the lowest post-WWII record of 1.58 in 1966.[31] Due to extensive media coverage of the "1.57 shock," in the 1990s, fertility decline became widely recognized as a social problem. The government expressed it as a problem strongly tied to issues related to an aging population, as attested by the mobilization of the term, *shōshi kōreika shakai*, "a society moving toward an aging population with a smaller number of children," which appeared in official documentation starting in the mid-1990s.[32] Along with this language, the government quickly established measures for declining fertility and an aging population. In 1994, it announced the five-year Angel Plan as the first policy package to tackle the issue of fertility decline, and it has updated the plan since then. Through the plan, the government launched a succession of programs that were designed to expand childcare services and make workplaces more family friendly. In 1989, the government formalized the Gold Plan, a ten-year strategy aimed to build infrastructure that would enable the state to provide health and welfare services for the elderly by 2000. However, the government was unable to implement many of the measures presented in the plan due to the financial pressure caused by the burst of the bubble economy in the early 1990s. Furthermore, these government policies, though intending to relieve family members of the burden of child and elderly care, in reality still relied on families as a source for care providers. At the same time, over the course of the decade, more families were unable to provide the required care, as labor became increasingly precarious and the family structure became diversified. Consequently, gaps between policies and realities widened further, which resulted in a further drop in the TFR.

Japan has inherited last century's demographic legacies. The TFR continued to decline, hitting the lowest-ever in 2005 at 1.26 children per woman. Since then, it has been gradually going up, though it is still well below the replacement level,[33] while the ratio of individuals sixty-five

---

[31] The dip in the TFR in 1966 was caused by young couples following the myth that, according to the Chinese zodiac, a girl born in the year of the fire horse (i.e., 1966) would be fiery tempered.

[32] Fujisaki, "Kea seisaku," 611.

[33] According to the latest statistics, the most recently available TFR is 1.45 in 2015. National Institute of Population and Social Security Research, "Population Statistics

years of age and older to the total population has been increasing at a phenomenal rate. The ratio was 17.4 at the end of 1999, and by 2020 it had risen to 28.4. Confronted with these demographic realities, the government has been expanding and diversifying social policies to tackle what it now categorically perceives as the twin problems of declining fertility and an aging population. The government enacted the Basic Law for Measures against the Declining Fertility Rate and the Law for Measures to Support Raising Next-Generation Children in 2003 and set up major outlines for measures in 2010 ("Vision for Children and Childcare") and 2015 ("Outline of Measures against Declining Fertility"). This time, policymakers tried to tackle work-related issues, such as long-term economic hardship among young people due to precarious labor, working long hours, low- and double-income households, and a lack of childcare provisions. These issues were acting as major disincentives to have children, so they stressed measures fostering a "work and life balance."[34]

For elderly care, the government enacted the Long-Term Care Insurance Law (LTCI) in 1997, and in 2000, introduced the mandatory insurance-based care system for the elderly, based on the infrastructure built under the Gold Plan and the eight welfare-related acts revised in the 1990s.[35] More recently, the government, recognizing the imminence of the "2025 vision" – the baby boomer cohort joining the age group of seventy-five years old and older and a contracting population of younger generations, who were previously expected to care for the elderly – established a number of measures to "mobilize" the elderly.[36] In 2005, it reformed the LTCI to shift its emphasis from care to prevention and independence for those with low levels of care needs. Since the early 2010s, the government has been promoting the "community-based integrated care system," which is attached to another, older scheme that aimed to build a large community of healthy elderly who could function as a peer-support group for their frailer peers. Linked to this, the Ministry of Health, Labour and Welfare collaborates with the public corporation National Silver Human Resources Center Association to secure further employment for the elderly. Since April 2021, the amended Elderly Persons Employment Stabilization Law has been in place, which encourages business owners to secure employment opportunities up to the age of

---

of Japan 2017, 4. Fertility,' www.ipss.go.jp/p-info/e/psj2017/PSJ2017.asp, accessed July 5, 2021.
[34] Moriizumi, "Kinnen ni okeru 'jinkō seisaku,'" 209.
[35] Mayumi Hayashi, *The Care of Older People*; Campbell, Ikegami, and Gibson, "Lessons from Public Long-Term Care Insurance," 87–95.
[36] Mayumi Hayashi, "Japan's Long-Term Care Policy for Older People," 11–21.

seventy. Through these social policy measures, the Japanese government has been clearly endorsing "active aging," a buzzword and policy framework that has been circulating globally since the early 2000s.[37]

For a long time, communities of population experts in Japan have concentrated their energy into studying policy-relevant topics. In addition to the activities of population experts mentioned above, since the end of WWII, the research of the IPP (today, it is the National Institute of Population and Social Security Research, IPPS) and the Population Association of Japan (PAJ) clearly shows affinity with the government's policy agenda concerning population issues.[38]

This affinity with policymaking, however, is a double-edge sword for the community of population scientists in Japan who wish to expand their field. On the one hand, over the post-WWII period, the proximity to the government permitted population science to thrive in Japan as policy science. On the other hand, in part due to the magnified role of population science in state politics, demography's potential to grow into an academically rigorous discipline was overlooked.[39] According to eminent demographer Atoh Makoto, the historical legacies covered in this book, such as the official endorsement of population/race policies and research (Chapter 4) during WWII and the over-concentration of efforts in policy-relevant birth control research immediately after the war (Chapters 5 and 6), were culpable; they had discouraged universities from investing their resources into population science.[40] All these factors made population studies a somewhat neglected subject in academia, while the Japanese government's recognition of its importance grew significantly over this period.

Yet, as argued in Chapter 6, post-WWII population policies and research in Japan were not solely shaped by the abovementioned domestic factors. In fact, as fertility in Japan declined, Japanese policymakers and population scientists increased the country's involvement in the transnational movement to contain the world population, which, from the mid-1960s, came to coalesce into the global politics of development and health promotion even further.[41]

---

[37] See WHO, 2002, https://extranet.who.int/agefriendlyworld/wp-content/uploads/2014/06/WHO-Active-Ageing-Framework.pdf.

[38] Nihon Jinkō Gakkai Sōritsu 50-shūnen Kinen Jigyō Iinkai, *Nihon jinkōgakkai 50-nenshi*, 149–209; Jinkō Mondai Kenkyūsho, ed., *Jinkō mondai kenkyūsho sōritsu gojusshūnen*, 3–8.

[39] For instance, in the past, the recognition of population scientists' efforts to produce policy relevant data and analysis overshadowed academic activities such as the Study Group on Demography (*Jinkōgaku Kenkyūkai*) launched by the eminent transwar population scientist Minami Ryūzaburō in 1958. Atoh, *Gendai jinkōgaku*, 13–14.

[40] Atoh, *Gendai jinkōgaku*, 4–5.

[41] Ogino, "Jinkō seisaku no sutorateji."

The Domestic and the Global
in the Science–Policy Nexus

In the 1960s, as the TFR in Japan dropped, birth control became less relevant within Japanese policymaking, and the government began to invest in international cooperation and development aids involving family planning.[42] Along with this, population experts who had established their careers via the policy-relevant birth control research in the 1950s channeled more of their energy into the transnational scientific exchange.

The Japanese government's interest in this field was influenced by a number of interlinked domestic and transnational factors. Within Japan, attempts to urge the government to participate in this field culminated in the 1960s, in part due to the campaign led by the health and family planning activist Kunii Chōjirō. In the 1950s, Kunii made himself known among the circle of family planning activists in Japan, but, foreseeing reduced demands for family planning, in the 1960s, he began to explore expanding his activism overseas. To convince influential politicians and economic leaders in Japan to join forces with him, in 1967, Kunii arranged for the American William Draper, a special advisor to the International Planned Parenthood Federation (IPPF) and a staunch proponent of global population control, to visit Japan. Draper met with eminent figures in business and politics, most importantly Kishi Nobusuke and Sasakawa Ryōichi, and stressed the importance of Japan's cooperation in the field of family planning and population in Asia. After Draper's visit, with the support of Kishi and Sasakawa, Kunii established the Japanese Organization for International Cooperation in Family Planning (JOICFP) in 1968, a nongovernmental organization approved by the Ministry of Foreign Affairs and the MHW and originally funded by the Health Center (*Hoken Kaikan*) and the IPPF.[43] Thereafter, the Japanese government began making donations to international organizations specializing in family planning and/or population issues, starting with the IPPF in 1969. By the mid-1970s, Japan had become one of the major donors within international cooperation on family planning, ranking in the top five after the United States, Sweden, West Germany, and the Netherlands.[44]

Japan's move was in line with the trend in the transnational population control movement described in Chapter 6, which merged with global

---

[42] Homei, "Between the West and Asia."

[43] Chōjirō Kunii, ed., *JOICFP no nijūnen* (Japanese Organization for International Cooperation in Family Planning, Inc., 1988).

[44] "Nihon no jinkō kazoku keikaku kyōryoku," n.d. c.1975, 1–9, NIPHL archives.

health promotion from the mid-1960s onward. Especially after the US government decided to fund population programs as part of its overseas development aid program, family planning became a staple in international health initiatives. In 1966, the United Nations General Assembly agreed to launch an organization that would help states develop population programs, and the United Nations Fund for Population Activities (UNFPA) was created in 1969. In 1967, the World Bank also began to fund population programs. Along with existing funding bodies, these international organizations became major donors in the field, fostering the integration of family planning with development aid health projects.

In this context, Japanese population experts and researchers who specialized in reproductive medicine, especially those in the IPP and IPH (Chapters 5 and 6), put more effort into transnational exchange. Muramatsu, for instance, actively attended international conferences and published in English to communicate his research to colleagues across the world. He also led international seminars on family planning that were organized by the JOICPF and taught about Japan's success with the state-led family planning program. As an institution, the PAJ organized and led a core symposium of the Eleventh Pacific Science Congress, held in Tokyo in 1966, titled "Population Problems of the Pacific." In the 1970s, the IPH and PAJ were heavily involved in organizing the Second Asian Population Conference that was held in Tokyo in 1972.[45] Finally, at the aforementioned Japan Population Conference, Japanese participants, including Muramatsu and Shinozaki, conversed with eminent figures in global population control, including Draper and Rafael Salas (Executive Director, UNFPA), who were invited for the occasion.

In the process of extending its network globally, Japanese population experts and activists also concentrated on forging regional ties in Asia. In the 1970s, the PAJ extended its international network by dispatching members to the Population Information Network to the United Nations Economic Commission for Asia and Far East (ECAFE; in 1974 it became the Economic and Social Commission for Asia and the Pacific, ESCAP).[46] Moreover, the Declaration of the Japan Population Conference called for the "promotion of increasing assistance for the developing countries in Asia ... and cooperation with other international organizations such as ... UNFPA and IPPF."[47] The preeminence of Asia

[45] Jinkō Mondai Kenkyūsho, ed., *Jinkō mondai kenkyūsho sōritsu gojusshūnen*, 7–8.
[46] Nihon Jinkō Gakkai Sōritsu 50-shūnen Kinen Jigyō Iinkai, *Nihon jinkōgakkai 50-nenshi*, 143–45.
[47] "Declaration of the 1974 Japan Population Conference (July 4)," in *Activities in Japan for the World Population Year 1974*, 48.

in Japanese population activities was directly compatible with the government's agenda to provide overseas development aid; in the early days of Japan's commitment to overseas development aid and international cooperation, Asia was without doubt the priority. This Japanese focus on Asia's population and development, both by the government and population experts, was informed by questions related to Japan's sovereignty that emerged as a result of Japanese colonialism and US-centric Cold War geopolitics. As much as the trajectory of Japanese sovereignty influenced the development of various medico-scientific population fields in Japan in the modern period, the globalization of Japanese population science and policy over the latter half of the twentieth century was spurred by Japan's perceived position in world politics, in the past and present.

Population, Development, and Social Orders

This story is not only about how politics influenced the globalization of Japanese population science and Japan's population management. It is also about how social orders realigned through the process of this globalization. Narratives inscribed in the Japanese science–policy nexus in favor of population control, which hitherto exposed and perpetuated preexisting social hierarchies domestically within Japan, extended globally and paved the way for the coproduction of social marginalization within Japan and global population control. For instance, the discourse associating fecundity with underdevelopment that informed Koya's experiment with foam tablets (Chapter 6) ultimately rendered villagers in Kajiya and Khanna not only as targets of Japanese and global population control but also fixed them within a lower social stratum. Similarly, in the 1960s, when "social development" surged as a policy agenda within Japan and the Japanese government embarked on providing overseas development aid, a Japanese expert studying population quality for social development portrayed the peripheral region of Tohoku – long associated with economic hardship and underdevelopment (Chapter 3) – as the "Tibet of Japan," attempting to stress the region's need for development.[48] The comparison of Tohoku with Tibet was buttressed by the chauvinistic progressivism inscribed in social science theories, including demographic transition theory (Chapter 5), which promoted the global developmentalist discourse that singled out Asia as a region in need of population control. By applying this globally circulating, ableist

---

[48] Sumiko Uchino, "Comparative Study of Tohoku and Kyushu Region Observed from Demographic Characteristics and Life Behavior," *Annual Reports of the Institute of Population Problems English Summary* 11, (1966): 82.

discourse to an "underdeveloped" region within Japan's internal borders, Japanese population experts and policymakers ultimately served to marginalize the region in post-WWII narratives of economic growth and social advancement, while also consolidating the status of Asia as a site of intervention in world politics. Through this, Japan's internal peripheries and Asia in general were simultaneously assigned the role of "recipients" of development assistance, occupying the lower end of the hierarchical relationship between "donors" and "recipients" that buttressed the global politics of development and population control.

The coproduction of domestic and global social orders was, in part, a byproduct of the Japanese aspiration to extricate the nation from the status of "mid-developed country" (chūshinkoku), a label some Japanese perceived as dishonorable because it allegedly described Japan's socio-economic regression after WWII. The byproduct of this coproduction was the eliding of the agency of the population control target. For example, consider Koya's policy-relevant birth control research discussed in Chapter 6. Through fieldwork, people – mostly married women – were asked to provide detailed information about the frequency of sexual intercourse, the use or nonuse of contraceptives, their menstrual history, and their experience of pregnancy and abortion. The information provided important data for the production of scientific knowledge, which was used to present the efficacy of the Japanese birth control pilot projects to an international audience. However, through the process of turning the information into scientific knowledge, the research participants' distinct sexual and reproductive experiences were translated to fit a "family line," a standardized – globally discernible – format, represented in numbers, chronological lines, and charts (see Figures 6.4 and 6.5).[49] Through this process, the messy and personal details of sexual experiences, which reflected research participants' agency but resisted numerical representation (such as the reasons why couples decided to stop using certain contraceptives or why people decided to discontinue their participation in the research), were stripped away to construct coherent datasets.

Needless to say, this was regarded as an essential component in the process of producing legible demographic knowledge for specialists, policymakers, and the general public. By the time Koya's team conducted the research, globally it had long been assumed as part of "modern" and "objective" science. However, modern data collection systems affected the construction of demographic facts, and they were far from value-free,

---

[49] "Kajiya Mura Page 1" Series III, Box 96, Folder 1577, Gamble Papers. The translation work, from Japanese to English, which was crucial for the development of globally circulated data, is one important element in the process.

inscribing Eurocentric and masculinist understandings of gender, class, civilization, and history onto them.[50] Japanese population studies followed the chauvinistic perspective embedded in this methodology, and as a consequence, they allowed little space for representing the agency of the "governed" individuals in the state-led population-governing endeavor.

One implication of the development of the science–policy nexus for the governing of a population covered in this book might be that policies informed by science failed to respond to the real needs of the governed population, whose sexual and reproductive behaviors were shaped by the micropolitics surrounding their everyday lives – the very details that population research got rid of in the process of making globally discernible, "scientific" knowledge. This book, relying chiefly on the sources that reflected the voices of elite, male bureaucrats, statesmen, doctors, and population experts, who were in the "governing" camp in the population-governing exercise, pointed to the possibility that this kind of scientific chauvinism might have shaped the ways in which the concept of the "governed" was constructed and acted upon – via the science that these elite men were keen to incorporate for the governing purposes.

Rather paradoxically, due to the expansiveness of the field's scope, the criticism of this chauvinism embedded in the interplay between the science of population and modern governance has come from the discipline of demography. This criticism peaked around the time when the International Conference on Population and Development took place in Cairo in 1994. Coinciding with the conference, which expanded the scope of policy discussion from one that focused on population control to a more holistic perspective that located family planning broadly in the arena of reproductive health rights and social development, feminist demographers such as Susan Watkins and Harriet Presser proclaimed that the science–policy nexus buttressing family planning in developing countries since the 1960s had been perpetuating a population discourse that overlooked the agency of the people who were on the receiving end of the initiatives.[51] In Japan, too, historical works incorporating a similar, critical approach to the analysis of the relationship between the discourse of population, governance, and the making of agency are quickly growing, in particular in works that study reproductive and family politics.[52]

---

[50] McCann, *Figuring the Population Bomb*.
[51] Presser, "Demography, Feminism, and the Science-Policy Nexus"; Susan Watkins, "If All We Knew About Women Was What We Read in Demography."
[52] Hiroshi Kojima, "Joshō jinkō, kazoku seisaku no gainen"; Kayo Sawada, *Sengo Okinawa no seishoku wo meguru poritikkusu*; Ogino, "Jinkō Seisaku No Sutorateji."

Population science and scientists critically informed this very relationship over the last century and a half, while determining the contours of modern Japan and the Japanese. In turn, population science today looks the way it does precisely because it was directly molded by the specific economic, social, and political conditions that constructed the self-understanding of Japan and the Japanese throughout the period. The science of population will continue to develop and shape – and be shaped by – statecraft as long as the population is implicated in socioeconomic problems that affect the nation's health and wealth. In Japan specifically, it will thrive by fulfilling its assigned role as a policy science by engaging with official efforts to curb the socioeconomic problems attached to the country's aging, low-fertility, and shrinking population. But a challenge lies ahead. One of the most pressing conundrums is: How does science adequately represent the population of Japan for policymaking without essentializing the category of "the Japanese population," especially in today's politics, which are increasingly exposed to globalization and multiculturalism? As the science–policy nexus continues to underpin the legal practice, social fabric, and national identities and experiences of people residing in the country, the science of population will be compelled to engage with this kind of question more than ever before.

Select Bibliography

Archives and Collections

Gamble, Clarence James Papers, 1920–70s, Center for the History of Medicine Francis A. Countway Library of Medicine, Boston, USA.

Irene B. Taeuber (1906–1974), Papers, 1912–1981, The State Historical Society of Missouri, Columbia, MO, USA.

Japan Center for Asian Historical Records (JACAR) digital archives, National Archives of Japan, Tokyo, Japan.

Minamoto Aiikukaikan Hozon Shiryō, Minamiarupusu-City, Yamanashi-Prefecture, Japan.

Rockefeller Foundations Archive, Sleepy Hollow, New York, USA.

The National Diet Library GHQ/SCAP Records, Tokyo, Japan.

National Institute of Public Health Library archives, Wako-shi, Saitama Prefecture, Japan.

The Tachi Bunko Archive, National Institute of Population and Social Security Research, Tokyo, Japan.

Official Publications

Hoken Eisei Chōsakai, "Hoken eisei chōsakai dai jūgo kai hōkokusho" (April 1931).

Hoken Eisei Chōsakai, "Hoken eisei chōsakai dai jūyonkai hōkokusho" (1930).

Hoken Eisei Chōsakai, "Hoken eisei chōsakai dai ikkai hōkokusho" (April 1917).

Jinkō Shokryō Mondai Chōsakai, "Jinkō mondai ni kansuru yoron" (January 1928).

Jinkō Shokryō Mondai Chōsakai, "Jinkō shokuryō mondai chōsakai jinkōbu tōshin setsumei" (April 1930).

Jinkō Mondai Kenkyūsho, ed., *Jinkō mondai kenkyūsho sōritsu gojusshūnen kinenshi* (1989).

Jinkō Mondai Shingikai, "Jinkō mondai shingikai dai 21 kai sōkai giji sokkiroku" (August 9, 1960).

Jinkō Mondai Shingikai, "Jinkō mondai shingikai dai 19 kai sōkai sokkiroku" (October 19, 1959).

Jinkō Mondai Shingikai, "Jinkō shishitsu kōjō taisaku ni kansuru ketsugi" (July 12, 1962).

Kokuritsu Kōshū Eisei In, ed., *Kokuritsu kōshū eisei in sōritsu gojusshūnen kinenshi* (1988).
Kokuritsu Kōshū Eisei In, *Kokuritsu kōshū eisei in sōritsu jūgo shūnen kinenshi* (1953).
Kōseirōdōshō (Ministry of Health, Labour and Welfare of the Government of Japan), *Heisei 28-nendo ban kōsei rōdō hakusho* (2016).
Kōseishō, *Kōsei hakusho (Shōwa 52-nendo ban)*. 1977. www.mhlw.go.jp/toukei_hakusho/hakusho/kousei/1977/, accessed December 12, 2019.
Kōseishō, *Kōsei hakusho (Shōwa 37-nendo ban)*. 1962. www.mhlw.go.jp/toukei_hakusho/hakusho/kousei/1962/, accessed December 12, 2019.
Kōseishō Jinkō Mondai Kenkyūsho, ed., *Jinkō mondai kenkyūsho no ayumi: 40-shūnen wo kinen shite* (Kōseishō Jinkō Mondai Kenkyūsho, 1979).
Kōseishō Gojūnenshi Henshū Iinkai, ed., *Kōseishō gojūnenshi* (Kōseishō Mondai Kenkyūkai and Chūō Hōki Shuppan, May 1988).
Naimushō, "Dainihon teikoku Naimushō tōkei hōkoku dai 1-kai" (1886).
Naimushō Eiseikyoku, *Nōson hoken eisei jicchi chōsa seiseki* (1929).
Naimushō Eiseikyoku, "Eiseikyoku nenpō Meiji 17-nen 7-gatsu – Meiji 20-nen 12-gatsu" (n.d., c.1887).
Naimushō ed., *Kokusei chōsa izen nihon jinkō tōkei shūsei 1 (Meiji 5-nen – 18-nen)*, vol. 1 (Tōyō shorin, 1992). Naimushō Eiseikyoku, "Eiseikyoku nenpō dai 6-ji" (July 1880).
Naimushō Eiseikyoku, "Eiseikyoku hōkoku dai 3-ji nenpō" (November 1877).
Naimushō Eiseikyoku, "Eiseikyoku hōkoku" (July 1877). Sōmushō Tōkeikyoku, "'Nihon tōkei nenkan' 120 kai no ayumi," accessed April 28, 2017, www.stat.go.jp/data/nenkan/pdf/120ayumi.pdf.
Sōmushō Tōkeikyoku, "Tōkei de miru anotoki to ima no. 3: Dai 1-kai kokuzei chōsaji (taishō 9-nen) to ima" (October 1, 2014), www.stat.go.jp/info/anotoki/pdf/census.pdf, accessed July 22, 2019.
Sōmushō Tōkeikyoku, "Nihon kindai tōkei no so 'sugi koji,'" accessed May 7, 2017, www.stat.go.jp/library/shiryo/sugi.htm.
Sōrifu Tōkeikyoku ed., *Sōrifu tōkeikyoku hyakunenshi shiryō shūsei*, vol. 2 (Sōrifu Tōkeikyoku, 1976).

Published Sources

Abe, Manami. "Meijiki no Osaka niokeru sanba seido no hensen." *Nihon ishigaku zasshi* 65, no. 1 (2019): 3–18.
Abe, Takashi, Shigeru Kawasaki, Atsushi Otomo, et al. "Chirigaku niokeru tōkei no riyō to kongo no kadai: 'Tōkei' wo meguru kan, gaku no renkei wo mezashite." *E-Journal GEO* 6, no. 1 (2011): 81–93, https://doi.org/10.4157/ejgeo.6.81.
Ackerman, Edward A. *Japan's Natural Resources and Their Relation to Japan's Economic Future.* Chicago: University of Chicago Press, 1953.
Ahluwalia, Sanjam, and Daksha Parmar. "From Gandhi to Gandhi: Contraceptive Technologies and Sexual Politics in Postcolonial India, 1947–1977." In *Reproductive States: Global Perspectives on the Invention and Implementation of Population Policy*, edited by Rickie Solinger and Mie Nakachi, 124–55. Oxford and New York: Oxford University Press, 2016.

Akami, Tomoko. "The Nation-State/Empire as a Unit of Analysis in the History of International Relations: A Case Study in Northeast Asia, 1868–1933." In *The Nation State and Beyond: Governing Globalization Processes in the Nineteenth and Early Twentieth Centuries*, edited by Isabella Löhr and Roland Wenzlhuemer, 177–208. Berlin and Heidelberg: Springer, 2013.

Ambaras, David Richard. *Japan's Imperial Underworlds: Intimate Encounters at the Borders of Empire*. Asian Connections. Cambridge: Cambridge University Press, 2018.

Ando, Yoshio, ed. *Showa keizaishi*. Tokyo: Nihon Keizai Shinbunsha, 1994.

Anesaki, Masahira. "History of Public Health in Modern Japan: The Road to Becoming the Healthiest Nation in the World." In *Public Health in Asia and the Pacific: Historical and Comparative Perspectives*, edited by Milton James Lewis and Kerrie L. Macpherson, 55–58. London: Routledge, 2011.

Aoki, Hidetora. *Osaka-shi sanba dantaishi*. Osaka: Osaka-shi Sanbakai, 1935.

Araragi, Shinzo. "The Collapse of the Japanese Empire and the Great Migrations: Repatriation, Assimilation, and Remaining Behind." In *The Dismantling of Japan's Empire in East Asia: Deimperialization, Postwar Legitimation and Imperial Afterlife*, edited by Barak Kushner and Sherzod Muminov, 66–84. London: Routledge, 2017.

Araragi, Shinzo, ed. *Teikoku igo no hito no idō: Posutokoroniarizumu to gurōbarizumu no kōsakuten*. Bensei Shuppan, 2013.

Asada, Kyōji. "Manshū nōgyō imin seisaku no ritsuan katei." In *Nihon teikokushugika no manshū imin*, edited by Manshū Iminshi Kenkyūkai, 3–107. Ryūkei Shosha, 1976.

Asano, Toyomi, ed. *Sengo nihon no baishō mondai to higashi ajia chiiki saihen*. Tokyo: Jigakusha Shuppan, 2013.

Asano, Toyomi. "Zentai no shikaku: Hikiage no tenkai to zaisan womeguru teikoku no butsuriteki kaitai to chiikiteki saihen." In *Sengo nihon no baishō mondai to higashi ajia chiiki saihen*, edited by Toyomi Asano, 1–27. Jigakusha Shuppan, 2013.

Asano, Toyomi. "Kokusai chitsujo to teikoku chitsujo womeguru nihon teikoku saihen no kōzō: Kyōtsūhō no rippō katei to hōteki kūkan no saiteigi." In *Shokuminchi teikoku nihon no hōteki tenkai*, edited by Toshihiko Matsuda and Toyomi Asano, 61–136. Shinzansha Shuppan, 2004.

Atoh, Makoto. *Gendai jinkōgaku shōshi kōrei shakai no kiso chishiki*. Tokyo: Nihon Hyōronsha, 2000.

Ayatoshi, Kure. *Riron tōkeigaku, jissai tōkeigaku*. Senshū Gakkō, 1890.

Ayatoshi, Kure, *Jissai tōkeigaku*. Senshū Gakkō, 1895.

Aydin, Cemil. "Japan's Pan-Asianism and the Legitimacy of Imperial World Order, 1931–1945." *Asia-Pacific Journal: Japan Focus* 6, no. 3 (March 2008): 1–33.

Azuma, Eiichiro. "Remapping a Pre-World War Two Japanese Diaspora: Transpacific Migration as an Articulation of Japan's Colonial Expansionism." In *Connecting Seas and Connected Ocean Rims*, 415–39, January 1, 2011.

Azuma, Eiichiro. "'Pioneers of Overseas Japanese Development': Japanese American History and the Making of Expansionist Orthodoxy in Imperial Japan." *Journal of Asian Studies* 67, no. 4 (November 2008): 1187–226.

Balasubramanian, Savina. "Motivating Men: Social Science and the Regulation of Men's Reproduction in Postwar India." *Gender & Society* 32, no. 1 (February 2018): 34–58.

Barnhart, Michael A. *Japan Prepares for Total War: The Search for Economic Security, 1919–1941*. Ithaca: Cornell University Press, 2013.

Bashford, Alison. *Global Population: History, Geopolitics, and Life on Earth*. New York: Columbia University Press, 2014.

Bashford, Alison. "Nation, Empire, Globe: The Spaces of Population Debate in the Interwar Years." *Comparative Studies in Society and History* 49, no. 1 (2007): 180–83.

Brecher, W. Puck. "Euraasians and Racial Capital in a 'Race War.'" In *Defamiliarizing Japan's Asia-Pacific War*, edited by W. Puck Brecher and Michael W. Myers, 207–26. Honolulu: University of Hawai'i Press, 2019.

Briggs, Laura. *Reproducing Empire: Race, Sex, Science, and U.S. Imperialism in Puerto Rico*. Berkeley: University of California Press, 2002.

Burns, Susan L. "Gender in the Arena of the Courts: The Prosecution of Abortion and Infanticide in Early Meiji Japan." In *Gender and Law in the Japanese Imperium*, edited by Susan L. Burns and Barbara J. Brooks, 81–108. Honolulu: University of Hawai'i Press, 2014.

Burns, Susan L. "Constructing the National Body: Public Health and the Body in Nineteenth-Century Japan." In *Nation Work: Asian Elites and National Identities*, edited by Timothy Brook and André Schmid, 17–49. Ann Arbor: University of Michigan Press, 2000.

Cai, Huiyu. *Taiwan in Japan's Empire-Building an Institutional Approach to Colonial Engineering*. Academia Sinica on East Asia. New York: Routledge, 2009.

Campbell, John. "The Old People Boom and Japanese Policy Making." *Journal of Japanese Studies* 5, no. 2 (1979): 321–57.

Campbell, John, Naoki Ikegami, and Mary Jo Gibson. "Lessons from Public Long-Term Care Insurance in Germany and Japan." *Health Affairs* 29, no. 1 (2010): 87–95.

Chan, Su-chuan. "Identification and Transformation of Plains Aborigines, 1895–1960: Based on the 'Racial' Classification of Household System and Census" [in Chinese]. *Taiwanshi yanjiu* 12, no. 2 (December 2005): 121–66.

Chapman, David. "Managing 'Strangers' and 'Undecidables': Population Registration in Meiji Japan." In *Japan's Household Registration System and Citizenship: Koseki, Identification and Documentation*, edited by David Chapman and Karl Jakob Krogness, 93–110. New York: Routledge, 2014.

Chapman, David. "Geographies of Self and Other: Mapping Japan through the Koseki." *The Asia-Pacific Journal: Japan Focus* 9, no. 29 (July 19, 2011). http://apjjf.org/2011/9/29/David-Chapman/3565/article.html, accessed June 1, 2022.

Chapman David, and Karl Jakob Krogness, eds. *Japan's Household Registration System and Citizenship: Koseki, Identification and Documentation*. New York: Routledge, 2014.

Chapman, David, and Karl Jakob Krogness. "The Koseki." In *Japan's Household Registration System and Citizenship: Koseki, Identification and Documentation*, edited by David Chapman and Karl Jakob Krogness, 1–18. New York: Routledge, 2014.

Chatani, Sayaka. "A Man at Twenty, Aged at Twenty-Five: The Conscription Exam Age in Japan." *The American Historical Review* 125, no. 2 (April 2020): 427–437.

Chen, Ching-Chih. "The Japanese Adaptation of the Pao-Chia System in Taiwan, 1895–1945." *The Journal of Asian Studies* 34, no. 2 (February 1975): 391–416.

Chin, Hsien-Yu, "Colonial Medical Police and Postcolonial Medical Surveillance Systems in Taiwan, 1895–1950s." *Osiris* 13, no. 1 (January 1998): 326–38.

Ching, Leo T. S. *Becoming "Japanese": Colonial Taiwan and the Politics of Identity Formation*. Berkeley: University of California Press, 2001.

Chuman, Ai. "Hoken eisei chōsakai hossoku eno michi: Nyūji shibōritsu modai no shiten kara." *Rekishigaku kenkyū*, no. 788 (2004): 16–26.

Chuman, Mitsuko. "Nagai Hisomu saikō: Yūseigaku keimō katsudō no shinsō wo saguru." In *Seimei no rinri*, edited by Kiyoko Yamazaki, 225–267. Fukuoka: Kyūshū Daigaku Shuppankai, 2013.

Chung, Yuehtsen Juliette. *Struggle for National Survival: Eugenics in Sino-Japanese Contexts, 1896–1945*. New York: Routledge, 2002.

Connelly, Matthew. *Fatal Misconception: The Struggle to Control World Population*. Cambridge, MA and London: The Belknap Press of Harvard University Press, 2008.

Cornell, L. L., and Akira Hayami. "The Shumon Aratame Cho: Japan's Population Registers." *Journal of Family History* 11, no. 4 (December 1986): 311–28.

Crook, Tom. *Governing Systems: Modernity and the Making of Public Health in England, 1830–1910*. Berkeley: California University Press, 2016.

Callahan, Karen Lee. "Dangerous Devices, Mysterious Times: Men, Women, and Birth Control in Early Twentieth-Century Japan." PhD diss., University of California, Berkeley, 2004.

Cullather, Nick. *The Hungry World: America's Cold War Battle Against Poverty in Asia*. Cambridge, MA and London: Harvard University Press, 2010.

Culver, Annika A. "Battlefield Comforts of Home." In *Defamiliarizing Japan's Asia-Pacific War*, edited by W. Puck Brecher and Michael W. Myers, 85–103. Honolulu: University of Hawai'i Press, 2019.

Daston, Lorraine. "Science Studies and the History of Science." *Critical Inquiry* 35, no. 4 (January 2009): 798–813.

Dear, Peter, and Sheila Jasanoff. "Dismantling Boundaries in Science and Technology Studies." *Isis* 101, no. 4 (2010): 759–74.

Desrosières, Alain, and Camille Naish. *The Politics of Large Numbers: A History of Statistical Reasoning*. Cambridge: Harvard University Press, 1998.

Dinmore, Eric G. "'Mountain Dream' or the 'Submergence of Fine Scenery'? Japanese Contestations over the Kurobe Number Four Dam, 1920–1970." *Water History* 6, no. 4 (December 2014): 315–40.

Dinmore, Eric G. "Concrete Results?: The TVA and the Appeal of Large Dams in Occupation-Era Japan." *The Journal of Japanese Studies* 39, no. 1 (January 2013): 10–12.

Dinmore, Eric G. "A Small Island Nation Poor in Resources: Natural and Human Resource Anxieties in Trans-World War II Japan." PhD diss., Princeton University, 2006.

Drixler, Fabian Franz. *Mabiki: Infanticide and Population Growth in Eastern Japan, 1660–1950*. Asia: Local Studies/Global Themes 25. Berkeley: University of California Press, 2013.

Doak, Kevin M. *A History of Nationalism in Modern Japan: Placing the People*. Leiden and Boston: Brill, 2007.

Donaldson, Peter J. *Nature Against Us: The United States and the World Population Crisis, 1965–1980*. Chapel Hill: University of North Carolina Press, 1990.

Dower, John W. *War without Mercy: Race and Power in the Pacific*. London: Faber, 1986.

Duara, Prasenjit. *Sovereignty and Authenticity: Manchukuo and the East Asian Modern*. New York: Rowman and Littlefield Publishers, 2004.

Endo, Masataka, *Koseki to kokuseki no kingendaishi: Minzoku, kettō, nihonjin*. Akashi Shoten, 2013.

Endo, Masataka. *Kindai nihon no shokuinchi tōchi ni okeru kokuseki to koseki*. Akashi Shoten, 2010.

Endoh, Toake. *Exporting Japan: Politics of Emigration toward Latin America*. Urbana: University of Illinois Press, 2009.

Fallwell, Lynne Anne. *Modern German Midwifery, 1885–1960*. London: Routledge, 2015.

Foucault, Michel. *The Foucault Reader*. Edited by Paul Pabinow. Penguin reprint. London: Penguin Books, 1991.

Francks, Penelope. *Rural Economic Development in Japan: From the Nineteenth Century to the Pacific War*. London and New York: Routledge, 2006.

Francks, Penelope. "Rice for the Masses: Food Policy and the Adoption of Imperial Self-Sufficiency in Early Twentieth-Century Japan." *Japan Forum* 15, no. 1 (January 2003): 125–46.

Frühstück, Sabine. *Colonizing Sex: Sexology and Social Control in Modern Japan*. Berkeley: University of California Press, 2003.

Fuchs, Stephen J. "Feeding the Japanese: Food Policy, Land Reform, and Japan's Economic Recovery." In *Democracy in Occupied Japan: The U.S. Occupation and Japanese Politics and Society*, edited by Mark E. Caprio and Sugita Yoneyuki, 26–47. London: Routledge, 2007.

Fuess, Harald. "Informal Imperialism and the 1879 'Hesperia' Incident: Containing Cholera and Challenging Extraterritoriality in Japan." *Japan Review*, no. 27 (2014): 103–40.

Fujime, Yuki. *Sei no rekishigaku: Kōshō seido, dataizai taisei kara baishun bōshihō, yūsei hogohō taisei e*. Fuji Shuppan, 1997.

Fujimoto, Hiro. "Women, Missionaries, and Medical Professions: The History of Overseas Female Students in Meiji Japan." *Japan Forum* 32, no. 2 (2020): 185–208.

Fujino, Yutaka. *Nihon fashizumu to yūsei shisō*. Kyoto: Kamogawa Shuppan, 1998.

Fujino, Yutaka. *Kōseishō no tanjō: Iryō wa fashizumu wo ikani suishin shitaka*. Kyoto: Kamogawa Shuppan, 2003.

Fujisaki, Hiroko. "Kea seisaku ga zentei tosuru kazoku moderu: 1970-nendai ikō no kosodate, kōreisha kaigo." *Shakaigaku hyōron* 64, no. 4 (2013): 605–24.

Fujitani, Takashi. *Race for Empire: Koreans as Japanese and Japanese as Americans during World War II*. Berkeley: University of California Press, 2011.

Garon, Sheldon. "Transnational History and Japan's 'Comparative Advantage.'" *Journal of Japanese Studies* 43, no. 1 (2017): 65–92.

Garon, Sheldon. *Molding Japanese Minds: The State in Everyday Life*. Princeton: Princeton University Press, 1997.

George, Timothy S. *Minamata: Pollution and the Struggle for Democracy in Postwar Japan*. Cambridge, MA: Harvard University Asia Center, 2001.

Germer, Andrea, Vera Mackie, and Ulrike Wöhr. *Gender, Nation and State in Modern Japan*. London: Taylor and Francis Group, 2014.

Gordon, Andrew. "Managing the Japanese Household: The New Life Movement in Postwar Japan." In *Gendering Modern Japanese History*, edited by Barbara Molony and Kathleen Uno, 423–51. Cambridge, MA: Harvard University Asia Center, 2005.

Gordon, Andrew. *The Evolution of Labor Relations in Japan: Heavy Industry, 1853–1955*. Cambridge, MA: Council on East Asian Studies, Harvard University Press, 1985.

Gosht, Arunabh. *Making It Count: Statistics and Statecraft in the Early People's Republic of China*. Princeton: Princeton University Press, 2020.

Greenhalgh, Susan. *Just One Child: Science and Policy in Deng's China*. Berkeley: University of California Press, 2008.

Hanes, Jeffrey E. *The City as Subject: Seki Hajime and the Reinvention of Modern Osaka*. Twentieth-Century Japan. Berkeley: University of California Press, 2002.

Hanscom, Christopher P., and Dennis C. Washburn, eds. *The Affect of Difference: Representations of Race in East Asian Empire*. Honolulu: University of Hawai'i Press, 2016.

Harrison, Mark. "Health, Sovereignty and Imperialism: The Royal Navy and Infectious Disease in Japan's Treaty Ports." *Social Science Diliman* 14, no. 2 (2018): 49–75.

Hartmann, Heinrich, and Ellen Yutzy Glebe. *The Body Populace Military Statistics and Demography in Europe before the First World War*. Cambridge: MIT Press, 2019.

Hartmann, Heinrich, and Corinna R. Unger, eds. *A World of Populations: Transnational Perspectives on Demography in the Twentieth Century*. New York: Berghahn Books, 2014.

Haruna, Nobuo. *Jinkō, shigen, ryōdo: Kindai nihon no gaikō shisō to kokusai seijigaku*. Chikura Shobo, 2015.

Haruyama, Meitetsu. "Meiji kenpō taisei to Taiwan tōchi." In *Iwanami kōza kindai nihon to shokuminchi 4 tōchi to shihai no ronri*, edited by Oe Shinobu, Kyoji Asada, Taichiro Mitani, Kenichi Goto, Hideo Kobayashi, Soji Takahashi, Masahiro Wakabayashi, and Minato Kawamura, 31–50. Iwanami Shoten, 1993.

Havens, Thomas R. H. "Kato Kanji (1884–1965) and the Spirit of Agriculture in Modern Japan." *Monumenta Nipponica* 25, no. 3/4 (1970): 249–66.

Hayami, Akira. "Koji Sugi and the Emergence of Modern Population Statistics in Japan: The Influence of German Statistics." *Reitaku Journal of Interdisciplinary Studies* 9, no. 2 (2001): 1–10.

Hayashi, Mayumi. "Japan's Long-Term Care Policy for Older People: The Emergence of Innovative 'Mobilisation' Initiatives Following the 2005 Reforms." *Journal of Aging Studies* 33 (2015): 11–21.

Hayashi, Mayumi. *The Care of Older People: England and Japan, A Comparative Study.* London: Pickering & Chatto, 2013.

Hayashi, Reiko. "Perception and Response to the Population Dynamics – on Fertility (Pre-War Period)" [in Japanese]. *Jinkō mondai kenkyū* 73, no. 4 (December 2017): 270–282.

Hein, Laura E. *Reasonable Men, Powerful Words: Political Culture and Expertise in Twentieth-Century Japan.* Berkeley and London: University of California Press, 2004.

Higami, Emiko. *Kindai Osaka no nyūji shibō to shakai jigyō.* Osaka: Osaka daigaku shuppankai, 2016.

Higasa, Kenta. "Sugi Kōji hakase to Meiji ishin no tōkei (7)." *Tōkeigaku zasshi,* no. 624 (June 1938): 22–34.

Hikita, Yasuyuki. "Daitoakyōeiken ni okeru tōsei keizai." In *"Teikoku" nihon no gakuchi dai 2 kan "teikoku" no keizaigaku,* edited by Shin'ya Sugiyama, vol. 2, 257–302. Iwanami Shoten, 2006.

Hiraoka, Koichi. "1980-nendai ikō no nihon niokeru shakai hoshō no seido kaikaku to seisaku tenkai." *Shakai seisaku kenkyū* 10, (2010): 23–28.

Hiroshima, Kiyoshi. "Horon sengo nihon no jinkō seisaku no hensen." In *Jinkō seisaku no hikakushi,* edited by Kojima and Hiroshima, 301–13. Nihon Keizai Hyōronsha, 2019.

Hiroshima, Kiyoshi. "Gendai nihon jinkō seisaku shi shōron 2: kokumin yūseihō ni okeru jinkō no shitsu seisaku to ryō seisaku." *Jinkō mondai kenkyū,* no. 160 (October 1981): 61–77.

Hiroshima, Kiyoshi. "Gendai nihon jinkō seisakushi shōron: Jinkō shishitsu gainen wo megutte (1916–1930)." *Jinkō mondai kenkyū,* no. 154 (April 1980): 46–61.

Hodgson, Dennis. "Demography as Social Science and Policy Science." *Population and Development Review* 9, no. 1 (1983): 1–34.

Hoffmann, David L. "Mothers in the Motherland: Stalinist Pronatalism in Its Pan-European Context." *Journal of Social History* 34, no. 1 (September 2000): 35–54.

Homei, Aya. "Between the West and Asia: 'Humanistic' Japanese Family Planning in the Cold War." *East Asian Science, Technology and Society* 10, no. 4 (December 2016): 445–67. https://doi.org/10.1215/18752160-3149695.

Homei, Aya. "The Science of Population and Birth Control in Post-War Japan." In *Science, Technology, and Medicine in the Modern Japanese Empire,* edited by David G. Wittner and Philip C. Brown, 227–43. London: Routledge, 2016.

Homei, Aya. "Midwife and Public Health Nurse Tatsuyo Amari and a State-Endorsed Birth Control Campaign in 1950s Japan." *Nursing History Review* 24, no. 1 (January 2016): 41–64.

Homei, Aya. "Midwives and the Medical Marketplace in Modern Japan." *Japanese Studies* 32, no. 2 (2012): 275–93.

Homei, Aya. "Birth Attendants in Meiji Japan: The Rise of the Biomedical Birth Model and a New Division of Labour." *Social History of Medicine* 19, no. 3 (2006): 407–424.

Homei, Aya. "Sanba and Their Clients: Midwives and the Medicalization of Childbirth in Japan." In *New Directions in History of Nursing: International Perspectives,* edited by Barbara Mortimer and Susan McGann, 68–85. London: Routledge, 2005.

Hopper, Helen M. *A New Woman of Japan: A Political Biography of Katō Shidzue*. Boulder: Westview Press, 1996.

Horikawa, Yuri. "Senji dōin seisaku to kikon josei rōdōsha: Senjiki ni okeru josei rōdōsha no kaisōsei wo meguru ichikōsatsu." *Shakai seisaku* 9, no. 3 (2018): 128–40.

Hoshino, Noriaki. "Racial Contacts across the Pacific and the Creation of Minzoku in the Japanese Empire." *Inter-Asia Cultural Studies* 17, no. 2 (April 2016): 186–205.

Hotta, Eri. *Pan-Asianism and Japan's War 1931–1945*. 1st ed. New York: Palgrave Macmillan, 2007.

Howell, David Luke. *Geographies of Identity in Nineteenth-Century Japan*. Berkeley: University of California Press, 2005.

Huang, Yu-ling. "The Population Council and Population Control in Postwar East Asia." Rockefeller Archive Center Research Reports Online: Rockefeller Archive Center, 2009. www.rockarch.org/publications/resrep/huang.pdf, accessed June 1, 2022.

Hyun, Jaehwan. "Racializing Chōsenjin: Science and Biological Speculations in Colonial Korea." *East Asian Science, Technology and Society* 13, no. 4 (December 2019): 489–510.

Iacobelli, Pedro. *Postwar Emigration to South America from Japan and the Ryukyu Islands*. London: Bloomsbury Academic, 2017.

Ichikawa, Tomoo. "Kindai nihon no kaikōchi ni okeru densenbyō ryūkō to gaikokujin kyoryūchi: 1879-nen 'Kanagawa-ken chihō eiseikai niyoru korera taisaku." *Shigaku zasshi*, no. 117 (June 2008): 1–38.

Igarashi, Yoshikuni. *Homecomings: the Belated Return of Japan's Lost Soldiers*. New York: Columbia University Press, 2016.

Iki, Hiroshi. "Meiji, Taisho-ki no maisō kyokashō ni miru yamai to shibō nenrei." *Nihon ishigaku zasshi* 45, no. 2 (1999): 246–47.

Ippo, Tsukada. *Kokka sōdōinhō no kaisetsu*. Shūhōen shuppanbu, 1938.

Ipsen, Carl. *Dictating Demography the Problem of Population in Fascist Italy*. Cambridge: Cambridge University Press, 1996.

Ishibashi, Yasumasa. "Mobilizing Structures in Manchuria Agricultural Emigration in Imperial Era: Idea and Practice of Kanji Kato as a 'Mediator'" [in Japanese]. *Korokiumu*, no. 6 (June 2011): 111–34.

Ishii, Akiko. "Statistical Visions of Humanity: Toward a Genealogy of Liberal Governance in Modern Japan." PhD diss., Cornell University, 2013.

Ishii, Ryoichi. *Population Problems and Economic Life in Japan*. Chicago: University of Chicago Press, 1937.

Ishii, Futoshi. "Posuto jinkō tenkanki no shibō dōkō." In *Posuto jinkō tenkanki no nihon*, edited by Sato and Kaneko, 91–109. Hara Shobo, 2016.

Ishikawa, Akira, and Tsukasa Sasai. "Gyōsei kiroku ni motozuku jinkō tōkei no kenshō." *Jinkō mondai kenkyū* 64, no. 4 (December 2010): 23–40.

Ishimura, Yoshinori, and Sakura Ishimura. "'Mizushina Shichisaburō' nōto (oboegaki): Hokkaido niokeru sangaku kōsō kansoku, kishō kansoku gyōsei oyobi tōkei kyōiku, Taiwan kokō chōsa wo chūshin toshite." *Takushoku daigaku ronshū* 2, no. 3 (July 1994): 143–95.

Ishizaki, Shoko. *Kingendai nihon no kazoku keisei to shusshōjisū: Kodomo no kazu wo kimetekita mono wa nanika*. Akashi Shoten, 2015.

Ishizaki, Shoko. "Meijiki no shussan wo meguru kokka seisaku." *Rekishi hyōron*, no. 600 (April 2000): 39–53.

Ishizaki, Shoko. "Kindai nihon no sanji chōsetsu to kokka seisaku." *Sogo joseishi*, no. 15 (1998): 15–32.

Ishizaki, Shoko. "Nihon no dataizai no seiritsu." *Rekishi hyōron*, no. 571 (November 1997): 53–70.

Ittmann, Karl, Dennis D. Cordell, and Gregory Maddox. *The Demographics of Empire: The Colonial Order and the Creation of Knowledge*. Athens: Ohio University Press, 2010.

Iwata, Masami, and Akihiko Nishizawa. *Poverty and Social Welfare in Japan*. Japanese Society Series. Melbourne: Trans Pacific Press, 2008.

Iwata, Shigenori. "*Inochi" wo meguru kindaishi: Datai kara jinkō ninshin chūzetsu e*. Yoshikawa Kobunsha, 2009.

Izumi, Takahide, ed. *Nihon kingendai jinmei jiten 1868–2011*. Igaku Shoin, 2012.

Jackson, Terrence. *Network of Knowledge: Western Science and the Tokugawa Information Revolution*. Honolulu: University of Hawai'i Press, 2016.

Jasanoff, Sheila. "The Idiom of Coproduction." In *States of Knowledge: The Co-Production of Science and the Social Order*, edited by Sheila Jasanoff, 1–12. Abingdon and Oxon: Taylor and Francis Group, 2004.

Jasanoff, Sheila. *The Fifth Branch: Science Advisors as Policymakers*. London and Cambridge, MA: Harvard University Press, 1990.

Jinkō Daijiten Henshū Iin, ed. *Jinkō daijiten*. Heibonsha, 1957.

Jinkōgaku Kenkyūkai, ed. *Gendai jinkō jiten*. Hara Shobo, 2010.

Johnston, William D. "Buddhism Contra Cholera: How the Meiji State Recruited Religion against Epidemic Disease." In *Science, Technology, and Medicine in the Modern Japanese Empire*, edited by David G. Wittner and Philip C. Brown, 62–78. London: Routledge, 2016.

Johnston, William. "Cholera and the Environment in Nineteenth-Century Japan." *Cross-Currents: East Asian History and Culture Review* 8, no. 1 (2019): 105–38.

Jones, Mark A. *Children as Treasures: Childhood and the Middle Class in Early Twentieth Century Japan*. Cambridge, MA and London: Harvard University Press, 2010.

Kamita, Seiji. "*Konketsuji" no sengoshi*. Seikyusha, 2018.

Kanazu, Hidemi. "Edo sankasho ni mirareru seishokuron: 'Umu shintai' towa dareno shintai ka." *Nihon shisōshi kenkyūkai kaihō*, no. 20 (2003): 152–64.

Kanazu, Hidemi, and Marjan Boogert. "The Criminalization of Abortion in Meiji Japan." *U.S.-Japan Women's Journal*, no. 24 (2003): 35–58.

Kaneko, Ayumi. "Nēshon to jitsugaku: 'Keimō' to 'gesaku' no kōten." In *Meiji, taisho-ki no kagaku shisoshi*, edited by Kanamori Osamu, 13–64. Keiso Shobo, 2017.

Kaneko, Tsutomu. *Edo jinbutsu kagakushi: "Mou hitotsu no bunmei kaika" wo tazunete*. Chuokoron-Shinsha, 2005.

Kasza, Gregory J. "War and Welfare Policy in Japan." *The Journal of Asian Studies* 61, no. 2 (May 2002): 417–35.

Kawai, Masashi. *Mirai no nenpyō: Jinkō genshō nihon de korekara okiru koto*. Kodansha, 2017.

Kawanishi, Kousuke. *Daitōa kyōwaken: Teikoku nihon no nanpō taiken.* Kodansha, 2016.

Kawauchi, Atsushi. "Jinkō to tohoku: Senjiki kara sengo niokeru tohoku 'kaihatsu' tono kanren de." In *Tohoku chihō "kaihatsu" no keifu: Kindai no sangyō seisaku kara higashi nihon daishinsai made,* edited by Takenori Yamamoto, 1–17. Akashi Shoten, 2015.

Kanamori, Osamu, ed. *Meiji, Taisho-ki no kagaku shisōshi.* Keiso Shobo, 2017.

Kanamori, Osamu, ed. *Showa zenki no kagaku shisōshi.* Keiso Shobo, 2011.

Kano, Mikiyo. *Onna tachi no jūgo.* Chikuma Shobo, 1995.

Kasahara, Hidehiko. *Nihon no iryō gyōsei.* Keiō Gijuku Daigaku Shuppankai, 1999.

Kasahara, Hidehiko, and Kazutaka Kojima. *Meijiki iryō, eisei gyōsei no kenkyū: Nagayo Sensai kara Gotō Shinpei e.* Kyoto: Mineruva Shobo, 2011.

Kashihara, Hiroki. *Meiji no gijutu kanryō.* Chuokoron-Shinsha, 2018.

Kato, Michiya. "Hidden from View?: The Measurement of Japanese Interwar Unemployment" [in Japanese]. *Annual Research Bulletin of Osaka Sangyo University,* no. 1 (December 2008): 77–103.

Kato, Shigeo. "Senjiki nihon no kagaku to shokuminchi, teikoku." *Rekishi hyōron* 832, (August 2019): 36–46.

Kawakami, Hajime. *Jinkō mondai hihan.* Sobunkaku, 1927.

Kim, Hoi-eun. *Doctors of Empire: Medical and Cultural Encounters between Imperial Germany and Meiji Japan.* Toronto: University of Toronto Press, 2016.

Kim, Sang-Hyun. "Science and Technology: National Identity, Self-Reliance, Technocracy and Biopolitics." In *The Palgrave Handbook of Mass Dictatorship,* edited by P. Corner and J. H. Lim, 81–97. London: Palgrave Macmillan, 2016.

Kimura, Naoko. *Shussan to seishoku wo meguru kōbō: Sanba, josanpu dantai to sankai no 100-nen.* Otsuki Shoten, 2013.

Kingsberg Kadia, Miriam. *Into the Field: Human Scientists of Transwar Japan.* Stanford: Stanford University Press, 2020.

Kitaoka, Shin'ichi. *Gotō Shinpei.* Chuokoron-sha, 1988.

"Ko Tachi Minoru shochō no ryakureki to gyōseki." *Jinkō mondai kenkyū* 123, (July 1972): 44–62.

Ko, Sunho. "Managing Colonial Diets: Wartime Nutritional Science on the Korean Population, 1937–1945." *Social History of Medicine* 34, no. 2 (2021): 592–610.

Kobayashi Deckrow, Andre. "São Paulo as Migrant-Colony: Pre-World War II Japanese State-Sponsored Agricultural Migration to Brazil." PhD diss., Columbia University, 2019.

Kojima, Kazutaka. *Nagayo Sensai to naimushō no eisei gyōsei.* Keio Gijuku Daigaku Shuppankai, 2021.

Koikari, Mire. *Gender, Culture, and Disaster in Post-3.11 Japan.* London: Bloomsbury Publishing, 2020.

Koikari, Mire. *Pedagogy of Democracy: Feminism and the Cold War in the U.S. Occupation of Japan.* Philadelphia: Temple University Press, 2008.

Kojima, Hiroshi, and Kiyoshi Hiroshima, eds. *Jinkō seisaku no hikakushi: Semegiau kazoku to gyōsei.* Nihon Keizai Hyōronsha, 2019.

Kojima, Hiroshi. "Joshō jinkō, kazoku seisaku no gainen, bunseki wakugumi, hikakushi." In *Jinkō seisaku no hikakushi: Semegiau kazoku to gyōsei*, edited by Kiyoshi Hiroshima and Hiroshi Kojima, 1–27. Nihon Keizai Hyouronsha, 2019.

Kondo, Kyoko. "Kokudo keikaku to jinkō no shiten no hensen." *Tōkei* 62, no. 12 (December 2011): 17–26.

Kovner, Sarah. *Occupying Power: Sex Workers and Servicemen in Postwar Japan*. Stanford: Stanford University Press, 2012.

Kramm, Robert. *Sanitizing Sex: Regulating Prostitution, Venereal Disease, and Intimacy in Occupied Japan, 1945–1952*. Berkeley: University of California Press, 2017.

Kratoska, Paul H., ed. *Asian Labor in the Wartime Japanese Empire: Unknown Histories*. London: Routledge, 2005.

Kratoska, Paul H. "Labor Mobilization in Japan and the Japanese Empire." In *Asian Labor in the Wartime Japanese Empire: Unknown Histories*, edited by Paul H. Kratoska, 3–21. London: Routledge, 2005.

Kubo, Hidebumi. *Nihon no kazoku keikaku shi: Meiji, Taisho, Showa*. Shadan Hōjin Nihon Kazoku Keikaku Kyōkai, 1997.

Kuniaki, Makino. *Senjika no keizagakusha: Keizaigaku to sōryokusen*. Chuokoron-Shinsha, 2020.

Kunii, Chōjirō, ed. *JOICFP no nijūnen*. Japanese Organization for International Cooperation in Family Planning, Inc., 1988.

Kurasawa, Aiko. *Shigen no sensō: "Daitōa kyōeiken" no jinryū, butsuryū*. Iwanami Shoten, 2012.

Kurihara, Jun. "The National Census, Family Registrations and the Extraordinary Taiwan Census of 1905" [in Japanese]. *Tokyo joshi daigaku hikaku bunka kenkyūsho kiyō* 65 (2004): 33–77.

Kushner, Barak, and Sherzod Muminov, eds. *The Dismantling of Japan's Empire in East Asia: Deimperialization, Postwar Legitimation and Imperial Afterlife*. London: Routledge, 2017.

Lam, Tong. *A Passion for Facts: Social Surveys and the Construction of the Chinese Nation-State, 1900–1949*. Asia Pacific Modern 9. Berkeley: University of California Press, 2011.

Latham, Michael E. *The Right Kind of Revolution: Modernization, Development, and U.S. Foreign Policy from the Cold War to the Present*. Ithaca: Cornell University Press, 2011.

Latour, Bruno. *Reassembling the Social: An Introduction to Actor-Network-Theory*. Oxford: Oxford University Press, 2007.

Latour, Bruno. *Laboratory Life: The Construction of Scientific Facts*. Princeton: Princeton University Press, 1986.

Lee, Sujin. "Differing Conceptions of 'Voluntary Motherhood': Yamakawa Kikue's Birth Strike and Ishimoto Shizue's Eugenic Feminism." *U.S.-Japan Women's Journal* 52 (2017): 3–22.

Lee, Sujin. "Problematizing Population: Politics of Birth Control and Eugenics in Interwar Japan." PhD diss., Cornell University, 2017.

Lim, Sungyun. *Rules of the House: Family Law and Domestic Disputes in Colonial Korea*. Berkeley: University of California Press, 2018.

Lin, Pei-Hsin. "The Unfolding and Significance of the Temporary Taiwan Household Investigation in Japanese Taiwan (1905–1915)" [in Chinese]. *Cheng Kung Journal of Historical Studies* 45, (December 2013): 87–128.

Lo, Ming-cheng Miriam. *Doctors within Borders: Profession, Ethnicity, and Modernity in Colonial Taiwan.* Berkeley: University of California Press, 2002.

López, Raúl Necochea. "Gambling on the Protestants: The Pathfinder Fund and Birth Control in Peru, 1958–1965." *Bulletin of the History of Medicine* 88, no. 2 (2014): 344–71.

Low, Morris, ed. *Building a Modern Japan: Science, Technology, and Medicine in the Meiji Era and Beyond.* New York: Palgrave Macmillan, 2005.

Löwy, Ilana. "Defusing the Population Bomb in the 1950s: Foam Tablets in India." *Studies in History and Philosophy of Science Part C: Studies in History and Philosophy of Biological and Biomedical Sciences* 43, no. 3 (September 2012): 583–93.

Löwy, Ilana. "'Sexual Chemistry' before the Pill: Science, Industry and Chemical Contraceptives, 1920–1960." *The British Journal for the History of Science* 44, no. 2 (June 2011): 245–74.

Liu, Michael Shiyung. *Prescribing Colonization: The Role of Medical Practices and Policies in Japan-Ruled Taiwan, 1895–1945.* Ann Arbor, MI: Association for Asian Studies, 2009.

Lu, Sidney Xu. *The Making of Japanese Settler Colonialism: Malthusianism and Trans-Pacific Migration, 1868–1961.* Cambridge: Cambridge University Press, 2019.

Lu, Sidney Xu. "Eastward Ho! Japanese Settler Colonialism in Hokkaido and the Making of Japanese Migration to the American West, 1869–1888." *The Journal of Asian Studies* 78, no. 3 (2019): 521–47.

Lu, Sidney Xu. "Japanese American Migration and the Making of Model Women for Japanese Expansion in Brazil and Manchuria, 1871–1945." *Journal of World History* 28, no. 3 (2017): 439–40.

Lu, Sidney Xu. "Colonizing Hokkaido and the Origin of Japanese Trans-Pacific Expansion, 1869–1894." *Japanese Studies* 36, no. 2 (May 2016): 251–74. https://doi.org/10.1080/10371397.2016.1230834.

Lu, Sidney Xu. "Good Women for Empire: Educating Overseas Female Emigrants in Imperial Japan, 1900–45." *Journal of Global History* 8, no. 3 (November 2013): 436–60.

Mackie, Vera C. *Feminism in Modern Japan: Citizenship, Embodiment, and Sexuality.* Cambridge: Cambridge University Press, 2003.

McClain, James L., and Osamu Wakita, eds. *Osaka: The Merchant's Capital of Early Modern Japan.* Ithaca: Cornell University Press, 1999.

Mathias, Regine. "Women and the War Economy in Japan." In *Japan's War Economy*, edited by Erich Pauer, 65–84. London and New York: Routledge, 1999.

Matsubara, Yoko. "Nihon – Sengo no yūsei hogohō toiu nano danshuhō." In *Yūseigaku to ningen shakai: Seimei kagaku no seiki wa doko e mukaunoka*, edited by Shohei Yonemoto, Yoko Matsubara, Jiro Nudeshima, and Yasutaka Ichinokawa, 169–236. Kodansha, 2000.

Matsubara, Yoko. "Nihon ni okeru yūsei seisaku no keisei." PhD diss., Ochanomizu University, 1998.

Matsubara, Yoko. "The Enactment of Japan's Sterilization Laws in the 1940s: A Prelude to Postwar Eugenic Policy." *Historia Scientiarum* 8, no. 2 (1998): 187–201.

Matsuda, Yoshiro. "Formation of the Census System in Japan: 1871–1945 – Development of the Statistical System in Japan Proper and Her Colonies." In *Historical Demography and Labor Markets in Prewar Japan*, edited by Michael Smitka, 100–24. New York: Garland, 1998.

Matsumura, Hiroyuki. "'Kokubō kokka' no yūseigaku: Koya Yoshio wo chūshin ni." *Shirin* 83, (2000): 102–32.

Matsumura, Janice. "Unfaithful Wives and Dissolute Labourers: Moral Panic and the Mobilisation of Women into the Japanese Workforce, 1931–45." *Gender & History* 19, no. 1 (2007): 78–100.

Matsusaka, Yoshihisa Tak. *The Making of Japanese Manchuria, 1904–1932*. Cambridge: Harvard University Asia Center, 2003.

McCann, Carole R. *Figuring the Population Bomb: Gender and Demography in the Mid-Twentieth Century*. Seattle: University of Washington Press, 2017.

Mikuriya, Takashi. *Seisaku no sōgō to kenryoku: Nihon seiji no senzen to sengo*. Tokyo Daigaku Shuppankai, 1996.

Mikuriya, Takashi. "The National Land Planning and the Politics of Development" [in Japanese]. *The Annuals of Japanese Political Science Association* 46 (1995): 57–76.

Miller, James A. "Betting with Lives: Clarence Gamble and the Pathfinder International." Population Research Institute, July 1, 1996. www.pop.org/betting-with-lives-clarence-gamble-and-the-pathfinder-international/ accessed May 22, 2022.

Mimura, Janis. "Economic Control and Consent in Wartime Japan." In *The Palgrave Handbook of Mass Dictatorship*, edited by Paul Corner and Jie-Hyun Lim, 157–69. London: Palgrave Macmillan, 2016.

Mimura, Janis. *Planning for Empire: Reform Bureaucrats and the Japanese Wartime State*. Ithaca: Cornell University Press, 2011.

Mimura, Janis. "Japan's New Order and Greater East Asia Co-Prosperity Sphere: Planning for Empire." *The Asia-Pacific Journal: Japan Focus* 9, no. 3 (2011): 1–12.

Mimura, Janis. "Technocratic Visions of Empire: Technology Bureaucrats and the 'New Order' for Science-Technology." In *The Japanese Empire in East Asia and Its Postwar Legacy*, edited by Harald Fuess, 97–118. Munich: Indicium Verlag GmbH, 1998.

Misaki, Yuko. "Jūrai kaigyō joi ni tsuite no ichi kōsatsu." *Nihon ishigaku zasshi* 65, no. 3 (September 2019): 301–13.

Mizuno, Hiromi. *Science for the Empire: Scientific Nationalism in Modern Japan*. Stanford: Stanford General, 2009.

Miyamoto, Kyoko. "Shimane-ken ni okeru kindai sanba seido unyō nikansuru kenkyū." *Shakai bunka ronshū* 11 (March 2015): 37–54.

Miyamoto, Kyoko. "Meiji-ki karano josanpu shoku no hatten to nyūji shibō no kanren: Shimane-ken no baai." *Shakai igaku kenkyū* 31, no. 2 (2014): 93–105.

Miyakawa, Tadao. *Tōkeigaku no nihonshi: Chikoku keisei eno negai*. Tokyo Daigaku Shuppankai, 2017.

Molony, Barbara. "Equality Versus Difference: The Japanese Debate over 'Motherhood Protection,' 1915–50." In *Japanese Women Working*, edited by Janet Hunter, 123–48. London and New York: Routledge, 1993.

Moore, Aaron Stephen. *Constructing East Asia: Technology, Ideology, and Empire in Japan's Wartime Era, 1931–1945*. Stanford: Stanford University Press, 2013.

Mori, Kenji. "The Development of the Modern Koseki." In *Japan's Household Registration System and Citizenship: Koseki, Identification and Documentation*, edited by David Chapman and Karl Jakob Krogness, 59–75. New York: Routledge, 2014.

Mori, Yasuo. *"Kokka sōdōin" no jidai: Hikaku no shiza kara*. Nagoya: Nagoya Daigaku Shuppankai, 2020.

Moriizumi, Rie. "Kinnen ni okeru 'jinkō seisaku': 1990-nendai ikō no shōshika taisaku no tenkai." In *Jinkō seisaku no hikakushi: semegiau kazoku to gyōsei*, edited by Hiroshi Kojima and Kiyoshi Hiroshima, 197–221. Nihon Keizai Hyōronsha, 2019.

Morita, Makiko. "Zenchi taiban no ichijikken." *Josan no shiori*, no. 40 (September 1899): 226–27.

Morita, Yūzō. *Tōkei henreki shiki*. Nihon Hyōronsha, 1980.

Morris-Suzuki, Tessa. "Beyond Racism: Semi-Citizenship and Marginality in Modern Japan." *Japanese Studies* 35, no. 1 (2015): 67–84.

Morris-Suzuki, Tessa. *Borderline Japan: Foreigners and Frontier Controls in the Post-War Era*. Cambridge: Cambridge University Press, 2010.

Morris-Suzuki, Tessa. "Ethnic Engineering: Scientific Racism and Public Opinion Surveys in Midcentury Japan." *Positions: East Asia Cultures Critique* 8, no. 2 (November 2000): 499–529.

Morris-Suzuki, Tessa. *Re-Inventing Japan: Time, Space, Nation*. Armonk, NY: M.E. Sharpe, 1998.

Morris-Suzuki, Tessa. "Debating Racial Science in Wartime Japan." *Osiris* 13, no. 1 (January 1998): 354–75.

Morris-Suzuki, Tessa. "Becoming Japanese: Imperial Expansion and Identity Crises in the Early Twentieth-Century." In *Japan's Competing Modernities: Issues in Culture and Democracy*, edited by Sharon Minichiello and Gail Bernstein, 157–80. Honolulu: University of Hawai'i Press, 1998.

Murakoshi, Kazunori. "Meiji, Taisho, Showa zenki ni okeru shizan tōkei no shinraisei." *Jinkōgaku kenkyū*, no. 49 (June 2013): 1–16.

Murphy, Michelle. *The Economization of Life*. Durham: Duke University Press, 2017.

Murphy, Michelle. "Economization of Life: Calculative Infrastructure of Population and Economy." In *Relational Architectural Ecologies Architecture, Nature and Subjectivity*, edited by Peg Rawes, 139–55. Florence: Taylor and Francis, 2013.

Myers, Ramon H. "Creating a Modern Enclave Economy: The Economic Integration of Japan, Manchuria, and North China, 1932–1945." In *The Japanese Wartime Empire, 1931–1945*, edited by Peter Duus, Ramon H. Myers, and Mark R. Peattie, 136–70. Princeton: Princeton University Press, 1996.

Nagayama, Sadanori. "Nihon no kanchō tōkei no hatten to gendai." *Nihon tokeigakkaishi* 16, no. 1 (1986): 101–9.

Nakamura, Ellen. "Ogino Ginko's Vision: 'The Past and Future of Women Doctors in Japan' (1893)." *U.S.-Japan Women's Journal*, no. 34 (2008): 3–18.

Nakamura, Miri. *The Monstrous Bodies: The Rise of the Uncanny in Modern Japan*. Cambridge: Harvard University Asia Center, 2015.

Nakanishi, Yasuyuki. "Takata Yasuma no jinkō riron to shakaigaku." *Keizai ronsō* 140, no. 5–6 (1987): 59–63.

Nakano, Ryoko. "Uncovering Shokumin: Yanaihara Tadao's Concept of Global Civil Society." *Social Science Japan Journal* 9, no. 2 (2006): 187–202.

Nakano, Satoshi. *Japan's Colonial Moment in Southeast Asia 1942–1945: The Occupiers' Experience*. Abingdon: Routledge, 2019.

Nakao, Katsumi. "Taihoku teikoku daigaku bunsei gakubu no dozoku jinshugaku kyōshitsu niokeru fīrudo wāku." In *Teikoku nihon to shokuinchi daigaku*, edited by Tetsuya Sakai and Toshihiko Matsuda, 221–50. Yumani Shobo, 2014.

Nakayama, Izumi. "Moral Responsibility for Nutritional Milk: Motherhood and Breastfeeding in Modern Japan." In *Moral Foods*, edited by Angela Ki Che Leung, Melissa L. Caldwell, Robert Ji-Song Ku, and Christine R. Yano, 66–88. Honolulu: University of Hawai'i Press, 2020.

Nakayama, Shigeru, Kunio Goto, and Hitoshi Yoshioka, eds. *A Social History of Science and Technology in Contemporary Japan*. 4 vols. Melbourne: Trans Pacific Press, 2001–06.

Nanta, Arno. "Physical Anthropology and the Reconstruction of Japanese Identity in Postcolonial Japan." *Social Science Japan Journal* 11 (2008): 29–47.

National Institute of Population and Social Security Research. "Population Projections for Japan (2016–2065): Summary Population Statistics." (2017), www.ipss.go.jp/pp-zenkoku/e/zenkoku_e2017/pp_zenkoku2017e.asp, accessed May 22, 2022.

Nihon Jinkō Gakkai, ed. *Jinkō daijiten*. Baifukan, 2002.

Nihon Jinkō Gakkai Sōritsu 50-shūnen Kinen Jigyō Iinkai. *Nihon jinkōgakkai 50-nenshi*. Nihon Jinkō Gakkai, 2002.

Nihon Kagakushi Gakkai, ed. *Kagakushi jiten*. Maruzen, 2021.

Norgren, Christiana A. E. *Abortion before Birth Control: The Politics of Reproduction in Postwar Japan*. Princeton: Princeton University Press, 2001.

Nishikawa, Mugiko. *Aru kindai sanba no monogatari: Noto, Takeshima Mii no katari yori*. Toyama: Katsura Shobo, 1997.

Nomura, Akihiro. "Shokuminchi ni okeru kindaiteki tōchi ni kansuru shakaigaku: Gotō Shinpei no Taiwan tōchi wo megutte." *Kyoto shakaigaku nenpō* 7 (1999): 1–24.

Oakley, Deborah. "American-Japanese Interaction in the Development of Population Policy in Japan, 1945–52." *Population and Development Review* 4, no. 4 (1978): 617–43.

Obayashi, Michiko. *Yamamoto Senji to haha Tane: Minshū to kazoku wo aishita hankotsu no seijika*. Domesu Shuppan, 2012.

Obayashi, Michiko. "Sengo nihon no kazoku keikaku fukyū katei nikansuru kenkyū." PhD diss., Ochanomizu University, 2006.

Obayashi, Michiko. *Josanpu no sengo.* Keiso Shobo, 1989.

O'Bryan, Scott. *The Growth Idea: Purpose and Prosperity in Postwar Japan.* Honolulu: University of Hawai'i Press, 2009.

Ochiai, Emiko. *Kindai kazoku to feminizumu.* Keiso Shobo, 1989.

Ogata, Masakiyo. *Nihon sanka gakushi.* Kyoto: Maruzen, 1919.

Ogawa, Keiko. "Seiyō kindai igaku no dōnyū to sanba no yōsei." In *Umisodate to josan no rekishi: Kindaika no 200-nen wo furikaeru,* edited by Chiaki Shirai, 26–46. Igaku Shoin, 2016.

Ogino, Miho. "Jinkō seisaku no sutoratejī 'umeyo fuyaseyo' kara kazoku keikaku e." In *Tekuno/baio poritikkusu: Kagaku iryō gijutsu no ima,* edited by Kaoru Tachi, 145–59. Sakuhinsha, 2008.

Ogino, Miho. *"Kazoku keikaku" eno michi: Kindai nihon no seishoku wo meguru seiji.* Iwanami Shoten, 2008.

Oguma, Eiji. *"Nihonjin" no kyōkai: Okinawa, Ainu, Taiwan, Chōsen, shokuminchi shihai kara fukki undō made.* Shin'yosha, 1998.

Oguma, Eiji. *Tan'itsu minzoku shinwa no kigen: "Nihonjin" no jigazō no fukei.* Shin'yosha, 1995.

Oguma, Eiji. "Tsumazuita junketsu shugi: Yūseigaku seiryoku no minzoku seisakuron." *Jyōkyō* 5, no. 11 (1994): 38–50.

Oguri, Shiro. *Chihō eisei gyōsei no sōsetsu katei.* Iryo Tosho Shuppansha, 1981.

Ohashi, Ryuken. *Nihon no tōkeigaku.* Kyoto: Horitsu Bunkasha, 1965.

Ohmi, Kenichi. "Mizushima Haruo ra no shokuminchi seimeihyō kenkyū ni miru dainiji sekai taisen zen, senchū no igaku kenkyū saikō." *Nihon kenkō gakkai zasshi* 86, no. 5 (September 2020): 209–223.

Ohmi, Kenichi. "Dainijisekaitaisen izen no wagakuni niokeru jinkō dōtai tōkei." *Nihon kōshū eisei zasshi* 51, no. 6 (2004): 452–60.

Oide, Harue. "Byōin shussan no seiritsu to kasoku: Seijōsan wo meguru kōbō to sanshihō seitei wo undō wo chūshin toshite." *Ningenkankeigaku kenkyū,* no. 7 (2006): 25–39.

Okamoto, Kiyoko. "Josanpu katsudō no rekishiteki igi: Meiji jidai wo chūshin ni." In *Nippon no josanpu Showa no shigoto,* edited by Reborn henshūbu, 169–90. Reborn, 2009.

Okamoto, Makiko. "Ajia taiheiyō sensō makki niokeru chōsenjin, taiwanjin sanseiken mondai." *Nihonshi kenkyū,* no. 401 (January 1996): 53–67.

Okazaki, Yoichi. "Tsuitō Shinozaki Nobuo hakase (tsuitōbun)" *Jinkōgaku kenkyū* 24, (June 1999): 74.

Okubo, Naomutsu and Yoshitoshi Misugi. *Nyūyōji hogo shishin.* Osaka: Osaka Nyūyōji Hogo Kyōkai, 1928.

Okumura, Makoto. "Tōhoku chihō kaihatsu no rekishi," *Toshi keikaku* 61, no. 2 (April 2012): 5–10.

Ono, Yoshiro, and Isao Somiya. "Meijiki nihon no kōshū eisei nikansuru jōhō kankyō." *Papers of the Research Meeting on the Civil Engineering History in Japan* 4, (1984): 41–48.

Osaka Mainichi Shinbun Jizendan. *Osaka Mainichi Shinbun Jizendan nijūnen-shi.* Osaka: Osaka Mainichi Shinbun Jizendan, 1931.

Ota, Motoko. *Kodakara to kogaeshi: Kinsei nōson to kazoku seikatsu to kosodate.* Fujiwara Shoten, 2007.

Ota, Tenrei. "Ota Ringu no hanseiki." *Gendai no me* 20, no. 9 (1979): 236–45.

Ota, Tenrei. *Nihon sanji chōsetsu hyakunenshi.* Ningen no Kagakusha, 1976.

Otsubo, Sumiko, and James R. Bartholomew. "Eugenics in Japan: Some Ironies of Modernity, 1883–1945." *Science in Context* 11, no. 3–4 (1998): 545–65.

Oyodo, Shoichi. *Gijutu kanryō no seiji sankaku: Nihon no kagaku gijutu gyōsei no makuaki.* Chuokoron-sha, 1997.

Palmer, Brandon. *Fighting for the Enemy: Koreans in Japan's War, 1937–1945.* Washington: University of Washington Press, 2013.

Park, Jin-kyung. "Interrogating the 'Population Problem' of the Non-Western Empire: Japanese Colonialism, the Korean Peninsula, and the Global Geopolitics of Race." *Interventions* 19, no. 8 (November 2017): 1112–31, https://doi.org/10.1080/1369801X.2017.1354961.

Park, Jin-kyung. "Husband Murder as the 'Sickness' of Korea: Carceral Gynecology, Race, and Tradition in Colonial Korea, 1926–1932." *Journal of Women's History* 25, no. 3 (2013): 116–40.

Park, Jin-kyung. "Corporeal Colonialism: Medicine, Reproduction, and Race in Colonial Korea." PhD diss., University of Illinois Urbana–Champaign, 2008.

Parmar, Inderjeet. *Foundations of the American Century: The Ford, Carnegie, and Rockefeller Foundations in the Rise of American Power.* New York: Columbia University Press, 2012.

Peattie, Mark R. "Japanese Attitudes toward Colonialism, 1895–1945." In *The Japanese Wartime Empire, 1931–1945*, edited by Peter Duus, Ramon H. Myers, and Mark R. Peattie, 80–127. Princeton: Princeton University Press, 1996.

Pflugfelder, Gregory M. "The Nation-State, the Age/Gender System, and the Reconstitution of Erotic Desire in Nineteenth-Century Japan." *The Journal of Asian Studies* 71, no. 4 (2012): 963–74.

Porter, Theodore M. *Trust in Numbers: The Pursuit of Objectivity in Science and Public Life.* Princeton: Princeton University Press, 1995.

Porter, Theodore M. *The Rise of Statistical Thinking 1820–1900.* Princeton: University Press, 1986.

Presser, Harriet. "Demography, Feminism, and the Science-Policy Nexus." *Population and Development Review* 23, no. 2 (1997): 295–331.

Prévost, Jean-Guy. *Total Science: Statistics in Liberal and Fascist Italy.* Montreal: McGill-Queen's University Press, 2009.

Reed, James. *The Birth Control Movement and American Society: From Private Vice to Public Virtue.* Princeton Legacy Library. Princeton: Princeton University Press, 1978.

Robertson, Jennifer. "Eugenics in Japan: Sanguinous Repair." In *The Oxford Handbook of the History of Eugenics*, edited by Alison Bashford and Philippa Levine, 430–48. Oxford: Oxford University Press, 2010.

Roebuck, Kristin A. "Orphans by Design: 'Mixed-Blood' Children, Child Welfare, and Racial Nationalism in Postwar Japan." *Japanese Studies* 36, no. 2 (2016): 191–212.

Roebuck, Kristin A. "Japan Reborn: Mixed-Race Children, Eugenic Nationalism, and the Politics of Sex after World War II." PhD diss., Columbia University, 2015.

Robertson, Thomas. *Malthusian Moment: Global Population Growth and the Birth of American Environmentalism*. New Brunswick: Rutgers University Press, 2012.

Rothermund, Dietmar. *The Global Impact of the Great Depression 1929–1939*. London: Taylor and Francis Group, 1996.

Saeki, Eiko. "Abortion, Infanticide, and a Return to the Gods: Politics of Pregnancy in Early Modern Japan." In *Transcending Borders*, edited by Shannon Stettner, Katrina Ackerman, Kristin Burnett, and Travis Hayl, 19–33. New York: Palgrave Macmillan, 2017.

Saeki, Riichirō. "Hamada Gen'tatsu sensei no omoide banashi: Hamada Gen'tatsu sensei no nijukkaiki wo shinobite." *Sanka to fujinka* 2, no. 2 (1934): 63–69.

Sakai, Naoki. "Ethnicity and Species/Radical Philosophy." *Radical Philosophy* 95, no. 1 (June 1999), www.radicalphilosophy.com/article/ethnicity-and-species, accessed May 22, 2022.

Sakai, Saburo. *Showa kenkyūkai: Aru chishikijin shūdan no kiseki*. Chuokoronsha, 1992.

Sakano, Toru. *Teikoku wo shiraberu: Shokuminchi fīrudo wāku no kagakushi*. Keiso Shobo, 2016.

Sakano, Toru. "Joron 'teikoku nihon' 'posto teikoku' jidai no fīrudowāku wo toinaosu." In *Teikoku wo shiraberu*, edited by Toru Sakano, 1–11. Keiso Shobo, 2016.

Sakano, Toru. *Teikoku nihon to jinrui gakusha: 1884–1952 nen*. Keiso Shobo, 2005.

Sakano, Toru, and Togo Tsukahara, eds. *Teikoku nihon no kagaku shisōshi*. Keiso Shobo, 2018.

Sams, Crawford F., and Zabelle Zakarian. *"Medic": The Mission of an American Military Doctor in Occupied Japan and Wartorn Korea*. Armonk, NY: M. E. Sharpe, 1998.

Sasaki, Toshiji, and Akinori Otagiri, eds. *Yamamoto Senji zenshū*. Chobunsha, 1979.

Sato, Jin. *"Motazaru kuni" no shigenron: Jizoku kanō na kokudo wo meguru mouhitotsu no chi*. Tokyo Daigaku Shuppankai, 2011.

Sato, Masahiro. *Kokusei chōsa nihon shakai no hyakunen*. Iwanami Shoten, 2015.

Sato, Masahiro. *Teikoku nihon to tōkei chōsa: Tōchi shoki Taiwan no senmonka shūdan*. Iwanami Shoten, 2012.

Sato, Masahiro. "Chōsa tōkei no keifu: Shokuminchi niokeru tōkei chōsa shisutemu," in *"Teikoku" nihon no gakuchi dai 6 kan kenkyū chiiki toshiteno ajia*, edited by Suehiro Akira, 181–204. Iwanami Shoten, 2006.

Sato, Masahiro. *Kokuzei chōsa to nihon kindai*. Iwanami Shoten, 2002.

Sato, Ryuzaburo, and Ryuichi Kaneko, eds. *Posuto jinkō tenkanki no nihon.* Hara Shobo, 2016.

Sato, Ryuzaburo, and Ryuichi Kaneko. "Entering the Post-Demographic Transition Phase in Japan: Its Concept, Indicators and Implications." Paper presented at the European Population Conference, Budapest, Hungary, June 25–28, 2014. https://epc2014.princeton.edu/papers/140662, accessed July 21, 2020.

Saito, Osamu. "Senzen nihon niokru nyūji shibō mondai to aiikuson jigyō." *Shakai keizaishigaku* 73, no. 6 (March 2008): 611–33.

Saito, Osamu. "Bosei eisei seisaku ni okeru chūkan soshiki no yakuwari: Aiikukai no jigyō wo chūshin ni." In *Senkanki nihon no shakai shudan to nettowāku: Demokurashī to chūkan dantai,* edited by Takenori Inoki, 359–79. NTT Shuppan, 2008.

Saito, Osamu. "Jinkō tenkan izen no nihon niokeru mortality: Patān to henka." *Keizai kenkyū* 43, no. 3 (July 1992): 248–67.

Saito, Osamu, and Masahiro Sato. "Japan's Civil Registration Systems Before and After the Meiji Restoration." In *Registration and Recognition: Documenting the Person in World History,* edited by Keith Derek Breckenridge and Simon Szreter, 113–35. Oxford: Oxford University Press, 2012.

Sakurada, Yuriko. "Senji ni itaru 'jinteki shigen' wo meguru mondai jōtai: Kenpei kenmin seisaku tōjō no haikei." *Nagano daigaku kiyō,* no. 9 (March 1979): 41–55.

Sasaki, Kahoru. "Meiji-ki niokeru Gunma-ken no sanba yōsei no hajimari." *Gunma kenritsu kenmin kenkō kagaku daigaku kiyō* 4 (March 2009): 1–11.

Sawada, Kayo. *Sengo Okinawa no seishoku wo meguru poritikkusu: Beigun tōchika no shusshōryoku tenkan to onna tachi no kōshō.* Otsuki Shoten, 2014.

Sawayama, Mikako. *Kindai kazoku to kosodate.* Yoshikawa Kobunkan, 2013.

Sawayama, Mikako. *Sei to seishoku no kinsei.* Keiso Shobo, 2005.

Saya, Makito. *Minzokugaku, Taiwan, kokusai renmei: Yanagita Kunio to Nitobe Inazō.* Kodansha, 2015.

Schaffner, Karen J., ed. *Eugenics in Japan.* Fukuoka: Kyushu Daigaku Shuppankai, 2014.

Schoen, Johanna. *Choice & Coercion: Birth Control, Sterilization, and Abortion in Public Health and Welfare.* Chapel Hill: University of North Carolina Press, 2005.

Schweber, Libby. *Disciplining Statistics: Demography and Vital Statistics in France and England, 1830–1885.* Durham: Duke University Press, 2006.

Sekiyama Naotarō. *Nihon jinkōshi.* Shikai Shobo, 1942.

Shapiro, Barbara J. *A Culture of Fact: England, 1550–1720.* Ithaca and London: Cornell University Press, 2000.

Shi, Mau-Shan, Chien-Heng Chen, Yen-Hao Huang, and Min-Shing Huang. "A Conversion of Population Statistics of Taiwan at the Sub-Provincial Layer: 1897–1943" [in Chinese]. *Renkouxuekan,* (2010): 157–202, https://doi.org/10.6191/jps.2010.4.

Sharpless, John. "World Population Growth, Family Planning, and American Foreign Policy." *Journal of Policy History* 7 (1995): 72–102.

Shibata, Yoichi. *Teikoku nihon to chiseigaku: Ajia taiheiyō sensōki ni okeru chiri gakusha no shisō to jissen.* Osaka: Seibunsha, 2016.

Shinmura, Taku. *Shussan to seishokukan no rekishi*. Hosei Daigaku Shuppankyoku, 1996.

Shiode, Hiroyuki. *Ekkyōsha no seijishi*. Nagoya: Nagoya Daigaku Shuppankai, 2015.

Shirai, Chiaki, ed. *Umisodate to josan no rekishi: Kindaika no 200-nen wo furikaeru*. Igaku Shoin, 2016.

Shimamura, Shiro. *Nihon tōkeishi gunzō*. Nihon Tōkei Kyōkai, 2009.

Shimoji, Lawrence Yoshitaka. *"Konketsu" to "nihonjin": Hāfu, daburu, mikkusu no shakaishi*. Seidosha, 2018.

Soeda, Yoshiya. *Seikatsu hogo seido no shakaishi [zōhoban]*. Tokyo Daigaku Shuppankai, 2014.

Solinger, Rickie, and Mie Nakachi, eds. *Reproductive States: Global Perspectives on the Invention and Implementation of Population Policy*. Oxford and New York: Oxford University Press, 2016.

Sreenivas, Mytheli. *Reproductive Politics and the Making of Modern India*. Washington, DC: University of Washington Press, 2021.

Steger, Brigitte. "From Impurity to Hygiene: The Role of Midwives in the Modernisation of Japan." *Japan Forum* 6, no. 2 (1994): 175–87.

Stichweh, Rudolf. "The Sociology of Scientific Disciplines: On the Genesis and Stability of the Disciplinary Structure of Modern Science." *Science in Context* 5, no. 1 (1992): 3–15.

Sugita, Naho. "Nihon niokeru jinkō shishitsu gainen no tenkai to shakai seisaku: senzen kara sengo e." *Keizaigaku zasshi* 116, no. 2 (2015): 59–81.

Sugita, Naho. "1950-nendai no nihon ni okeru jinkōgaku no kenkyū kyōiku taisei kakuritsu ni muketa ugoki ni tsuite." *Jinkōgaku kenkyū* (May 2019): 4–7.

Sugita, Naho. *"Yūsei," "yūkyō" to shakai seisaku: Jinkō mondai no nihonteki tenkai*. Kyoto: Horitsu Bunka Sha, 2013.

Sugita, Naho. *Jinkō, kazoku, seimei to shakai seisaku: Nihon no keiken*. Kyoto: Hōritsu Bunka Sha, 2010.

Sugita, Yoneyuki. "Toward a National Mobilization: The Establishment of National Health Insurance." In *Japan's Shifting Status in the World and the Development of Japan's Medical Insurance Systems*, edited by Yoneyuki Sugita, 93–125. Singapore: Springer, 2019.

Suh, Soyoung. *Naming the Local: Medicine, Language, and Identity in Korea since the 15th Century*. Cambridge, MA: Harvard University Asia Center Publications Program, 2017.

Suzuki, Zenji. *Nihon no yūseigaku*. Sankyo Shuppan Kabushikigaisha, 1983.

Szreter, Simon. "The Idea of Demographic Transition and the Study of Fertility Change: A Critical Intellectual History." *Population and Development Review* 19, no. 4 (1993): 659–701.

Tachi, Minoru. "Japan's Population To-Day." *Japan Planned Parenthood Quarterly* 1, no. 1 (1950): 3–5.

Takagi, Shugen. "Akiyoshi Mizukuri to tōkeigaku." *Kansai daigaku keizai ronshū* 19, no. 1 (April 1969): 1–17.

Takahashi, Masuyo. "'Taiwan tōkei kyōkai zasshi' sōmokuji kaidai." Hitotsubashi University Research Unit for Statistical Analysis in Social Sciences, May 2005.

Takahashi, Masuyo. "Meijiki wo chūshin nimita nihon no jinkō tōkei shiryō nitsuite (keizai tōkei tokushū)." *Keizai shiryō kenkyū* 14 (June 1980): 14–31.

Takaoka, Hiroyuki. *Sōryokusen taisei to "fukushi kokka": Senjiki nihon no "shakai kaikaku" kōsō*. Iwanami Shoten, 2011.

Takaoka, Hiroyuki. "Senji no jinkō seisaku." In *Kazoku kenkyū no saizensen jinkō seisaku no hikakushi: Semegiau kazoku to gyōsei*, edited by Hiroshi Kojima and Kiyoshi Hiroshima, 101–25. Nihon Keizai Hyōronsha, 2019.

Takase, Masato. "1890nen–1920nen no wagakuni no jinkō dōtai to jinkō seitai." *Jinkōgaku kenkyū*, no. 14 (May 1991): 21–34.

Takeda, Hiroko. *The Political Economy of Reproduction in Japan: Between Nation-State and Everyday Life*. London and New York: RoutledgeCurzon, 2005.

Takehara, Kazuo. "Meiji shoki no eisei seisaku kōsō: 'Naimushō eiseikyoku zasshi' wo chūshin ni." *Nihon ishigaku zasshi* 55, no. 4 (2009): 509–20.

Takeno, Manabu. "Jinkō mondai to shokuminchi: 1920, 20 nendai no Karafuto wo chūshin ni." *Keizaigaku kenkyū* 50, no. 3 (December 2000): 117–32.

Takeuchi-Demirci, Aiko. *Contraceptive Diplomacy: Reproductive Politics and Imperial Ambitions in the United States and Japan*. Stanford: Stanford University Press, 2018.

Takeuchi-Demirci, Aiko. "Birth Control and Socialism: The Frustration of Margaret Sanger and Ishimoto Shizue's Mission." *Journal of American-East Asian Relations* 17, no. 3 (2010): 257–80.

Takeuchi, Aiko. "The Transnational Politics of Public Health and Population Control: The Rockefeller Foundation's Role in Japan, 1920s–1950s." Rockefeller Archive Center (RAC) Research Reports Online (2009). https://rockarch.org/publications/resrep/takeuchi.pdf, accessed July 16, 2010.

Tama, Shinnosuke. *Sōsenryoku taiseika no manshū nōgyō imin*. Yoshikawa Kobunkan, 2016.

Tama, Yasuko. "Jutai chōsetsu (bāsu kontorōru) to botai hogohō." In *Umi sodate to josan no rekishi: Kindaika no 200-nen wo furikaeru*, edited by Chiaki Shirai, 110–35. Igaku Shoin, 2016.

Tama, Yasuko. *"Kindai kazoku" to bodī poritikkusu*. Kyoto: Sekai Shisosha, 2006.

Tama, Yasuko. *Boseiai toiu seido: Kogoroshi to chūzetsu no poritikkusu*. Keiso Shobo, 2001.

Tamai, Kingo, and Naho Sugita. "Nihon niokeru jinkō no 'ryō' 'shitsu' gainen to shakai seisaku no shiteki tenkai: Ueda Teijirō kara Minoguchi Tokijirō e." *Keizaigaku zasshi* 3, no. 1 (September 2015): 25–40.

Tamanoi, Mariko. "Knowledge, Power, and Racial Classifications: The 'Japanese' in 'Manchuria.'" *The Journal of Asian Studies* 59, no. 2 (May 2000): 248–76.

Terazawa, Yuki. *Knowledge, Power, and Women's Reproductive Health in Japan, 1690–1945*. New York: Palgrave Macmillan, 2018.

Terazawa, Yuki. "The State, Midwives, and Reproductive Surveillance in Late Nineteenth- and Early Twentieth-Century Japan." *US-Japan Women's Journal* 24 (2003): 59–81.

The Population Knowledge Network. *Twentieth Century Population Thinking: A Critical Reader of Primary Sources*. Abingdon, Oxon, and New York: Routledge, 2015.

Thompson, Malcom. "Foucault, Fields of Governability, and the Population–Family–Economy Nexus in China." *History and Theory* 51, no. 1 (2012): 42–62.

Tipton, Elise K. "Defining the Poor in Early Twentieth-Century Japan." *Japan Forum* 20, no. 3 (October 2008): 361–82.

Tipton, Elise K. "Birth Control and the Population Problem." In *Society and the State in Interwar Japan*, edited by Elise K. Tipton, 42–62. London and New York: Routledge, 1997.

Tipton, Elise K. "Birth Control and the Population Problem in Prewar and Wartime Japan." *Japanese Studies* 14, no. 1 (1994): 54–64.

Tomida, Hiroko. "The Controversy over the Protection of Motherhood and its impact upon the Japanese Women's Movement." *European Journal of East Asian Studies* 3, no. 2 (2004): 243–71.

Tomita, Akira. "1905-nen rinji Taiwan kokō chōsa ga kataru Taiwan shakai: Shuzoku, gengo, kyōiku wo chūshin ni." *Nihon Taiwan gakkaihō* 5, (May 2003): 87–106.

Toyoda, Maho. "Sengo nihon no bāsu kontorōru undō to Kurarensu Gyanburu: Dai 5 kai kokusai kazoku keikaku kaigi no kaisai wo chūshin ni." *Jendā shigaku* 6 (2010): 55–70.

Ts'ai, Hui-Yu. "Shaping Administration in Colonial Taiwan, 1895–1945." In *Taiwan under Japanese Colonial Rule, 1895–1945: History, Culture, Memory*, edited by Binghui Liao and Dewei Wang, 99–104. New York: Columbia University Press, 2006.

Tsu, Timothy Y. "Japanese Colonialism and the Investigation of Taiwanese 'Old Customs.'" In *Anthropology and Colonialism in Asia and Oceania*, edited by Jan Van Bremen and Akitoshi Shimizu, 197–218. London: Routledge, 1999.

Tsuboi, Hideto, ed. *Sengo nihon wo yomikaeru*. Volume 4 of *Jendā to seiseiji*. Kyoto: Rinsen Shoten, 2019.

Tsurumi. Yūsuke. *Seiden Gotō Shinpei*. Fujiwara Shoten, 2004.

Uchida, Jun. *Brokers of Empire: Japanese Settler Colonialism in Korea, 1876–1945*. Cambridge: Harvard University Press, 2011.

United Nations Department of Economic and Social Affairs. "World Population Prospects 2019." https://population.un.org/wpp/Maps/, accessed May 22, 2022.

Wada, Mayumi. "Osaka Mainichi Shinbun Jizendan no nyūyōji hogo katsudō to katei eno shien: Muryō josan jigyō to hoiku gakuen no sōsetsu wo chūshin ni." *Himeji daigaku kyōiku gakubu kiyō*, no. 11 (2018): 171–74.

Walker, Brett. *Toxic Archipelago: A History of Industrial Disease in Japan*. Seattle: University of Washington Press, 2010.

Watkins, Susan. "If All We Knew About Women Was What We Read in Demography, What Would We Know?" *Demography* 30, no. 4 (1993): 551–77.

Wang, Tiejun. "Kindai nihon bunkan kanryō seido no nakano Taiwan sōtokufu kanryō." *Chūkyō hōgaku* 45, no. 1–2 (2010): 97–300.

Waswo, Ann. "Japan's Rural Economy in Crisis." In *The Economies of Africa and Asia in the Inter-War Depression*, edited by Ian Brown, 115–36. London: Routledge, 1989.

Waswo, Ann, and Yoshiaki Nishida. *Farmers and Village Life in Twentieth-Century Japan*. London: RoutledgeCurzon, 2003.

Watanabe, Atsuko. *Japanese Geopolitics and the Western Imagination*. Cham: Palgrave Macmillan, 2019.

Watanabe, Tomoe. "Josanpu de nakereba ajienai shokugyōjō no taiken ninshiki nitsuite." In *Osaka no samba wa kataru taisetsu na osan no hanashi*. Osaka: Osaka Mainichi Shinbun Shakai Jigyōdan, 1936.

Watt, Lori. *When Empire Comes Home: Repatriation and Reintegration in Postwar Japan*. Cambridge, MA: Harvard University Press, 2009.

Weiner, Michael A. *Japan's Minorities: The Illusion of Homogeneity*. London and New York: Routledge, 1997.

Weiner, Michael A. *Race and Migration in Imperial Japan*. Sheffield Centre for Japanese Studies/Routledge Series. London: Routledge, 1994.

Williams, Rebecca. "Storming the Citadels of Poverty: Family Planning under the Emergency in India, 1975–1977." *The Journal of Asian Studies* 73, no. 02 (May 2014): 471–92.

Williams, Rebecca. "Rockefeller Foundation Support to the Khanna Study: Population Policy and the Construction of Demographic Knowledge, 1945–1953." Rockefeller Archive Center Research Reports Online. Rockefeller Archive Center, 2011. www.issuelab.org/resources/28011/28011.pdf, accessed April 21, 2012.

Wilson, Sandra. "Securing Prosperity and Serving the Nation: Japanese Farmers and Manchuria, 1931–33." In *Farmers and Village Life in Twentieth-Century Japan*, edited by Ann Waswo and Yoshiaki Nishida, 156–74. New York: Routledge, 2003.

Wilson, Sandra. "The 'New Paradise': Japanese Emigration to Manchuria in the 1930s and 1940s." *International History Review* 17, no. 2 (1995): 121–40.

Winkler, W. *A History of the International Statistical Institute 1885–1960*. Oxford: Blackwell, 1962.

Wittner, David G., and Philip C. Brown, eds. *Science, Technology, and Medicine in the Modern Japanese Empire*. London: Routledge, 2016.

Yabuuchi, Takeshi. *Nihon tōkei hattatsushi kenkyū*. Kyoto: Horitsu Bunka Sha, 1994.

Yabuuchi, Takeshi. "Nihon ni okeru minkan tōkei dantai no shōtan: 'Hyōki gakusha' to sono keifu." *Kansai daigaku keizai ronshū* 26, nos. 4–5 (January 1977): 585–623.

Yasukawa, Masaaki. *Yasashii jinkōgaku kyōshitsu*. Japanese Organization for International Cooperation in Family Planning, 1978.

Yamaguchi, Kiichi, Zenji Nanjo, Kazumasa Kobayashi, and Takao Shigematsu. *Seimeihyō kenkyū*. Tokyo: Kokon Shoin, 1995.

Yamaguchi, Toshiaki. "Kokka sōdōin kenkyū josetsu: Dai ichiji sekai taisen kara shigenkyoku no seiritsu made." *Kokka gakkai zasshi* 92, no. 3–4 (1979): 266–85.

Yamamuro, Shinichi. "Kokumin teikoku, nihon no keisei to kūkanchi." In *Kūkan keisei to sekai ninshiki*, edited by Shinichi Yamamuro, 19–76. Iwanami Shoten, 2006.

Yamamoto, Shunichi. *Nihon korera shi*. Tokyo Daigaku Shuppankai, 1982.

Yamamuro, Shinichi. *Manchuria under Japanese Dominion*. Philadelphia: University of Pennsylvania Press, 2006.

Yamazaki, Kiyoko. ed. *Seimei no rinri: Yūsei seisaku no keifu*. Fukuoka: Kyushu Daigaku Shuppankai, 2013.

Yao, Jen-to. "The Japanese Colonial State and Its Form of Knowledge in Taiwan." In *Taiwan Under Japanese Colonial Rule, 1895–1945: History, Culture, Memory*, edited by Binghui Liao and Dewei Wang, 37–61. New York: Columbia University Press, 2006.

Yasuda, Rihito. "Kindai nihon niokeru jinkō seisaku kōsō no ichi danmen (II)." *Kokusai bunkagaku*, no. 32 (March 2019): 155–79.

Yellen, Jeremy A. *The Greater East Asia Co-Prosperity Sphere: When Total Empire Met Total War*. Ithaca: Cornell University Press, 2019.

Yokoyama, Yuriko. *Meiji ishin to kinsei mibunsei no kaitai*. Yamakawa Shuppansha, 2005.

Yokota, Yoko. *Gijutsu karamita nihon eisei gyōsei shi*. Kyoto: Koyo Shobo, 2011.

Yokoyama, Takashi. "Yūseigakushi ni okeru nihon minzoku eisei gakkai no ichi." *Nihon kenkō gakkaishi* 86, no. 5 (2020): 197–208.

Yokoyama, Takashi. *Nihon ga yūsei shakai ni narumade: Kagaku keimō, media, seishoku no seiji*. Keiso Shobo, 2015.

Yoshinaga, Naoko. "The Modernization of Childbirth and the Indoctrination of Motherhood in Prewar Japan: The 'Aiiku-Son' Project of Imperial Gift Foundation 'Aiiku-Kai'" [in Japanese]. *Tokyo daigaku daigakuin kyoikugaku kenkyuka kiyō* 37, (1997): 21–29.

Young, Louise. "When Fascism Met Empire in Japanese-Occupied Manchuria." *Journal of Global History* 12, no. 2 (July 2017): 274–96.

Young, Louise. *Japan's Total Empire: Manchuria and the Culture of Wartime Imperialism*. Berkeley: University of California Press, 1998.

Young, Louise. "Marketing the Modern: Department Stores, Consumer Culture, and the New Middle Class in Interwar Japan." *International Labor and Working-Class History*, no. 55 (April 1999): 52–70.

Yoshino, Koji. "Yutakana shakai no binbōron: Takata Yasuma to Kawakami Hajime." *Hitotsubashi kenkyū* 30, no. 3 (October 2005): 35–52.

Index

Printed in the United States
by Baker & Taylor Publisher Services